Soft Skills for Engineering

Other related titles:

You may also like

- PBME024 | Cuiñas | Design Thinking for Engineering: A practical guide | June 2023
- PBTE108 | Forsyth | The Art of Successful Business Communication | August 2008
- PBMT025 | Wellington | Effective Team Leadership for Engineers | May 2009

We also publish a wide range of books on the following topics:
Computing and Networks
Control, Robotics and Sensors
Electrical Regulations
Electromagnetics and Radar
Energy Engineering
Healthcare Technologies
History and Management of Technology
IET Codes and Guidance
Materials, Circuits and Devices
Model Forms
Nanomaterials and Nanotechnologies
Optics, Photonics and Lasers
Production, Design and Manufacturing
Security
Telecommunications
Transportation

All books are available in print via https://shop.theiet.org or as eBooks via our Digital Library https://digital-library.theiet.org.

IET MANUFACTURING SERIES 31

Soft Skills for Engineering

Methodologies, social skills and course development

Edited by
Iñigo Cuiñas and Itziar Goicoechea

The Institution of Engineering and Technology

About the IET

This book is published by the Institution of Engineering and Technology (The IET).

We inspire, inform and influence the global engineering community to engineer a better world. As a diverse home across engineering and technology, we share knowledge that helps make better sense of the world, to accelerate innovation and solve the global challenges that matter.

The IET is a not-for-profit organisation. The surplus we make from our books is used to support activities and products for the engineering community and promote the positive role of science, engineering and technology in the world. This includes education resources and outreach, scholarships and awards, events and courses, publications, professional development and mentoring, and advocacy to governments.

To discover more about the IET please visit https://www.theiet.org/.

About IET books

The IET publishes books across many engineering and technology disciplines. Our authors and editors offer fresh perspectives from universities and industry. Within our subject areas, we have several book series steered by editorial boards made up of leading subject experts.

We peer review each book at the proposal stage to ensure the quality and relevance of our publications.

Get involved

If you are interested in becoming an author, editor, series advisor, or peer reviewer please visit https://www.theiet.org/publishing/publishing-with-iet-books/ or contact author_support@theiet.org.

Discovering our electronic content

All of our books are available online via the IET's Digital Library. Our Digital Library is the home of technical documents, eBooks, conference publications, real-life case studies and journal articles. To find out more, please visit https://digital-library.theiet.org.

In collaboration with the United Nations and the International Publishers Association, the IET is a Signatory member of the SDG Publishers Compact. The Compact aims to accelerate progress to achieve the Sustainable Development Goals (SDGs) by 2030. Signatories aspire to develop sustainable practices and act as champions of the SDGs during the Decade of Action (2020–2030), publishing books and journals that will help inform, develop, and inspire action in that direction.

In line with our sustainable goals, our UK printing partner has FSC accreditation, which is reducing our environmental impact to the planet. We use a print-on-demand model to further reduce our carbon footprint.

Published by The Institution of Engineering and Technology, London, United Kingdom

The Institution of Engineering and Technology (the "**Publisher**") is registered as a Charity in England & Wales (no. 211014) and Scotland (no. SC038698).

First published 2025

The Institution of Engineering and Technology
Futures Place,
Kings Way, Stevenage,
Herts, SG1 2UA,
United Kingdom

www.theiet.org

British Library Cataloguing in Publication Data
A catalogue record for this product is available from the British Library

ISBN 978-1-83724-017-3 (hardback)
ISBN 978-1-83724-018-0 (PDF)

Typeset in India by MPS Limited

Cover image: Image – Laughing female engineer leading project discussion: Thomas Barwick/DigitalVision via Getty images

Contents

About the editors

Iñigo Cuiñas is a professor at the Department of Signal Theory and Communications, Universidade de Vigo, where he teaches courses on Remote Sensing and the Social Links of Engineering. His main research interest focuses on radio wave propagation within complex environments, environmental aspects of radio frequency systems, and innovation in education (virtual reality, design thinking, or project/challenge-based learning). He received his degree in Telecommunication Engineering and his PhD degree from Universidade de Vigo, Spain, in 1996 and 2000, respectively.

Itziar Goicoechea is an associate professor in the Department of Engineering Design, Universidade de Vigo, where she teaches courses on Project Management. She has 15 years' experience in project management, working in private companies (automotive, construction, and shipping sectors). Her research focuses on project management applications in various contexts, applying both predictive and agile methodologies. She received her degree and PhD in Industrial Engineering from the Universidade de Vigo, Spain, in 1996 and 2006, respectively.

Chapter 1
Introduction

Iñigo Cuiñas[1] *and Itziar Goicoechea*[2]

Employers today are increasingly recognizing the significance of soft skills, and students are well aware of this crucial fact. Upon graduation, students are expected to possess a firm grasp of the technical skills directly related to their completed courses, whether at the Bachelor's or Master's level. However, there is another realm of skills that can profoundly impact securing initial employment and, more importantly, advancing one's professional journey.

The importance of developing soft skills cannot be overstated, as they have the potential to shape the future of current and aspiring university students. Providing training in these essential skills becomes a pivotal aspect of their overall development. A comprehensive handbook outlining strategies for enhancing the most sought-after skills serves as a valuable resource for both students and lecturers, facilitating the integration of these vital competencies into traditional university programs.

In 2016, during a PLA (Peer Learning Activity) titled "Developing future skills in higher education," and organized by the Working Group on the Modernization of Higher Education (WG-MHE) of the European Commission, a framework on soft skills was developed. These PLA seminars bring together ministries and HEI representatives to share experiences and help identify common areas for further work and cooperation at the European level [1]. This framework, as defined by the EU Skills panorama [2] and The Occupational Information Network (O*NET) [3], represents the core of the contents of this book.

More recently, and also with the support of the European Commission, other frameworks have been published focusing on more specific soft skills:

- LifeComp, on personal, social, and learning to learn competencies [4];
- DigComp, related to digital competencies [5];
- GreenComp, which focuses on sustainability competency [6]; and
- EntreComp, for promoting entrepreneurship [7].

[1]Departamento de Teoría do Sinal e Comunicacións, Universidade de Vigo, atlanTTic, Spain
[2]Departamento de Deseño na Enxeñaría, Universidade de Vigo, Spain

These new paradigms are also included in the different chapters, linked to the fundamental collection from O*NET [3].

Education in Engineering has been traditionally focused on providing technical knowledge for future engineers. Thus, they are generally good professionals with a strong background in both basic sciences, mainly Mathematics and Physics, and in specific topics related to each engineering branch. Now, with this book, professors and lecturers have a tool to support the integration of soft skills within engineering academic programs.

By recognizing the significance of soft skills and equipping students with the tools to cultivate them, universities can empower their graduates to thrive in the competitive professional landscape. This handbook aims to serve as a roadmap for unlocking and developing these highly sought-after skills, benefiting both individuals embarking on their careers and institutions committed to nurturing well-rounded and successful professionals.

1.1 Future skills in the European higher education area

The EU Skills panorama and The Occupational Information Network (O*NET) classify soft skills into three categories: methodological, social, and cognitive. As this is the framework that supports the entire handbook, the following subsections will be devoted to defining the considered soft skills, which are the individual aims of most of the chapters.

1.1.1 Methodological skills

This group of skills is oriented toward developing capacities used to allocate resources efficiently. The included skills are time management, decision-making, problem solving, learning strategies, planning, and digital skills, and they are briefly explained as follows:

- Time management: managing one's own time and the time of others.
- Decision-making: considering the relative costs and benefits of potential actions to choose the most appropriate one.
- Problem solving: identifying complex problems and reviewing related information to develop and evaluate options and implement solutions.
- Learning strategies: selecting and using training/instructional methods and procedures appropriate for the situation when learning or teaching new things.
- Planning: developing specific goals and plans to prioritize, organize, and accomplish the work.
- Digital skills: involving the confident and critical use of information and communication technology (ICT) for working and learning in the knowledge society.

1.1.2 Social skills

Social skills are the developed capacities used to work with people to achieve goals, including skills such as teamwork, interpersonal communication, conflict management and negotiation, and intercultural understanding.

- Teamwork: skills required to work well as part of the team.
- Interpersonal communication: skills to express ideas and views clearly, confidently, and concisely in speech, writing, and body language.
- Conflict management and negotiation: handling complaints, settling disputes, resolving grievances and conflicts, or otherwise negotiating with others.
- Intercultural understanding: valuing and engaging with diverse cultures in ways that recognize commonalities and differences, create connections with others, and cultivate mutual respect.

1.1.3 Cognitive skills

Cognitive skills are those used in the process of obtaining and understanding new knowledge through thought, reflection, experience, and the senses. They include analytical, creative, critical, and reflective thinking.

- Analytical thinking: analyzing information and using logic to address work-related issues and problems.
- Creative thinking: developing, designing, or creating new applications, ideas, relationships, systems, or products, including artistic contributions.
- Critical thinking: using logic and reasoning to identify the strengths and weaknesses of alternative solutions, conclusions, or approaches to problems.
- Reflective thinking: processes of analyzing and making judgments about what has happened.

1.2 DigComp

The Digital Competence Framework for Citizens (DigComp) is an initiative supported by the European Commission to provide a common understanding of the significance of digital competencies. This is the first step to developing such competencies and creating the basis for measuring them. They are closely aligned with the European Union's efforts to promote digitization across the entire population. In fact, the objective of the EU is to reach 80% of the population with basic digital skills and about 20 million specialists in Information and Communication Technology by 2030 [5].

The European Skills Agenda supports the promotion of digital skills among the entire population as one of the pillars of future policies in education and wellness. Until now, the use of DigComp has been related to different purposes, mainly in the context of employment, education and training, and lifelong learning. This obviously fits with the purpose of this handbook, as including these skills in engineering academic programs should be a future and present concern. In fact, engineers must receive a high-level evaluation of their digital skills when obtaining their diplomas.

This framework is a response to the Council Recommendation of 2018: "Digital competence involves the confident, critical, and responsible use of, and engagement with, digital technologies for learning, at work, and for participation in society. It includes information and data literacy, communication and collaboration, media literacy, digital content creation (including programming), safety (including digital well-being and competencies related to cybersecurity), intellectual property-related questions, problem solving, and critical thinking." [8]. Including a chapter devoted to "Digital Thinking" represents the contribution of this book to this common effort.

Assuming that competencies are a combination of knowledge, skills, and attitudes, DigComp is structured around five areas, which link to the soft skills considered throughout this book, and are analyzed in the following paragraphs:

- Information and data literacy: this includes locating and retrieving digital information, judging the relevance of both the source and the content, and managing digital data. Thus, this first area connects to soft skills such as learning strategies, analytical thinking, creative thinking, and, of course, digital skills.
- Communication and collaboration: this area relates to how we communicate and collaborate using digital services, how we interact with other people or institutions, and the management of our reputation. This area is linked to skills such as interpersonal communication, teamwork, conflict management and negotiation, intercultural understanding, and again, digital skills.
- Digital content creation: this third area focuses on the creation and editing of digital content, connecting digital information to other bodies of knowledge while taking care of intellectual property rights, and efficient interaction with computer systems. The soft skills involved in such an area could be time management, planning, decision-making, creative thinking, and digital skills.
- Safety: while it might intuitively be interpreted as the protection of devices, data, and privacy in digital environments, this fourth area goes further, including the protection of physical and psychological health, awareness of digital technologies for social inclusion and well-being, as well as the environmental impact of these digital activities. This broad focus needs to be supported by skills such as problem solving, critical thinking, reflective thinking, and, once more, digital skills.
- Problem solving: in this fifth area, problem solving is aimed at digital problems or situations to be resolved in digital environments. This includes the use of digital technologies and staying up to date with digital knowledge. Therefore, the soft skills involved are obviously problem solving,

as well as learning strategies, analytical thinking, decision-making, and digital skills.

1.3 GreenComp

The European sustainability competence framework (GreenComp) is one of the actions within the European Green Deal aimed at promoting learning about environmental sustainability in the European Union. GreenComp identifies competencies to be included in education programs to create citizens with empathy, responsibility, and care for the planet and public health [6]. This EU initiative can be incorporated into engineering academic programs, making its analysis relevant in this book.

GreenComp is organized into four competence areas:

- Embodying sustainability values: this includes valuing sustainability, supporting fairness, and promoting nature. Analytical thinking, learning strategies, and intercultural understanding can be identified as the related soft skills.
- Embracing complexity in sustainability: this second area is devoted to systems thinking, critical thinking, conflict management and negotiation, and problem framing. Among the skills included in this book, critical thinking, problem solving, and digital skills are linked to this area.
- Envisioning sustainable futures: the competencies included here are futures literacy, adaptability, and exploratory thinking. Decision-making and creative thinking are skills that have important implications in this area.
- Acting for sustainability: political agency, collective action, and individual initiative are the competencies considered in this area. Reflective thinking, teamwork, interpersonal communication, time management, and planning are soft skills related to these concepts.

1.4 LifeComp

The European Framework for Personal, Social and Learning to Learn Key Competence (LifeComp) includes a set of competencies applicable to all spheres of life, to be acquired through formal, informal, and non-formal education. The objective is to help citizens thrive in the 21st century, providing them with a key competence set identified by the Council Recommendation on Key Competencies for Lifelong Learning [8]. LifeComp has nine competencies grouped into three areas [4]. They are outlined in the following lines and connected to the soft skills in this book that best help in developing each competency.

- Personal area, including self-regulation, flexibility, and well-being. They are connected to creative thinking, reflective thinking, problem solving, time management, and planning, among the soft skills involved.

- Social area, encompassing empathy, communication, and collaboration. They link to skills such as interpersonal communication, teamwork, conflict management and negotiation, and intercultural understanding.
- Learning to learn area, featuring growth mindset, critical thinking, and managing learning as the key competencies. Learning strategies, decision-making, digital skills, critical thinking, and analytical thinking are related soft skills.

1.5 EntreComp

The European Entrepreneurship Competence Framework aims to develop the competencies needed to act on opportunities and ideas, to work with others, to manage dynamic careers, and to shape the future for the common good. The created value is found in both financial and cultural or social spheres. In other words, it seeks to foster an entrepreneurial mindset in every aspect of life [7].

This framework is adaptable to different situations, and it contains 15 entrepreneurship competencies, grouped into three main areas:

- Ideas and opportunities. This area includes five competencies, such as spotting opportunities, creativity, vision, valuing ideas, and ethical and sustainable thinking.
- Resources. The competencies are self-awareness and self-efficacy, motivation and perseverance, mobilizing resources, financial and economic literacy, and mobilizing others.
- Into action. Competencies such as taking the initiative, planning and management, and coping with uncertainty. Ambiguity and risk, working with others, and learning through experience are the core of this area.

EntreComp also defines a timeline of progression levels in order to help individuals identify their starting point and to define a coherent pathway to achieve full development in entrepreneurship skills. These progression levels begin at the foundation stage (relying on support from others) and advance toward the expert level (driving transformation, innovation, and growth), passing through intermediate (building independence) and advanced (taking responsibility) stages.

1.6 Relations among the different considered frameworks

In order to help organize the contributions of various European Union competence frameworks discussed in previous sections and the soft skills explained throughout this book, Table 1.1 summarizes the relationships among them.

The cross-referenced information compiled in Table 1.1 allows us to see that each of the competence frameworks analyzed takes into account all soft skills developed through this book. This reinforces the importance of such a collection of skills, and even more, of incorporating them into the engineering curricula.

Table 1.1 Relationship between areas at each competence framework and soft skills

		Dig Comp					Green Comp				Life Comp			Entre Comp		
		Data literacy	Communication, collab.	Digital content creation	Safety	Problem solving	Embodying values	Embracing complexity	Envisioning futures	Acting	Personal	Social	Learning to learn	Ideas and opportunities	Resources	Into action
Methodological	Time management			■						■	■				■	
	Problem solving				■	■		■			■			■		
	Decision-making			■		■			■				■	■		■
	Learning strategies	■				■	■						■		■	■
	Planning			■						■	■			■		■
	Digital skills	■	■	■	■	■		■					■		■	
Social	Interpersonal comm.		■							■		■		■	■	
	Teamwork		■							■		■			■	■
	Conflict management		■					■				■			■	■
	Intercultural understand.		■				■					■		■		
Cognitive	Analytical thinking	■				■	■						■	■		
	Reflective thinking				■					■	■			■	■	■
	Critical thinking				■			■					■			■
	Creative thinking	■		■					■		■			■		

1.7 Incorporating soft skills in engineering programs

This book is intended for lecturers and students to gain knowledge about the importance of soft skills and the characteristics of each of them. Thus, the editors have selected a taxonomy of these skills and organized the book in a way that all of those included are analyzed and explained. As the book is designed as a manual to aid in training soft skills, some activities are suggested throughout the chapters, and they are indicated within gray boxes in between the text, close to the elements to be reinforced.

The book is organized into chapters, most of which are individually devoted to each of the considered soft skills. The first chapters focus on the reason why soft skills should be introduced in relation to the learning outcomes of different engineering academic programs, from various perspectives: the development of individuals, the advancement of professional careers, the companies' demands, and many others.

After these introductory chapters, each of the 14 considered soft skills (methodological, social, and cognitive) is introduced, explained, and developed along the core of the book. As mentioned, these descriptions and considerations are accompanied by exercises and activities that can be proposed to students to train and improve their soft skills through reflection.

Finally, the last chapters are devoted to how to incorporate these learning outcomes within the academic curriculum: course syllabus, assessment tips, and links to the digital world.

References

[1] ET2020 Working Group on Modernization of Higher Education, "Developing future skills in higher education", *Peer Learning Activity (PLA), European Commission*, Directorate-General for Education and Culture, Brussels, February 25–26, 2016.

[2] "*EU Skills panorama*", https://www.cedefop.europa.eu/en/tools/skills-intelligence. [Accessed 31 Jul 2024].

[3] T. C. Burgoyne, and M. C. Reeder, "O*NET® Analyst Ratings of Occupational Skills: Analysis Cycle 24 Results", *Final Report*, November 7, 2023.

[4] A. Sala, Y. Punie, V. Garkov, and M. Cabrera, "LifeComp. The European Framework for Personal, Social and Learning to Learn Key Competence", *EUR 30246 EN*, Joint Research Centre, Publications Office of the European Union, 2020. doi:10.2760/302967.

[5] R. Vuorikari, S. Kluzer, and Y. Punie, "DigComp 2.2. The Digital Competence Framework for Citizens", *EUR 31006 EN*, Joint Research Centre, Publications Office of the European Union, 2022. doi:10.2760/115376.

[6] G. Bianchi, U. Pisiotis, M. Cabrera, Y. Punie, and M. Bacigalupo, "GreenComp. The European sustainability competence framework", *EUR 30955 EN*, Joint Research Centre, Publications Office of the European Union, 2022. doi:10.2760/13286.

[7] M. Bacigalupo, P. Kampylis, Y. Punie, and G. Van den Brande, "EntreComp: The Entrepreneurship Competence Framework", *EUR 27939 EN, Joint Research Centre, Publications Office of the European Union*, 2016. doi:10.2791/593884.

[8] EU Council Recommendation on Key Competences for Life-long Learning, 22 May 2018. ST 9009 2018 INIT.

Part I

Context

Chapter 2
Why soft skills?

Bogdan Wlodarczyk[1]

Today's dynamic world is one of ubiquitous technological development and global access to technological advances that are visible at every turn. Thanks to technologies such as cloud computing, big data analytics, and the Internet of Things, it is possible to access any information at any time from anywhere in the world. Technology has changed the environment, and technological progress has created a modern society in which information and access to it have become a basic good and an integral factor of development. Technological development is not limited to modern companies, autonomous machines, digital networks, or new communication technologies. At its core, it is driven by human capital—specifically, a company's employees—who possess both hard skills (such as the technical expertise needed to implement and operate technology) and soft skills essential for effective communication and collaboration. These interpersonal abilities are particularly crucial to the success and functioning of modern organizations.

2.1 Soft skills in industry 4.0

Competencies determine the qualifications of people in certain professions and are a kind of guarantee of good work and success. Only people with specific and high competencies can be professionals.

The socio-economic development of mankind takes place in three stages, which A. Toffler [1] defined as waves of civilization development. The first wave is the agrarian revolution, associated with the acquisition of agricultural skills and the spread of sedentary lifestyles. The second wave is the industrial revolution, initiated by the invention of the steam engine, electricity, new means of transportation and mass communication, and the creation of mass production. The invention of the computer ushered in the third wave, the post-industrial revolution, associated with the use of automated machinery and equipment, unrestricted access to information, and the shift from mass to individualized production. The third wave of human socio-economic evolution is also considered a continuation of the industrial

[1]Department of Finance, Faculty of Economic Sciences, University of Warmia and Mazury in Olsztyn, Poland

revolution through the creation of automated production based on flexible production systems and smart factories with cyber-physical production systems, where information is transmitted via the Internet. The changes taking place in industry as a result of the third wave are also referred to as the third and fourth industrial revolutions (Industry 3.0 and Industry 4.0).

The direction of change in the era of the Fourth Industrial Revolution is clearly emphasized and indicated by various institutions, including those of the European Union. In the European Digital Agenda, the European Commission emphasized that information and communication technologies "stimulate value creation and economic growth." This means that industry increasingly needs open and interoperable solutions to exploit these technologies across all sectors. EU-funded programs will support industry initiatives to create standards and open platforms for new products and services. The Commission will strengthen activities that bring stakeholders together through joint research programs in areas such as the "Internet of the Future," including the "Internet of Things," and in key technologies in the information and communication sector [2].

But Industry 4.0 and its enterprises are all about human capital, an essential element that is the glue between technology and reality. Every modern organization builds its present and future on people. Each is also unique in the resources and capabilities it possesses that set it apart from its competitors.

Thus, there is little doubt today that regardless of the size of the organization, its structure, and other attributes, it is its people who constitute its most valuable strategic capital: capital of qualifications, knowledge, skills, experience, personality, and value. Human capital is considered the most important part of intellectual capital, which, together with structural capital (process capital), is the primary source of competitive advantage and value for any organization in today's knowledge-based economy. Therefore, people management is one of the most important elements of management. The most important and probably the most difficult, because people, like no other asset of an organization, are difficult to know, changeable, often irrational, and unpredictable, and at the same time endowed with the subjectivity and dignity of the human person.

In the past, technical knowledge helped to absorb industry-specific issues that were needed on a daily basis. Engineers were research-oriented specialists who carefully analyzed all aspects of a case before making a decision. They followed policies and procedures according to a plan of action that was refined to the last detail. The most important thing for them was the quality of their work. These competencies of a specialized and technical nature, i.e., specific skills and knowledge (hard competencies), are the absolute minimum for the functioning of a company employee. In the era of the challenges of Industry 4.0, in addition to specialized knowledge, soft competencies are expected from employees and workers as an important area of functioning for the implementation of tasks and cooperation. However, the definitional analysis of competencies is not obvious either. It is often pointed out that before the term "competence" was used, it was written about skills, abilities, aptitudes, qualifications, and work potential. Undoubtedly, the basic components of competence are knowledge and skills—mentioned by all authors dealing with the issue of competence. In addition, the diagnosis of both individual and organizational competencies is an integral part of competence management.

The term “soft skills” refers to an individual’s mental capacity and ability to cope with social situations. They relate to the ability to manage oneself, i.e., the ability to plan, anticipate, cope with stress, manage time, as well as the ability to motivate oneself, assertiveness, creativity, or emotional intelligence. The second area relates to functioning in society—these competencies are the ability to work in a team as well as to manage others—motivate them, set goals for them, inspire or persuade them to be right, and resolve conflicts that arise.

These topics are worth training, and these skills are worth deepening in order to bring tasks and projects to the expected results. The keyword is also openness. Today’s experts cannot be confined to their narrow specialization. They must be able to understand concepts and ideas from other fields. They must be open to change, both in their field and in the team in which they work. Communication skills, including intercultural communication and the use of virtual tools, are also becoming increasingly important. Of course, an engineer’s authority is still primarily based on “hard,” technological knowledge, but the importance of its modern segments is growing.

Enterprise 4.0 employees are outgoing and active persons who enjoy variety, both in their interactions with people and in the tasks they perform. They have the ability to communicate highly technical and detailed information to others with enthusiasm and optimism, inspiring positive feelings about the ideas they are sharing. Pay close attention to detail and strives for perfection. Ensure high quality of work and adherence to standards. Adhere to policies and procedures.

Developing soft skills is not easy. In fact, it requires the presence in a person’s consciousness of such elements as understanding the purpose of one’s change, practicing, and having access to feedback on the changes made. Typical difficulties a person encounters after deciding to work on developing their soft skills include overcoming habitual reactions, struggling with their own resistance and defensiveness, examining their self-perception, and—which requires a great deal of effort and support—painstaking training in everyday social situations [3]. Soft skills are useful in any job and can be a determining factor in career success. In job advertisements, soft skills are often mentioned and presented in terms of personality traits: creativity, responsibility, self-reliance, recognition of one’s own strengths and weaknesses, high motivation, and resistance to stress.

The biggest difficulty in implementing Industry 4.0 solutions will not be acquiring the right technology, but the need to change the organizational culture and acquire new skills. Of course, the technological aspect is also of paramount importance. Companies are now moving from a phase of discovering and understanding the availability of data and its importance to a phase of extensive use of analytical tools. In addition, with the proliferation of digital ecosystems, the need to establish a sufficiently high level of digital trust, supported by transparency and non-repudiation, confirming the integrity and provenance of their own and third-party data, is becoming increasingly important. All in all, in high-tech companies, people are a resource of strategic importance in the management process.

Company employees, including, for example, an Industry 4.0 engineer, should no longer focus solely on their narrow specialization. They should be able to understand concepts and ideas from other fields. Be open to change, both in their field and in the team in which they work. New communication and social technologies require employees to work effectively, often in a virtual team, and often with people from different cultures. The team-based nature of an engineer's work requires highly developed interpersonal skills. Effective communication is particularly important, including the ability to persuade, motivate, and explain difficult and unpopular decisions. As before, in Industry 4.0, the basis of an engineer's authority will be the extent of their technical knowledge, with new areas of technical competence likely to gain in importance. These may include, for example, the integration of IT control systems, the programming of industrial robot assemblies, the integration of cloud-based analytical systems with local systems, cybersecurity, or the application of artificial intelligence algorithms to automation systems.

Automation and robotics have a direct impact on reducing the need for unskilled labor. In turn, the need to program and maintain complex jobs, equipment, and systems will create new jobs. Studies show that the jobs most likely to be replaced by systems and technologies are those that perform simple administrative tasks and, in the case of manufacturing companies, the positions of people who perform simple activities in the production process. On the other hand, positions and occupations such as project managers, operations managers, engineers, or doctors, as well as members of corporate boards, are reported to be the least susceptible to being substituted [4].

The vision of Industry 4.0 requires a constant willingness to innovate, to be creative, and thus to change the way goods are produced and managed. The management of human capital will depend on many factors and will require far-reaching changes: from the planning of training programs (at lower and higher levels), to changes in the nature of the work of those already employed (re-qualification, training), to organizational changes in companies (implementation of new technologies, communication networks that coordinate work inside and outside the company).

The increasingly widespread use of advanced technological solutions, the interconnectedness of the economy that affects the development of various forms of communication, global competitiveness, and the increasing and popular mobility of workers are trends and phenomena that affect—and often profoundly change—the way people live and work. These regularities, combined with a rapidly changing professional environment, mean new opportunities but also challenges for organizations, their employees, and, above all, their human resources managers. In the "factory of tomorrow," people will be at the center of an intelligent production system where technology will support their cognitive and physical abilities.

Enterprises 4.0 will need to hire employees endowed with a range of competencies, starting with those basic specialized ones (hard skills) as well as key social ones (soft skills). Employees who are professional, committed, easily networked, and eager to cooperate are the dream/needed personnel of today's enterprises. The key to creating and implementing an Industry 4.0 strategy in a company is understanding how customers are changing, what the competition is doing, what challenges the organization is facing, and what can cause a business and technological status quo.

The challenge now posed by the market and consumers will be realized if, together with advanced technology, the emphasis is placed on the development of soft competencies of the employee of the enterprise 4.0. It is these that can determine the market success of the enterprise, where a close-knit, well-functioning team of employees, able to work together, with a range of soft competencies, will function effectively and efficiently, even with the difficult challenges of implementing and operating the technology of Industry 4.0.

2.2 The importance of universities and the educational ecosystem

The Fourth Technological Revolution is overturning the established paradigms of market, production, labor, and consumption. It is also changing the paradigm of education: the understanding of its purpose and the ways of acquiring it. More rapidly changing labor market conditions require greater flexibility from the entire education system and a better connection between educational institutions and employers, NGOs and public institutions within a multi-element educational ecosystem. At the center of this ecosystem is an individual who has the cognitive and social competencies to navigate the ever-changing labor market, and who understands the need for continuous further education and the acquisition of new skills, and above all, knows how to do it.

In this view, formal educational institutions—including universities—should first and foremost shape cognitive and social competencies and create a base of knowledge and skills that graduates will be able to extend and expand with further skills required in the labor market, both through further stages of formal education (e.g., MBA courses or postgraduate studies), as well as non-formal or informal education (vocational micro courses, training courses and self-education through e-learning platforms, offered both for educational purposes or in a non-intentional way—in everyday life). In other words, the primary task of educational institutions is to develop the fundamental competencies that enable people to reach for further forms of education on their own, not necessarily planned as part of a course. The ultimate goal is the maximum personalization of education, that is, the adaptation of education to both the needs of the individual, the requirements of functioning in society, and the demands of the labor market.

An important role in the educational ecosystem is played by universities, networked and actively cooperating with other entities. The gap between the expectations of students and the needs of the environment for the competencies of the future and the current educational results is alarming. This is true of most tertiary education systems around the world. In knowledge-based economies, the traditional role of the university is changing, as it ceases to be the proverbial "ivory tower" with a monopoly on the production and dissemination of knowledge, becoming one of many, though invariably a significant component of an educational ecosystem focused on continuous learning. Stimulated by the technological revolution, the change in the model of knowledge and skills acquisition and the emergence of competitive educational models mean that, all over the world, universities are facing new challenges to adapt to the changing expectations of students and employers.

The problem is that there is a massive gap between the beliefs of those working in higher education about the level of competence of graduates and the opinion of employers about it. We put forward the thesis that a higher education institution wishing to educate the competencies of the future must itself assimilate them and incorporate them into the basis of its operations.

The university of the future actively engages with the broader educational ecosystem: this means moving away from the traditional university model to an engaged model in which learning and teaching are aligned with academic standards, but at the same time meet the expectations of the university environment and jointly create new knowledge, skills, and attitudes. The primary challenge for universities is to change the way they educate students so that they acquire key competencies for the job market, while being prepared to function in a world where constant change in conditions is the norm.

Modern developments are moving in the direction of increasing the flexibility of programs and changing the forms of teaching students: moving away from the rigid division into semesters and years of study, teaching and applying theory to practice, and putting more emphasis on involving students in practical project work.

The universities' response to students' expectations is also to expand their educational offerings through cooperation with external entities, e.g., organizing intensive courses to improve specific skills or competencies in cooperation with specific companies or NGOs, involving the possibility of obtaining a certificate. Competency needs, expressed by companies, could be more quickly reflected in study programs if universities had closer and more frequent relations with the environment. These relations should include: intersectoral exchanges (hiring researchers at companies, institutions or foundations, and employees of these organizations at universities); research cooperation between students and the environment as part of theses or classes; broader participation of universities in commissioned research, joint publications with business or in the commercialization of knowledge and technology.

2.3 How is the university of the future?

Creative. Tirelessly seeks new ways to shape the competencies of the future in students; it is committed to tradition, but is not afraid of didactic and organizational innovations. Boldly introduces alternative teaching models based on new technologies, joins in the creation of new methods of skill certification (so-called "nano-degrees"), and highly personalizes the learning experience.

Collaborative. Understands its role in the broader educational ecosystem and in the context of the lifelong learning requirement. Eagerly and actively enters into alliances with employers to expand and add a practical dimension to educational offerings. In cooperation with businesses, it creates an attractive and accessible educational offer for employees who are changing their skills, and is thus a key component of lifelong learning.

Managing. Motivates and develops its staff, lecturers, and researchers not only in terms of content, but also in terms of the competencies of the future, so that they can successfully instill them in students.

Socializing. Provides a rich and open environment to develop social and emotional competencies in direct contact with lecturers and other students.

Digitized. Uses new technologies to:

- collecting and analyzing data on the quality of scientific and research and teaching activities through automated information systems, which allows authorities to respond quickly and flexibly to emerging challenges and adapt teaching offerings to the needs of the changing labor market;
- changing the way of teaching (e.g., increasing the role of e-learning and communication platforms);
- changes in the study experience: efficient handling of student and employee affairs.

Solving complex problems. First and foremost—how to combine the traditions of university education with the expectations of students and employers in terms of educational offerings; how to educate a flexible employee, able to continuously improve his qualifications and competencies, who at the same time will be an enlightened and satisfied citizen?

Critical. Produces knowledge crucial to the development of digital society and economy; creates proposals for solutions to problems arising from changes in the labor market, for example, actively participates in discussions on the direction of social and economic development. Teaches theory in practice.

Entrepreneurial. Actively cooperates with the environment (business, public institutions, third sector); in cooperation with the environment, develops key competencies for entrepreneurial attitudes in its students.

2.4 Universities in the vanguard: good practices

Stanford University's response to the changes taking place is the Stanford 2025 project, based on four concepts that are revolutionizing university education. The open-loop concept focuses on changing the period of study from a uniform block into time-independent and repeatable phases of study that meet the student's needs in the labor market. The concept of one's own pace of learning (paced education) is designed to provide a personalized learning process and time, allowing the students to better understand their learning style and strengths. The concept of axis change (axis flip) puts interdisciplinary competencies in the spotlight at the expense of attachment to traditional academic disciplines. The final element is purpose-driven learning, which seeks to ensure that the student is engaged in the study process by creating a purpose and meaning to work on. Instead of declaring "I'm studying biology," a student should be able to say "I'm studying human biology to create solutions to fight world hunger."

At Arizona State University, USA's largest university, students work as part of multidisciplinary teams—their members may be studying engineering, law or management, as well as arts, communications, and pedagogy. Together, they carry out a practical project, looking for a solution to a problem submitted by a specific company, and then present it to the principal. The university also emphasizes creating an environment in which students can learn from their mistakes, "but without suffering the consequences that await them in the real world," nor does the university make it difficult to change majors if a student finds that the current one does not meet his or her expectations. The university has developed various strategies for including employers—from the business world and public institutions—in the process of educating students.

In 2017, The Netherlands Association of Universities, the Netherlands Association of Universities of Applied Sciences and SURF (an organization dedicated to ICT for education and research) formulated the Acceleration Agenda for innovation in education. It defines the strategic areas of action needed to accelerate innovation in the higher education sector, namely: better connection between higher education and the labor market, making higher education more flexible, and supporting the use of new technologies in teaching. Activities include supporting digital competency training as a means to better prepare students to enter the labor market, sharing e-learning materials, or supporting lecturers to prepare them to teach with more innovative methods. Implementation of the agenda is based on eight areas of activity, which are coordinated by researchers from various universities affiliated with the aforementioned associations.

The University of Warsaw is implementing a project called "Applied Thesis—Effective Professional Start," under which students can apply the knowledge they acquire in classes to practice. External partners—companies, public institutions, and non-governmental organizations—can submit research requests, and students of the humanities and social sciences, under the direction of a supervisor, prepare a thesis proposing a solution to a specific problem. "The project responds to the need to prepare graduates for the practical application of knowledge at work, but also to the growing demand for the exchange of data, knowledge, and competence between the scientific sector and business, public institutions and NGOs."

As a joint initiative of the European Commission, the Directorate General for Education and Culture and the OECD, an evaluation tool has been created through which universities can assess themselves in terms of entrepreneurship and innovation in their own educational environment. The evaluation covers eight scopes: leadership and management, organizational capacity, entrepreneurial teaching and learning, preparing and supporting entrepreneurs, digital transformation and capacity, knowledge exchange and collaboration, internationalization, and impact measurement.

2.5 Quality of competence education

The future of universities depends on investment in teaching and teaching staff. In theory, universities are oriented toward teaching the competencies of the future. According to the requirements, the plan for each course must include learning outcomes regarding the knowledge, skills, and competencies that students will derive from the course. However, these ambitious assumptions are crashing into teaching practice. The weakest link in the system for educating the competencies of the future is the educators themselves. The reasons for this state of affairs must be sought in the systemic and long-standing overload of didactics while undervaluing it in the evaluation of a research and teaching employee, held accountable mainly for publications and projects. The motivation and commitment of academics are directed to where the highest return is expected in the form of the highest possible university revenues. These, in turn, are predominantly based on the publication efficiency of academic units and researchers. Thus, there is no space—either institutional or even more so personal—for active teaching based on flexible adaptation of content to the needs of the labor market. Universities that make such efforts do so on a small scale and often under difficult organizational conditions. In turn, the lecturers who actually undertake such teaching often lack relevant experience, and therefore—in the eyes of students—are not a reliable source of education aimed at achieving the competencies of the future.

Today, the role of the lecturer is changing: from a master with access to elite knowledge to an intellectual guide with high didactic and social competencies. A university that wants to teach the competencies of the future must emphasize the development of its teaching staff in terms of these competencies. In addition to systemic solutions, such as valuing didactics in employee evaluation and reducing salaries to increase the quality of teaching, it is crucial to involve lecturers in the open educational system through contacts with businesses, NGOs, and public institutions.

Under conditions of rapid acceleration of technological, economic, and social changes, lecturers' knowledge becomes outdated faster, so it turns out to be all the more important to cooperate with the university environment. Universities must also consistently engage practitioners from the market to teach. This type of practice occurs, among others, at many other universities, which have the infrastructure and technology to use innovative methods in teaching, but instructors do not know how to use these opportunities. Their concerns include, for example, the use of materials from the Internet. This has prompted the creation of a system of expert consultations aimed at universities to support the development of individualized digitization strategies for universities.

Technological advances mean that the development of the economy will increasingly be based on knowledge and the ability to apply it in practice.

Meanwhile, many companies in Europe consider the lack of personnel with the right skills to be a barrier to investment; a growing number of companies also consider this obstacle to be the most important. Skills gaps can negatively affect companies' productivity and their ability to grow, innovate, and adapt technology. For workers, the lack of these "right skills" limits employment prospects and access to quality jobs. Also, the economy is increasingly knowledge-based, so an insufficient supply of workers with the right competencies is a challenge for companies competing in local and global markets and threatens growth potential.

2.6 Personal benefits of teaching soft skills

Soft competencies are defined as the skills a person acquires, both during the period of education and throughout the entire life. To determine the benefits that are associated with the teaching of these competencies, it is necessary to take into account the immediate benefits that accrue to the student during the period of study, and the long-term benefits that are already evident in adult life. Another element that emphasizes the importance of soft competencies is society's expectations of an individual's resourcefulness and the requirements created in the labor market. The competencies acquired by a young person benefit him in three main areas of his life: the professional area, interpersonal relations, and self-realization.

The formation of such competencies has an impact on the development of interpersonal and self-realization skills. It creates pro-social thinking, characterized by orientation to the capabilities and needs of other people. This creates a deeper ability to observe the changing behavior and feelings of others, evaluate situations, and give and receive messages.

Students also learn how to succeed, how to fail, and how to cope. An important part of the impact of acquired soft skills is the sphere related to self-realization. A person with high social competence is able to accurately read his own needs, communicate them, and strive to meet them. Such people are assertive, direct, and have a high plasticity to assimilate changes. As a result, they have a greater ability to control their emotions and low rates of anxiety and depression. In addition, self-knowledge and self-awareness give them greater ability to control their lives, self-development, make more accurate choices, and plan their career paths more accurately. A person who has a high degree of self-awareness can also effectively use the various self-motivation techniques necessary to better achieve goals. Such a person demonstrates internal consistency, is consistent, and is responsible for his actions.

Teaching soft skills also has benefits in the area of marketing, as the young person learns personal branding, negotiation, and cooperation. These are key skills in the area of building one's own image and positioning oneself in the job market. Personal branding is often a key element when looking for a job in a profession. It is a kind of self-presentation that can be practiced through various speeches or activities. Young persons learn to present themselves at their best and to exercise control over how others perceive them. Such behavior shapes a desirable personal identity, increases self-esteem, and brings material or social benefits related to the perception of our image by others.

Among the soft skills that are or should be taught are intercultural competencies, which prepare individuals to participate constructively and effectively in society not only as a citizen of a country but also as a citizen of the world. A young person learns interpersonal communication in different environments by listening attentively, speaking, and interacting with people from other cultures. Related to this is the ability to express and understand different points of view, widely referred to as tolerance. By activating multicultural programs, students have the opportunity to learn about the cultural heritage of other countries. Such skills can later be used in working for foreign companies or in domestic companies on international projects. Foreign language skills are also a valuable skill.

Going a rung higher in terms of career, it is advantageous to have competencies that enable a person to perform a managerial or executive function. In this case, the qualities valued by employers are leadership competencies, enabling the efficient implementation of the company's mission and vision, as well as managing the team of people entrusted to the person. Through the introduction of various simulations, workshops, and discussions that include modeling of behavior, a young person has the opportunity to develop skills that improve delegation of tasks, organization of individual's and the team's time, and change management. Recruitment specialists emphasize that a high level of soft skills is very important for managers and entrepreneurs, as it determines the quality of work performance and efficient management of employees.

The effective functioning of an individual in society depends on the soft skills he or she possesses. Educational institutions are an excellent environment for developing them, as students have the opportunity for a large number of social experiences, through which they receive information to verify their own behavior. The teaching of soft skills has widely demonstrated benefits for young people, both related to their functioning at school, in society, and later in professional life. Possession of such skills positively affects interpersonal and professional relationships, as well as self-realization. A young person with social skills is more confident in himself, his needs, and his abilities, so he can manage his life and career more efficiently.

It also affects relationships with others through the ability to read their emotions and respond appropriately. Teaching soft skills also contributes to improving interpersonal communication and shaping the ability to work in a team, manage time, and delegate and accept tasks. These are abilities that are essential and sought after in future employees. The above-mentioned benefits are only a part of the gains that a person with soft skills obtains. Therefore, educational systems are encouraged to introduce subjects and programs that support the development of these competencies.

The Recommendation of the European Parliament and the Council of the European Union of December 18, 2006, on key competencies for lifelong learning urges member states to develop key competencies as part of a lifelong learning strategy [5]. The annex to this Recommendation was the European Commission's document entitled "Key Competences for Lifelong Learning. European Reference Framework" [6]. Within the framework of this document, eight key competencies, which are a combination of knowledge, skills, and attitudes, were defined, among them also social competencies, as necessary for self-realization and personal development. Due to the development of modern technologies and the changes that have since taken place in society, there was a need to update this document. A proposal for a revised recommendation was adopted by the College of Commissioners in 2018 as part of a package on the future of learning, and subsequently adopted by the Council of the European Union. Among other things, the update took into account the increased importance of soft competencies, which play an important role in preparing for life in today's societies.

References

[1] Toffler A. *The Third Wave*. Warsaw: PIW; 2006. pp. 39–40.

[2] European Commission. Communication from the Commission to the European Parliament, the Council, the European Economic and Social Committee and the Committee of the Regions. A Digital Agenda for Europe. Brussels. 2010; pp. 27.

[3] Stewart D. (ed.). *The Practice of Management. How to manage yourself, others and the company*, Warsaw: PWE; 1994. p. 33.

[4] McKinsey Digital. *Industry 4.0 after the initial hype. Where manufacturers are finding value and how they can best capture it*. McKinsey & Company; 2016.

[5] European Commission. *Recommendation of the European Parliament and of the Council of December 18, 2006 on key competencies for lifelong learning (2006/962/EC)*. Brussels. 2006.

[6] European Commission. *Key Competences for Lifelong Learning. European Reference Framework*. Luxembourg: Office for Official Publications of the European Communities. 2007.

Chapter 3

What do companies and the labor market demand? What do universities offer?

Almudena González Costas[1] *and Itziar Goicoechea Castaño*[2]

3.1 Introduction

Soft skills are increasingly in demand in the labor market, and higher education institutions should incorporate them among the competencies they provide to graduates. But, which are the most valued soft skills among the complete catalogue covered in this book?

In 2018, as part of the *UNIFORS 2020* project, funded by the Erasmus+ program of the European Union, the project consortium launched a survey to identify the most in-demand soft skills among companies in six European countries: *Belgium, Greece, Poland, Portugal, Romania, and Spain*. The objective was to gain a deeper understanding of the competencies valued in the labor market and to assess potential skill gaps that could impact employability and workforce readiness.

Simultaneously, the project consortium launched a parallel survey in *Higher Education Institutions* to gather insights from students about which soft skills they considered most important for their professional development. Additionally, the survey examined how difficult it was for students to find training opportunities or courses that would help them acquire these essential skills. The findings aimed to bridge the gap between the expectations of employers and the perceptions of future professionals, fostering better alignment between academia and the labor market.

In 2024, the same survey was conducted again, allowing for a comparative analysis of results over time. This new assessment sought to evaluate the evolution of skill demands at the European level, detect emerging trends, and analyze the possible effects of the *COVID-19 pandemic in 2020* on workforce requirements and educational priorities. Given the significant disruptions caused by the pandemic, the study aimed to determine whether employers' expectations had shifted and whether students' awareness of soft skills had changed because of the evolving job market landscape.

[1]PhD Program on Creativity and Social and Sustainable Innovation, Universidade de Vigo, Spain
[2]Departamento de Deseño na Enxeñaría, Universidade de Vigo, Spain

Finally, to obtain the most accurate and up-to-date picture of the current labor market, the collected data was compared with public sources, such as the *real-time analysis of online job advertisements (OJAs)* conducted by *CEDEFOP* in collaboration with *EUROSTAT*. This integration of multiple data sources provided a broader perspective on labor market trends, highlighting which skills were actively sought by employers and how these demands aligned with the perceptions of students and educational institutions.

3.2 Methodology

The initial questionnaires, developed in 2018, were based on the ET2020 PLA Report "Developing future skills in higher education," published by the European Commission in 2016 [1]. The report categorizes the main skills that the European Commission identified as critical for higher education training into three main groups:

- **Cognitive skills**: analytical, critical, reflective, and creative thinking.
- **Methodological skills**: time management, problem-solving, decision-making, learning strategies, planning, and digital skills.
- **Social skills**: interpersonal communication, teamwork, conflict management and negotiation, and intercultural understanding.

Since the report itself did not contain the definition of the skills, the definitions provided were sourced from the EU Skills Panorama and The Occupational Information Network (O*NET) [2], developed under the sponsorship of the U.S. Department of Labor/Employment and Training Administration (USDOL/ETA).

Each skill was defined as in Section 1.1, and participants in the questionnaire were asked to rate:

- The *importance* of each skill on a scale from 0 to 5, where 0 meant "*I don't have an opinion*" and 5 meant "*It is very important.*" This rating was collected from both companies and students.
- The *difficulty* of finding courses to learn these skills was also on a scale from 0 to 5, where 0 meant "*I don't have an opinion*" and 5 meant "*They are very difficult to find.*"

The same survey was conducted in 2024 to obtain comparable results.

3.3 Main findings from 2018 surveys

3.3.1 Main findings from the 2018 survey for companies

Once the company surveys were conducted, the first step was to analyze the profile of the respondents to later draw meaningful conclusions from the collected data.

Regarding company size, approximately 35.70% of the companies surveyed are large enterprises (with 250 or more employees). As shown in Figure 3.1, the remaining companies form a balanced mix of medium-sized enterprises, small enterprises, and micro-enterprises.

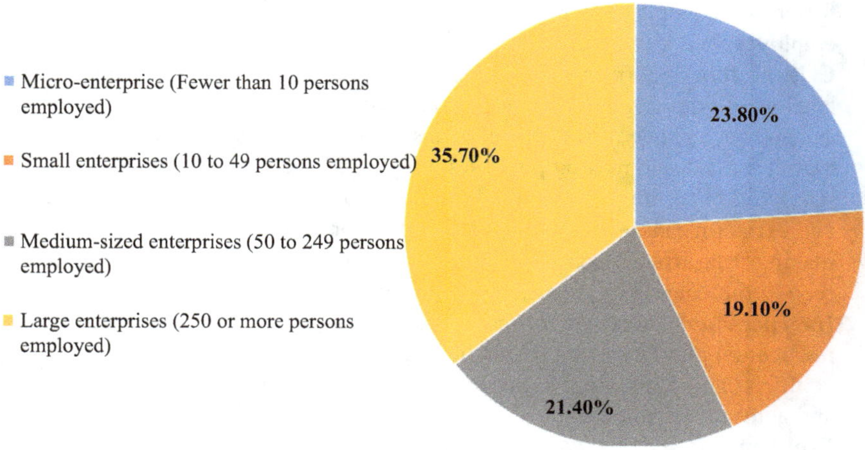

Figure 3.1 *Distribution of surveyed companies by size (%)*

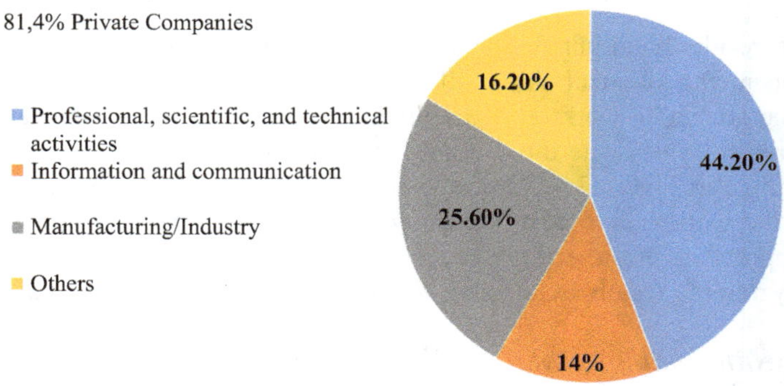

Figure 3.2 *Distribution of company sectors (%)*

Of the companies responding, 81.4% were private enterprises, distributed across various sectors, as shown in Figure 3.2. The distribution by sector was as follows: 44.2% of the companies surveyed were engaged in professional, scientific, and technical activities, followed by Information and Communication with 25.60%.

Additionally, nearly 80% of them had hired or offered internships to recent graduates or final-year university students in the past year. This is particularly relevant, as these companies have firsthand knowledge of the soft skills that newly graduated professionals bring to the workforce, giving them valuable insight into the actual skill set that new employees possess.

Analyzing the results, the five most in-demand soft skills among companies were as follows:

- **Teamwork** (*social skill*)—rated 4.13/5. Collaboration is a key factor in the success of most modern organizations. This skill is highly valued because it fosters a cooperative work environment, increases productivity, and ensures employees work toward shared goals.

- **Critical thinking** (*cognitive skill*)—rated 3.98/5. Companies are looking for employees who can independently evaluate information, challenge assumptions, and approach problems with a reasoned, systematic mindset. This skill is especially crucial for decision-making processes in dynamic and competitive industries.
- **Analytical thinking** (*cognitive skill*)—rated 3.96/5. The ability to analyze data and extract meaningful insights is essential for solving complex problems and making data-driven decisions. This skill is especially sought after in industries relying on technology, research, or strategic planning.
- **Interpersonal communication** (*social skill*)—rated 3.92/5. Effective communication plays a central role in building professional relationships, resolving conflicts, and facilitating clear and concise exchanges of ideas within teams and across departments.
- **Problem-solving** (*methodological skill*)—rated 3.85/5. Employers place significant value on employees who can identify and resolve issues efficiently. This skill enables individuals to find practical solutions to challenges, improve processes, and contribute to the overall success of the organization.

These results highlight that companies are prioritizing a blend of cognitive, social, and methodological skills, all of which are essential for fostering a productive and adaptable workforce. Employees who possess these skills are seen as better equipped to navigate the evolving demands of the modern labor market.

As shown in Figure 3.3, companies consistently rate the five most important skills for the workplace at approximately *four out of five points*, highlighting the critical importance of these skills in professional environments.

Analyzing the results by country, the same pattern is observed in all of them.

3.3.2 Main findings from the 2018 survey on university students' opinions: which soft skills are most important

Next, we will analyze the results of the surveys conducted among students from different countries. The demographic profile of the respondents is as follows: 42% of the respondents are pursuing technical degrees, as illustrated in Figure 3.4.

Regarding gender, there is a relatively balanced distribution, with 46% of the respondents being women and 54% men. Additionally, 56% of the respondents have had some form of work experience within companies, indicating a notable level of practical exposure to the workforce. Furthermore, 31% of the respondents have participated in Erasmus exchange programs at other universities, highlighting the international mobility and experience gained by a significant portion of the student population.

The results obtained are presented below. Regarding the skills that students consider most important for their future careers, the following are rated highly:

- **Problem-solving** (*methodological skill*)—rated *3.98/5*: This skill is seen as crucial for overcoming challenges and adapting to dynamic work environments.

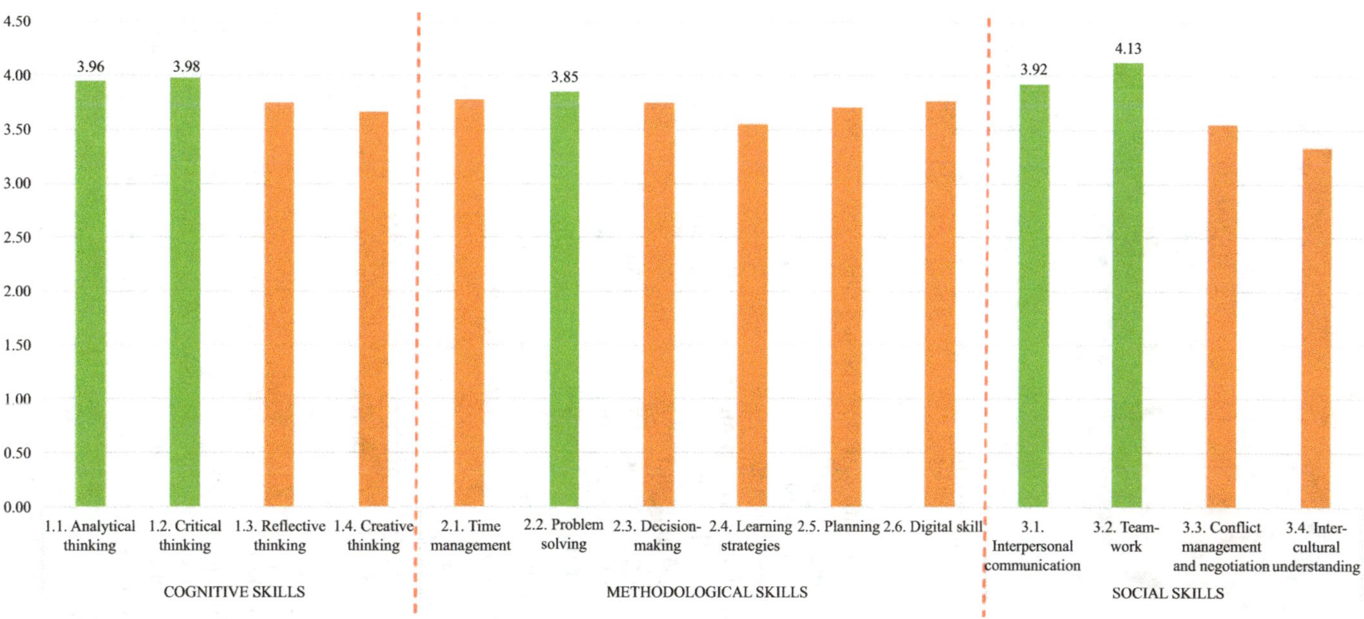

Figure 3.3 Comparison of the most valued soft skills by companies

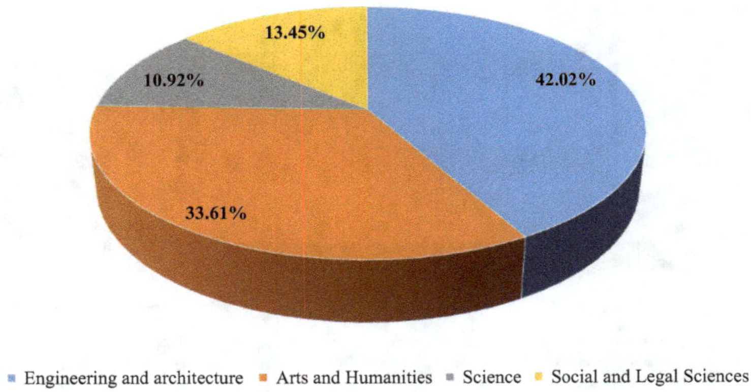

Figure 3.4 Distribution of study areas among respondents (%)

- **Teamwork** (*social skill*)—rated *3.97/5*: Collaboration and the ability to work effectively in teams are essential for success in the modern workplace.
- **Critical thinking** (*cognitive skill*)—rated *3.96/5*: The capacity to analyze, evaluate, and make reasoned decisions is highly valued by students for their professional development.
- **Interpersonal communication** (*social skill*)—rated *3.91/5*: Strong communication skills are important for building relationships and conveying ideas clearly in professional settings.
- **Decision-making** (*methodological skill*)—rated *3.91/5*: The ability to make informed and effective decisions is seen as a key competence for managing tasks and responsibilities in a career.

These results reflect the students' awareness of the importance of these skills for their future employability and career success.

3.3.3 Main findings from the 2018 survey on university students' opinions: access to soft skills training

Now, analyzing the students' opinions regarding the difficulty of acquiring these skills or their ability to find courses for learning them, the students indicated the following (Figure 3.5):

- **Creative thinking** (*cognitive skill*)—rated *3.14/5*: Students find it moderately challenging to find courses or training opportunities focused on developing creative thinking, which they consider essential for innovation in their future careers.
- **Critical thinking** (*cognitive skill*)—rated *3.06/5*: Although critical thinking is highly valued by students, they report some difficulty in locating specialized courses to help them sharpen this skill, indicating a gap between demand and availability.
- **Decision-making** (*methodological skill*)—rated *3.05/5*: Students feel that courses aimed at enhancing decision-making skills are somewhat hard to come

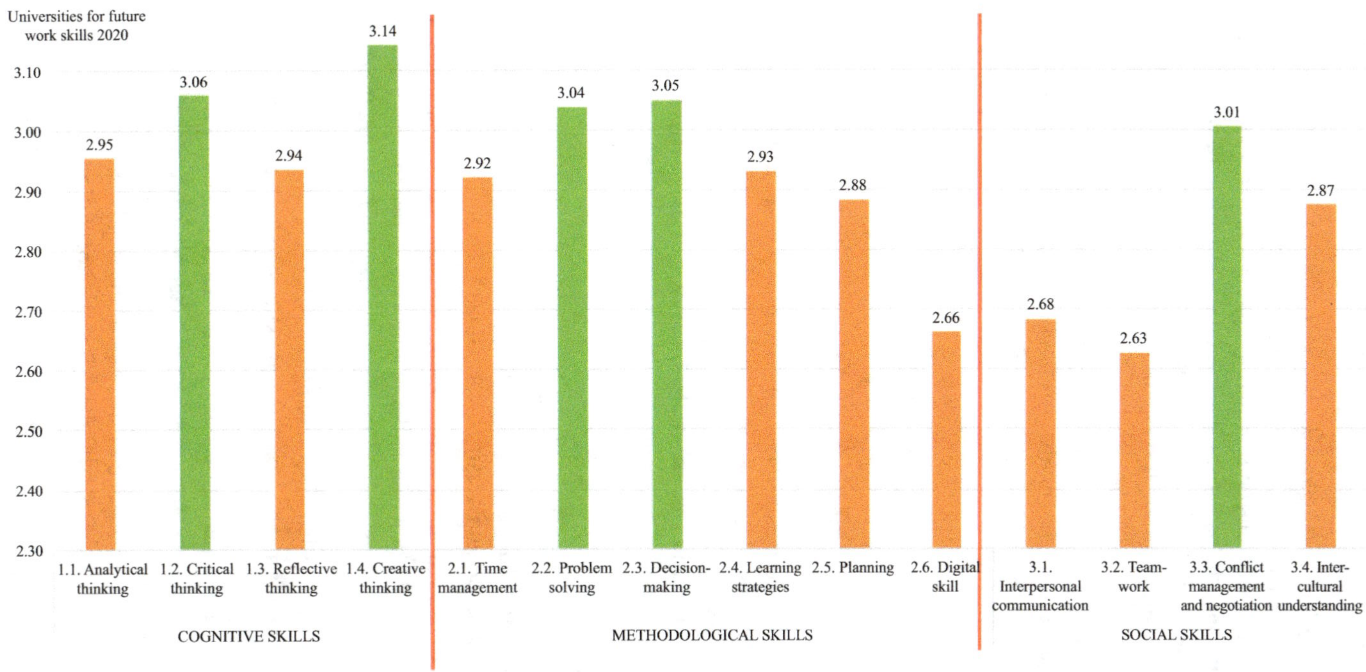

Figure 3.5 Student perception of difficulty in acquiring or finding training courses for key soft skills

by, even though they recognize their importance in both academic and professional settings.

- **Problem solving** (*methodological skill*)—rated *3.04/5*: While problem-solving is frequently emphasized as a key skill, students encounter challenges in finding structured courses that focus on practical, hands-on problem-solving strategies.
- **Conflict management** (social skill)—rated *3.01/5*: This skill, which is crucial for managing workplace relationships and navigating challenges within teams, is perceived as difficult to develop through available courses, despite its importance for career success.

These results suggest that while students recognize the importance of these soft skills for their professional growth, they face barriers when it comes to accessing relevant training programs or courses that could help them acquire these competencies effectively.

3.4 Main findings from 2024 surveys

3.4.1 Main findings from the 2024 survey on companies' perception

The surveys directed to companies in 2024 were programmed with the same content as those conducted in 2018. They were sent through the same communication channels across several countries. Data collection took place over a period of 3 months, from September to December 2024. Below is a summary of the top five soft skills most in demand by companies:

- **Teamwork** (*social skill*)—rated 4.47/5.
- **Analytical thinking** (*cognitive skill*)—rated 4.27/5.
- **Critical thinking** (*cognitive skill*)—rated 4.23/5.
- **Digital skills** (*methodological skill*)—rated 4.13/5.
- **Creative thinking** (*cognitive skill*)—rated 4.00/5.

The list reflects current labor market trends, emphasizing the need for a balance between social and cognitive skills. There is a noticeable increase in the demand for cognitive skills, as creative thinking has risen in its rating from 3.5 to 4 compared to the 2018 surveys.

Teamwork continues to lead the ranking, underscoring the importance of collaboration in work environments, while digital skills stand out as a necessity in an increasingly technological world—particularly emphasized by the impact of the pandemic. It is noteworthy that companies place greater value on cognitive skills such as analytical, critical, and creative thinking, suggesting that professionals must combine structured analysis with innovation to excel. This multifaceted approach is essential for addressing the dynamic challenges of today's labor market.

After analyzing the collected data, the following conclusions can be drawn:

- A noticeable increase in the value companies assign to soft skills is observed when compared to the surveys conducted in 2018, prior to the pandemic. This implies that companies are now more aware of the importance of these skills. Specifically, the average rating has increased by 7%.

- The most in-demand soft skill for companies continues to be *Teamwork* (the rating increased from 4.13 to 4.47).
- In the 2024 surveys, *Digital skills* appear within the top five most demanded soft skills, ranking fourth. This is a significant shift when compared to the pre-pandemic surveys in 2018, where it did not appear at all.
- Among the five most in-demand skills, three are cognitive skills, which account for 60% of the total. This highlights the growing importance of abilities related to analytical, critical, and creative thinking in today's workplace.
- The preference for cognitive skills indicates that companies increasingly value professionals who can analyze information in a structured way, solve complex problems, and deliver innovative solutions. This emphasis reflects a labor market that seeks to adapt to the complexity and rapid pace of technological, social, and economic changes, where strategic thinking and creativity are essential to staying competitive.

Furthermore, this trend underscores the need to develop and strengthen these skills, both through academic training and continuous professional development, to meet the demands of a globalized and technologically advanced market.

3.4.2 Main findings from the 2024 survey on university students' opinions: which soft skills are most important

The 2024 survey on university students' perceptions of soft skills highlights a strong preference for cognitive and methodological skills, with problem-solving emerging as the most valued skill, rated at 4.10 out of 5. This suggests that students recognize the importance of structured approaches to overcoming challenges, a critical competency in both academic and professional settings.

Closely following, critical thinking (4.06) and analytical thinking (4.01) reinforce the emphasis on cognitive abilities, indicating that students highly value the capacity to evaluate information and make reasoned decisions. Reflective thinking (3.93) also ranks prominently, suggesting an appreciation for self-assessment and continuous learning.

On the social skills front, interpersonal communication and teamwork received equal ratings (3.91), demonstrating their perceived importance in collaborative environments. While slightly lower than cognitive skills, these competencies remain essential for workplace success, particularly in team-based and interdisciplinary contexts.

Overall, the results reflect a balanced recognition of both individual cognitive abilities and collaborative social skills, with problem-solving standing out as the most crucial for university students. These insights can inform educational strategies aimed at strengthening students' preparedness for future professional challenges.

- **Problem-solving** (*methodological skill*)—rated *4.10/5.*
- **Critical thinking** (*cognitive skill*)—rated *4.06/5.*

- **Analytical thinking** (*cognitive skill*)—rated *4.01/5*.
- **Reflective thinking** (*cognitive skill*)—rated *3.93/5*.
- **Interpersonal communication** (*social skill*)—rated *3.91/5*.
- **Teamwork** (*social skill*)—rated *3.91/5*.

3.4.3 Main findings from the 2024 survey on university students' opinions: access to soft skills training

The results of the 2024 survey on university students' access to soft skills training reveal that while these competencies are recognized as important, the availability of training opportunities varies across different skill areas.

Conflict management received the highest rating (3.35/5), suggesting that some training is available but may not be widespread. Given the increasing need for effective communication and problem resolution in professional environments, this result highlights the potential for expanding such programs.

Cognitive and methodological skills, including creative thinking (3.12), decision-making (3.11), and time management (3.06), received moderate ratings, indicating that students may have access to some training but likely feel there is room for improvement. Notably, critical thinking and reflective thinking were rated the lowest (both at 3.05), despite being highly valued in professional and academic settings. This gap suggests that while these skills are deemed essential; however, structured training opportunities may be limited.

Overall, the findings point to a need for enhanced training programs that address these critical soft skills, particularly in areas such as decision-making, time management, and cognitive flexibility. Strengthening these training opportunities could better equip students for both academic success and future workplace challenges.

- **Conflict management** (*social skill*)—rated *3.35/5*.
- **Creative thinking** (*cognitive skill*)—rated *3.12/5*.
- **Decision making** (*methodological skill*)—rated *3.11/5*.
- **Time management** (*methodological skill*)—rated *3.06/5*.
- **Critical thinking** (*cognitive skill*)—rated *3.05/5*.
- **Reflective thinking** (*cognitive skill*)—rated *3.05/5*.

3.5 Main findings: comparison of surveys with public data to analyze the current labor market

The European Centre for the Development of Vocational Training (CEDEFOP), in collaboration with Eurostat, has created the Skills-OVATE [3] digital tool, which offers detailed information about job and skills demand based on OJAs in 32 European countries. Since not all job portals provide stable and comparable data, the tool only analyzes the OJAs published on the EURES portal, and other job offers are excluded from the analysis.

The skills are displayed in two classification systems, ESCO version 1 and O*NET. For consistency with previous research, all skills are displayed here

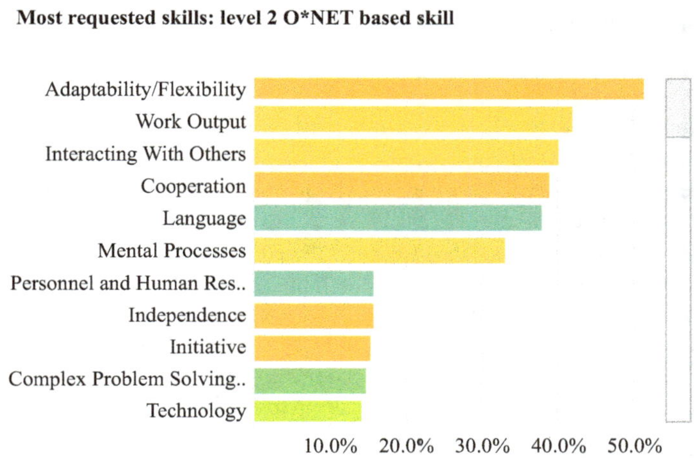

Figure 3.6 *Most requested skills 2 O+NET-based skill classification. Source: CEDEFOP.*

according to the O*NET classification. The data presented cover 15.480.991 OJAs from the last quarter of 2023 (Q4-2023) through the third quarter of 2024 (Q3-2024).

Figure 3.6 summarizes the most requested skills at the European level, calculated by counting the number of times each skill appears in job offers relative to the total number of published offers.

According to the available data, 51.4% of the OJAs published during that period included the adaptability/flexibility skill as a requirement, making it the most demanded skill. The second skill, work output, includes several level 3 O*NET skills such as interacting with computers or handling and moving objects, which are more linked to work activities than to soft skills.

The third skill, interacting with others, which appears in 40.2% of the OJAs, includes other level 3 O*NET skills such as communication with people from outside the organization and developing and building teams, both of which are soft skills.

Cooperation skills are required in 38.9% of the OJAs, while mental processes, which include level 3 O*NET skills such as thinking creatively, analyzing data, making decisions, and solving problems, are listed in 33.2% of the OJAs.

Independence, at 15.8%, and complex problem solving, at 14.8%, complete the list of the 11 most required skills at the European Level.

At first glance, it is clear that soft skills are important in the job market, with flexibility and adaptability being the most important ones.

When comparing the available data with the results from the surveys launched in 2018 and 2024, we can confirm the growing importance of soft skills for companies along with the changes on the labor market like the rise in the demand for digital skills or cognitive skills; that are also perceived as crucial to enter the labor market by university students.

3.6 Conclusions and recommendations

3.6.1 Key findings

Since Adaptability/Flexibility was not included in the series of skills assessed in the company surveys, nor did it appear in the European Commission document, its importance for the surveyed companies cannot be directly evaluated. However, the increasing relevance of digital skills is confirmed by the data available at CEDEFOP, which highlights the growing demand for job-related skills and the digital competencies associated with them. This rise can be partially explained by the digitalization process in companies, which was accelerated during the COVID-19 pandemic, as well as by ongoing technological developments.

Advances in Artificial Intelligence require a workforce that is not only digitally capable but also possesses cognitive skills to implement AI tools in their daily tasks from a critical and creative perspective. Businesses place high importance on cognitive abilities, including critical, analytical, and creative thinking, indicating that professionals need to combine innovation with structured analysis to succeed. The significant number of OJAs that involve cognitive abilities like creative thinking or data analysis—33.2% of all published OJAs—supports the survey findings on the growing relevance of cognitive skills.

The results of the business survey, along with the proportion of OJAs indicating that cooperation and teamwork are necessary for the role, further emphasize the significance of these abilities. This is confirmed by students' perception of the importance of social skills. Social skills remain essential in the job market, and according to the student survey, conflict resolution is the social skill for which it is hardest to find specialized training.

3.6.2 Proposals to bridge the gap between companies and universities

Based on the findings from the 2018 and 2024 surveys on university students' opinions about soft skills and access to training, several proposals can be made to strengthen the collaboration between universities and companies, ultimately ensuring that students are better equipped with the necessary skills for the professional world.

1. **Enhanced Soft Skills Training Programs**: Universities should prioritize the integration of training programs focused on essential soft skills such as conflict management, creative thinking, decision-making, and time management. Given the relatively moderate ratings for these skills, universities can collaborate with companies to design and implement specialized workshops or courses that cater to the practical needs of the workforce. In Chapter 18, a proposal for a course sheet is described.
2. **Industry-Driven Curriculum Development**: To bridge the gap between the theoretical knowledge students gain and the real-world demands, universities can partner with companies to co-develop curriculum content. This

collaboration would ensure that the courses offered are aligned with current industry trends and needs, particularly in areas like critical and reflective thinking, which receive lower ratings in terms of access to training.

3. **Internships and Work-Based Learning**: Expanding opportunities for internships, co-op programs, and other forms of work-based learning would allow students to gain practical experience in applying soft skills in real-life scenarios. These experiences would help students develop key competencies like teamwork, communication, and problem-solving, while simultaneously giving companies a chance to mold prospective employees.
4. **Mentorship and Networking Opportunities**: Companies can play a key role in fostering mentorship programs where experienced professionals guide students in developing essential soft skills. These programs would not only help students gain valuable insight into career paths but also offer hands-on support for developing skills such as conflict management and decision-making, which are crucial in professional environments.
5. **Feedback Loops Between Universities and Companies**: Establishing regular feedback mechanisms between universities and companies would help ensure that the skills students are developing align with industry expectations. Companies could provide direct feedback on students' performance during internships or on collaborative projects, offering insights on how universities can refine their soft skills training programs.

By implementing these proposals, universities and companies can work together to create a more cohesive educational experience that equips students with the soft skills most valued in the workplace, ensuring a smoother transition from academia to professional life.

References

[1] European Commission, Directorate-General for Education and Culture, "Developing future skills in higher education", *ET2020 – Peer Learning Activity (PLA)*, Brussels, 25–26 February 2016.

[2] The Occupational Information Network (O*NET). Available from https://www.onetonline.org/ [Accessed 28 Jan 2025].

[3] *Occupations and skills*. CEDEFOP. Available from: https://www.cedefop.europa.eu/en/tools/skills-online-vacancies/occupations/skills [Accessed 29 Jan 2025].

Part II

Methodological skills

Chapter 4
Time management

Anna Laska-Leśniewicz[1]

Time management is an essential skill that holds great significance in both personal and professional spheres. The ability to effectively manage time allows individuals to prioritize tasks, allocate resources efficiently, and achieve optimal productivity. In today's fast-paced world, time management becomes even more crucial for meeting goals and adhering to deadlines. By mastering time management, individuals can minimize procrastination, eliminate inefficiencies, and maintain a healthy work-life balance. Effective time management not only enhances productivity but also reduces stress, fostering a sense of control over their schedule and allocating adequate time for relaxation and self-care. It enables individuals to set realistic goals, break them down into actionable steps, and stay focused on their priorities. Moreover, time management cultivates discipline, self-motivation, and a sense of accountability, all of which are essential for success in any endeavor. By developing strong time management skills, individuals can unlock their full potential, maximize their achievements, and lead a more fulfilling and balanced life.

4.1 Introduction

Time management plays an important role in professional and private life. In general, the term "time management" refers to the ability to use time effectively, focusing on making the most of available time. Time management is the process of planning and controlling consciously how time is spent on specific activities. It involves balancing various demands, setting priorities, and managing time wisely. In the context of learning, time management can be defined as the ability to plan study time and tasks, for example, scheduling dedicated periods for learning activities [1].

According to Encyclopaedia Britannica [2], time management is explained as self-management with an explicit focus on time. It involves making decisions about what tasks to prioritize, determining how much time to allocate to various activities, finding ways to perform tasks more efficiently, and identifying the most appropriate time for particular activities. Initially, the concept of time management

[1]Institute of Mechatronics and Information Systems, Lodz University of Technology, Poland

was used in the context of how employees use their work time and to increase their productivity. In the 1950s, managers introduced lists of "dos and don'ts," which were a useful tool for improving management practices.

Time management has become increasingly critical in today's fast-paced world. Never before has the efficient use of time been so essential. Over the years, many ideas and strategies have been proposed for effective time management [3–6]. At the core of these approaches lies the principle of prioritizing tasks and developing a strategic plan to achieve them.

Time management skills are techniques that help individuals to use their time effectively and efficiently. Time management skills involve strategic planning, efficient organization, and effective prioritization to optimize time utilization and to achieve set goals. Key time management skills include the following [7–9]:

- Planning and Organizing: Creating schedules, establishing objectives, and breaking down large tasks into manageable parts.
- Prioritization: Identifying and focusing on high-impact tasks.
- Time Estimation: Accurately assessing the time required to complete tasks.
- Time Blocking: Dividing the day into focused work blocks.
- Task Management: Employing tools and techniques to manage and track tasks effectively.
- Stress Management: Developing strategies to cope with stress, maintain focus, and stay on track.
- Self-Discipline: Avoiding distractions, adhering to schedules, and maintaining focus on priorities.
- Flexibility: Adjusting to unexpected changes and reprioritizing tasks as needed.

Mastering these skills can significantly increase productivity, reduce stress, and enhance the ability to achieve set goals.

4.2 Importance of time management skills

Time management skills have long been recognized as core competencies required by workers. The ability to adapt to disrupted workplaces and surroundings has been repeatedly tested in recent years, especially during the COVID-19 pandemic. In the "Future of Jobs Report 2023," time management was identified as a key skill among analytical and creative thinking, which are leading the way in shaping the workforce [10]. The report emphasized the importance of these skills in navigating the changing job landscape. What is more, the European Union's "Key Competences for Lifelong Learning" framework outlines eight key competencies, including personal, social, and learning-to-learn skills, which encompass time management [11]. These skills are crucial for personal development, professional success, and active citizenship. Harvard Business Review has also published plenty of articles emphasizing time management as a critical skill for leadership and productivity. The topic is interesting not only from an academic point of view but also highly relevant from a practical, day-to-day perspective. For instance, Vasundhara [12] reported that the phrase "manage time" garners a monthly search volume of 60,500 on Google (the number of times a keyword is searched for in a specific time frame), significantly surpassing searches for "make/earn more money," which stands at just 260.

From the perspective of university studies, the ability to organize time is one of the crucial skills for achieving set goals [13]. At a basic level, time management covers attending scheduled classes, events, and exams, as well as meeting deadlines for assignments. While studying, time management skills can be developed in the meantime and further transferred to different aspects of life [9]. In the following areas of academic activities, such skills are tested and improved: attending lectures, seminars, and tutorials; delivering oral presentations; preparing for exams and revisions; completing tasks and exercises with deadlines; conducting research; and engaging in personal development planning.

Time management competencies and their influence on various aspects have been widely investigated in many studies. Self-organization, based on time management competence, is among the behaviors that enable individuals to cope with the workload [14]. According to [9], the authors found that individuals apply time management practices more extensively as they gain experience with age and take on multiple roles. Moreover, higher time management skills are associated with better use of time and reduced procrastination. Several studies have identified a significant positive relationship between time management and academic achievement [1,15–18]. Furthermore, there is also a link between time management and motivation. Time management skills contribute to maintaining motivation [19] and have a positive impact on well-being. This is mainly achieved through the perception of better time control, which leads to improved job satisfaction, reduced stress and anxiety, and better psychological state [9,20].

4.3 Useful methods in time management

Several time management methods and techniques can be applied in both personal and professional life. The most well-known methods include:

- Pomodoro technique,
- Eisenhower Matrix,
- Eat that Frog,
- Time blocking,
- Rapid Planning Method (RPM),
- Pareto rule (80/20 Rule), and
- SMART goal system.

This section provides descriptions of these techniques along with some lesser-known methods.

4.3.1 Pomodoro technique

The Pomodoro technique was developed by Francesco Cirillo over 30 years ago [21]. The method was created as a consequence of the author's own problems with

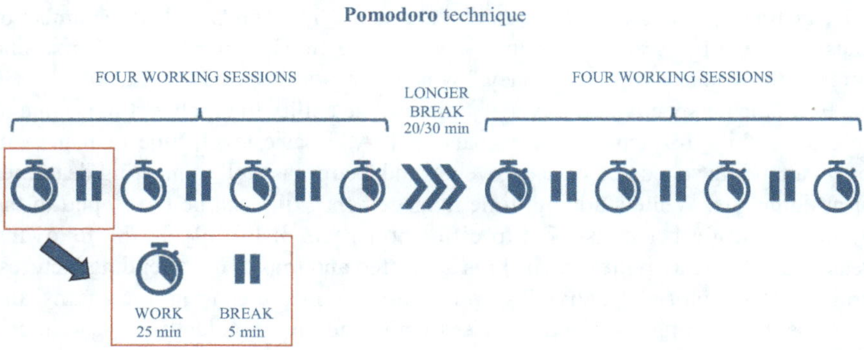

Figure 4.1 The steps of the Pomodoro technique

concentration while studying. As a university student preparing for a sociology exam, Cirillo observed that he was easily getting distracted and focusing on the learning material was a great challenge for him. Implementing shorter, concentrated study periods allowed him to improve concentration, productivity, and efficiency over time. By trial and error, he created an optimal work plan consisting of intervals known as "*pomodoros*" (named after the Italian word for "tomato," inspired by a tomato-shaped kitchen timer he used). The technique involves working in focused intervals, taking short breaks between each *pomodoro*, and incorporating a longer break after every four pomodoros.

The Pomodoro method is based on a concept of working in time intervals. Each working session lasts 25 minutes, and it is followed by a 5-minute break. After every four work sessions, a longer break of 20 or 30 minutes is recommended. The steps of the Pomodoro technique are illustrated in Figure 4.1.

Nowadays, there are several apps that are based on the idea of the Pomodoro technique, such as Flora, Fauna, Focus Keeper, Toggl, and others.

4.3.2 The Eisenhower Matrix

The Eisenhower Matrix, also known as the Time Management Matrix, is a time-management framework designed to help prioritize a list of tasks or agenda items. The matrix was created under the inspiration of the 34th President of the United States—Dwight Eisenhower, who was renowned for his productivity. His famous quote, "What is important is rarely urgent, and what is urgent is rarely important" serves as the foundation of this matrix. Stephen Covey presented Eisenhower's concept as a simple tool, a matrix, in his book "The 7 Habits of Highly Effective People" [22].

In the Eisenhower Matrix, there are only two factors taken into consideration: urgency and importance. Evaluating these two factors, daily tasks can be divided into four categories:

1. **Urgent and Important:** tasks that require immediate attention and completion.
2. **Not Urgent and Important**: tasks that should be scheduled on a calendar.

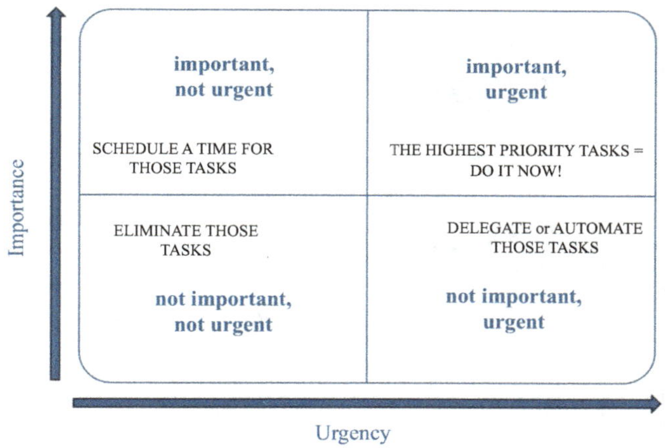

Figure 4.2 The Eisenhower Matrix

3. **Urgent and Unimportant:** tasks that can be delegated to someone else.
4. **Not Urgent and Unimportant**: tasks that can be deleted.

A graphical representation of this matrix is shown in Figure 4.2.

Urgent tasks require immediate action. These matters arise unexpectedly and demand action as soon as possible. While it is impossible to avoid those tasks entirely, spending too much time on them can produce stress and result in eventual burnout. On the other hand, important tasks are related to long-term goals and values, contributing to a sense of empowerment and self-esteem.

To implement the Eisenhower Matrix, each task should be labeled as either "important" or "urgent." The highest priority tasks will get both labels—"important" and "urgent." In contrast, the lowest priority tasks will have no labels. Applying the Eisenhower Matrix to prioritize tasks allows individuals to avoid the "mere urgency" trap and focus more on what is truly important to them.

4.3.3 Eat that frog

The "Eat that frog" strategy was popularized by Brian Tracy [23]. The method is said to be inspired by a quote attributed to a famous writer, Mark Twain "Eat a live toad the first thing in the morning, and nothing worse will happen to you the rest of the day." The "eat the frog" metaphor encourages tackling the most challenging task of the day first. This approach helps minimize anxiety and procrastination, making the rest of the day feel more manageable.

The "Eat that frog" method helps to identify difficult tasks and tackle them directly. In this framework, the frog represents an important and challenging task. The basic idea is to identify one major task (the frog) and complete the task, first thing in the morning ("eating it"). If you have two challenging tasks, the method suggests dealing with the more difficult one first (rule no. 1, Figure 4.3). The second rule of the method is about starting "eating the frog" as early as possible.

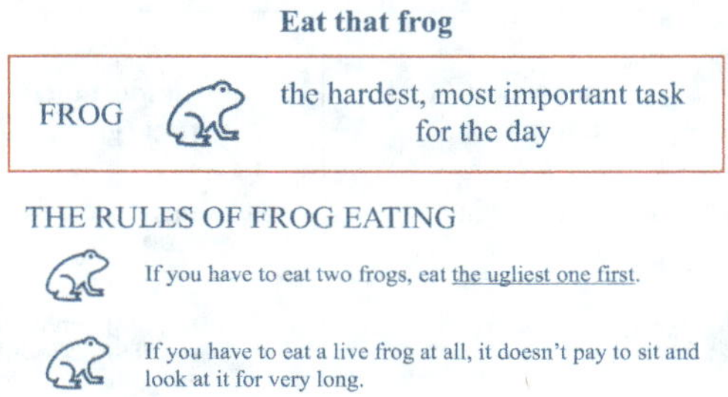

Figure 4.3 The rules of Eat that frog technique

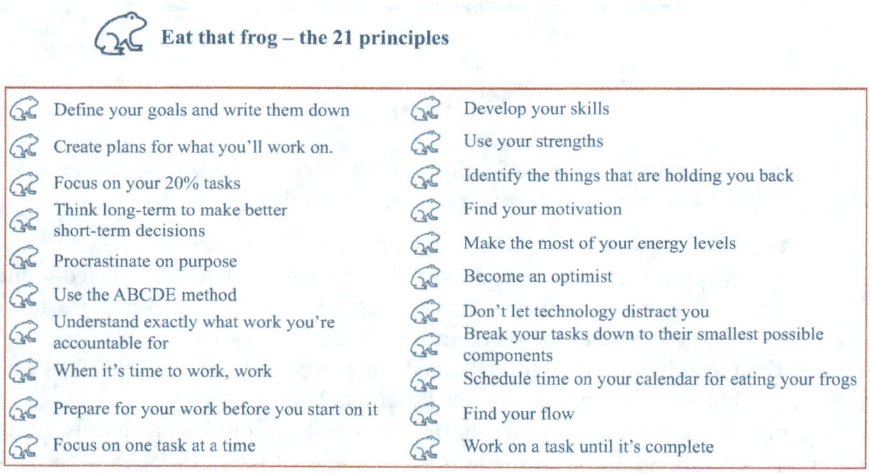

Figure 4.4 The 21 principles of the "eat that frog" method

This ensures that the hardest task is out of the way, allowing for greater focus and productivity throughout the rest of the day.

The 21 principles were proposed by Brian Tracy in his book [23], Figure 4.4 presents all of them.

"Eat that frog" method is similar to another one (not described in this chapter)—the MIT method (Most Important Tasks), which is based on identifying the most important tasks and focusing on the two or three most important tasks for the day.

4.3.4 Time blocking

Time blocking is a method that boosts productivity, enhances focus, and increases organization by categorizing tasks and providing a clear structure. Figure 4.5 illustrates a comparison between an unorganized, casual day and a day organized using the time blocking method.

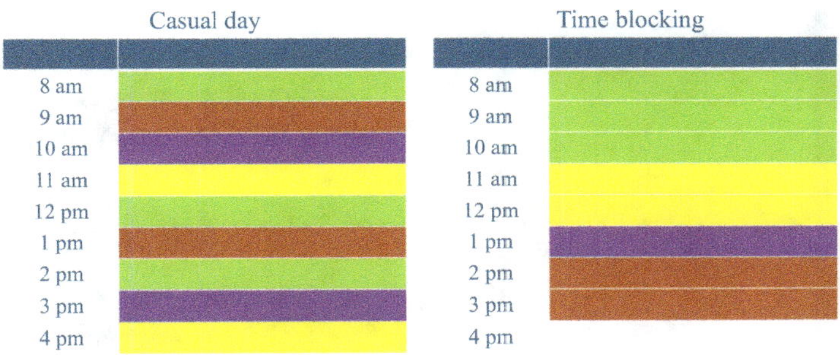

Figure 4.5 Time blocking method (right) in comparison to a non-organized multi-tasking day (left)

Time blocking occurs in various options [24]:

- **Casual time blocking**: implementing a time-blocking system where each time slot is allocated to a single, dedicated task or activity.
- **Task Batching:** grouping similar tasks together within specific time blocks to minimize the mental strain caused by frequent task switching.
- **Day Theming**: implementing a "theme day" system by assigning specific types of tasks or projects to designated days of the week.
- **Time Boxing**: working on a task for a set amount of time and aiming to complete as much as possible within that timeframe.

The time blocking method has two main rules. The first one is to use visual scheduling of time blocks on the calendar to avoid interruptions and overlapping tasks. The second rule is about Grouping similar tasks into one concentrated block of time. To apply the method, the following steps should be taken:

1. List the tasks.
2. Prioritize the tasks.
3. Estimate the time needed for each task.
4. Block out time in the calendar.
5. Work with the schedule.

Time blocking helps maximize energy cycles and facilitates entering a state of deep work that involves intense and uninterrupted focus on a cognitively demanding task.

4.3.5 Rapid planning method

RPM was created by Tony Robbins, a famous American coach, book author, and public speaker. RPM is [25]:

- Result-oriented,
- Purpose-driven,
- Massive action plan.

Figure 4.6 Key steps in the RPM process

The following key questions are directly linked to the RPM method:

- WHAT: What is the final GOAL?
- WHY: What are the reasons and motives for doing that?
- HOW: What needs to be done to accomplish this? What is the plan?

The RPM is a flexible system that should be adjusted to personal needs and preferences. The key steps are shown in Figure 4.6. The process starts with capturing all thoughts, ideas, and tasks [26]. Afterward, a long-term vision for each area of life should be created. The vision is linked to a sense of purpose, as understanding the "why" behind the goals provides intrinsic motivation, according to the method's creator.

4.3.6 Pareto rule (80/20)

The Pareto rule (also known as 80/20 rule) was developed by Italian economist Vilfredo Pareto. It suggests that approximately 80% of outcomes result from 20% of causes (Figure 4.7). This highlights the imbalance between inputs and outputs. For example, according to this rule, 80% of the total profit is generated by 20% of the product categories, or 80% of the maintenance expenses are incurred by 20% of the machines. In terms of time management, the Pareto principle says that 20% of daily tasks contribute to 80% of achievements. This means that it is crucial to focus on certain tasks that contribute to success. The key element here is to identify and prioritize crucial tasks, allocate time more efficiently, and accomplish more with less effort [27].

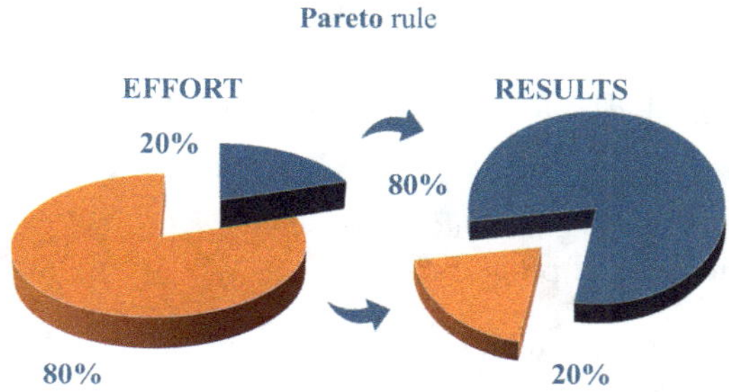

Figure 4.7 The Pareto rule—80/20 rule

By applying the Pareto rule, by identifying and prioritizing the 20% of tasks with the highest potential impact, individuals can optimize their time and enhance their overall productivity. This approach emphasizes focusing on the most critical activities while delegating or eliminating less impactful tasks.

4.3.7 SMART goal system

The SMART goal system was originally developed by George T. Doran and described in his article "There is a SMART Way to Write Management Goals and Objectives" [28]. The SMART stands for Specific, Measurable, Achievable, Realistic, and Time-bound (Figure 4.8). It means that each goal should possess those characteristics. The SMART goal system is a valuable tool for defining objectives and may improve time management skills by focusing on attainable goals with a clear timeline.

A specific (S) goal means that it is well-defined, clear, and unambiguous [29]. The "5 W questions" (also known as "5Ws") can help in creating a specific goal. The 5Ws are:

1. Who: Who is involved in this goal?
2. What: What do I want to accomplish?
3. Where: Where is this goal to be achieved?
4. When: When do I want to achieve this goal?
5. Why: Why do I want to achieve this goal?

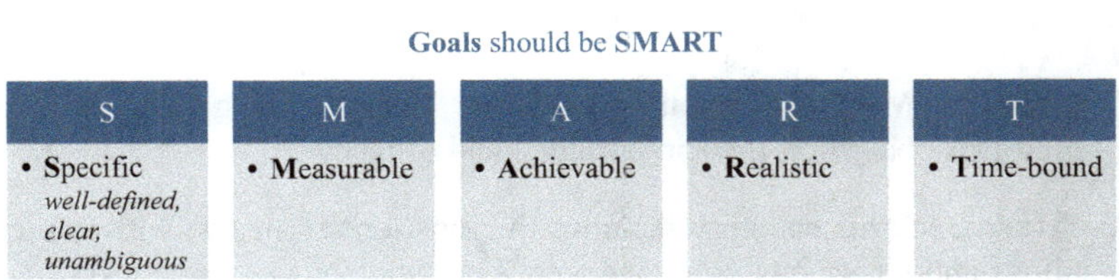

Figure 4.8 The SMART goal system—five components

The letter "M" stands for "measurable." There should be some criteria to measure the progress of a goal and to check if the activities undertaken are leading toward the desired goal. The following questions will be helpful:

1. How many/how much?
2. How will I know if the goal is accomplished?
3. What is my indicator of progress?

A SMART goal must be Achievable (or Attainable). An achievable goal means that it is possible to accomplish it. The right question to ask is, *How can I achieve it?* Achievability relates to proper analysis of available resources and capabilities, along with additional research. For example, *Have others successfully achieved this goal before?*

The next goal attribute is Realistic or Relevant (R). A realistic goal means that it can be accomplished, taking into account the available resources and time. A relevant goal is one that seems worthwhile and aligned with overall objectives.

The letter "T" stands for "time-bound," so each goal should be time-constrained, with a clear start and finish date. Setting the deadline is crucial to keep motivation at a high level.

Setting SMART goals can be beneficial because it improves time management, boosts productivity, and promotes focus. The key issue is to avoid unrealistic deadlines that cause stress and lead to burnout. With a balanced approach to meeting deadlines, the feeling of being overwhelmed will not occur. Moreover, the positive energy of success can motivate further achievements and trials.

4.3.8 Other methods

A list of methods of time management is extensive. Among other methods mentioned in the literature, the following ones are notable: ALPEN method, ABCDE technique, salami slicing method, Swiss cheese method, among others.

The **ALPEN method** was created by German economist and time management expert, Professor Lothar J Seiwert [30]. The acronym comes from the German word for "the Alps" (Alpen). This technique is designed to make efficient use of working hours with minimal effort through effective prioritization and buffer time. The ALPEN acronym contains the essentials of the method and is explained in Figure 4.9.

The **ABCDE Method** is another technique for prioritizing tasks. It involves categorizing tasks based on their importance and urgency:

- **A—Must Do:** This category includes the most critical tasks, with immediate deadlines and significant consequences if not completed. These tasks demand immediate attention.

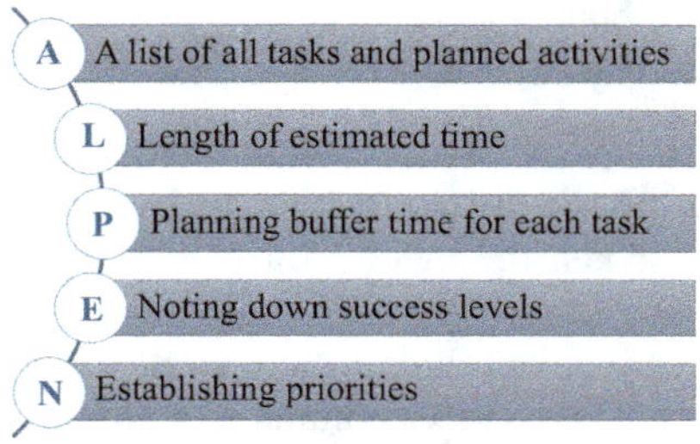

Figure 4.9 The ALPEN method elements from its acronym

- **B—Should Do:** This category contains tasks that are important but do not have immediate deadlines. These tasks contribute to long-term goals and should be addressed promptly.
- **C—Could Do:** This category involves tasks that are nice to do but not essential. These tasks can be postponed or even eliminated if time constraints arise.
- **D—Delegate:** Tasks in this category can be assigned to someone else.
- **E—Eliminate:** This category includes tasks that are neither important nor urgent and should be discarded to avoid wasting time and energy.

Two additional methods are based on a comparison with food: the salami slicing and Swiss cheese methods.

The **Salami Slicing Method** is a time management technique that consists of dividing a large task into smaller, more manageable pieces. Overwhelming tasks are broken down into smaller chunks, similar to slicing salami. By tackling one small slice at a time, you can reduce feelings of overwhelm. Completing these smaller tasks boosts motivation and reduces procrastination.

Similar benefits can also be attributed to another method called the **Swiss Cheese Method**. This method focuses on making small, incremental progress on a large or overwhelming task. The task is broken down into smaller and more manageable activities, allowing for action within a short timeframe. These "holes" (similar to those in Swiss cheese) represent small steps toward the larger goal.

4.4 Time management tools

Currently, time management tools can be categorized as either traditional or digital. Traditional tools are well-known, having been used for many years, and they are still applicable today.

Time management tools that belong to the "traditional category" are:

- Paper planners and calendars,
- Paper notebooks,
- Bullet journals,

- Whiteboards and corkboards to visually organize tasks and goals,
- Checklists and to-do lists,
- Paper agendas (e.g., meeting agendas).

They are all used to physically write down tasks, deadlines, and notes.

"Digital" tools have taken the market for over a decade. While the methods and techniques are the same as with "traditional" time management tools, digital options are often more convenient due to the prevalence of smartphones, laptops, and a digital environment. There is a vast number of applications available, making it impossible to list them all. However, these tools can be categorized according to their specific use:

- Calendar Apps help to schedule appointments, set reminders, and plan every single day,
- Task Management Apps are used to organize tasks, set priorities, and track progress,
- Time Tracking Apps allow users to monitor how time is spent and identify time-wasting activities,
- Note-Taking Apps are used to capture ideas, notes, and important information,
- Team Collaboration and Mind Mapping Tools help visualize complex information and brainstorm ideas.

Table 4.1 presents examples of the applications that support time management in both personal and professional life.

In university and project practice, various types of timelines are commonly used. These timelines allow for visualizing all tasks simply and clearly, which is especially helpful for larger projects that include multiple subtasks.

Many university and professional activities can be classified as a type of project—a temporary endeavor undertaken to create a unique result, with specific objectives defined within certain time, cost, and resource limits [31]. In such cases, using a timeline provides a clear and concise overview of the project schedule.

The most popular types of timelines are:

- Horizontal timeline,
- Gantt chart timeline,
- Kanban timeline.

Table 4.1 The exemplary apps for specific use in time management

Specific use	**Popular applications**
Calendar apps	Google Calendar, Outlook Calendar, Apple Calendar
Task Management Apps	Trello, Asana, Todoist
Time Tracking Apps	Toggl, Clockify, RescueTime
Note-Taking Apps	Evernote, OneNote, Notion
Team Collaboration Tools including Mind Mapping Tools	MindMeister, Coggle, Miro

A **timeline** is a simple tool to provide a visual representation of a schedule, helping to organize and track all tasks, deadlines, and goals. An example of a timeline for a student project is presented in Figure 4.10.

Gantt chart timelines provide a quick overview of an entire project, including task end dates, key project milestones, task dependencies, and the team members assigned to each task. A basic version of the Gantt chart includes all tasks, their duration, and their start and end dates, as shown in Figure 4.11.

Kanban timeline is an extended version of the Kanban board that adds a time dimension, providing a more comprehensive view of work progress. It helps teams better plan, track, and manage their projects. The Kanban timeline may vary depending on the tools used and the specific features of the project. Figure 4.12 presents a Kanban board with key dates for the current stage of the exemplary project.

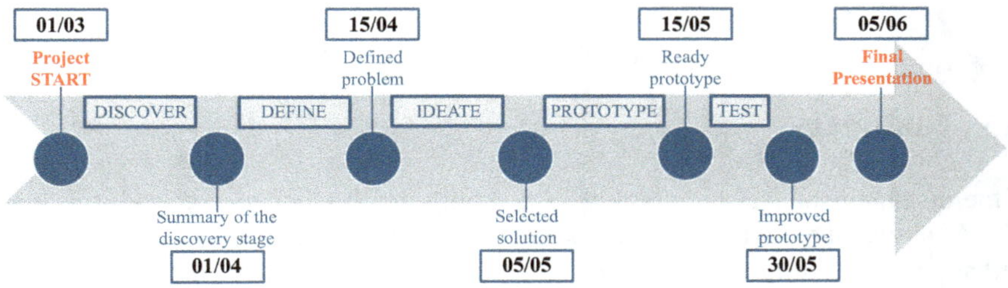

Figure 4.10 An example of a timeline for a student project based on the Double Diamond process

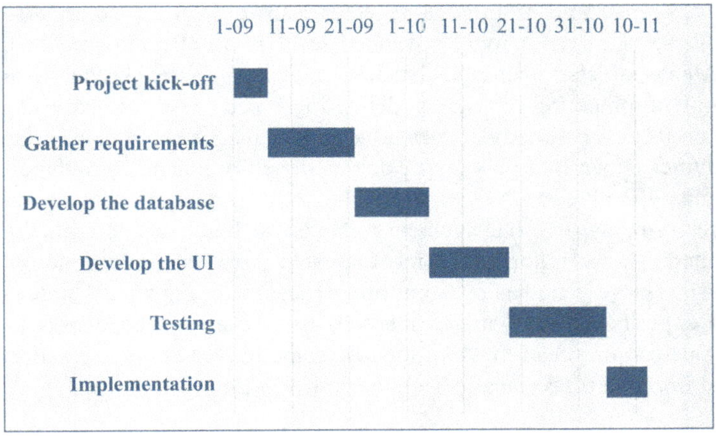

Figure 4.11 An example of a Gantt chart for an IT project

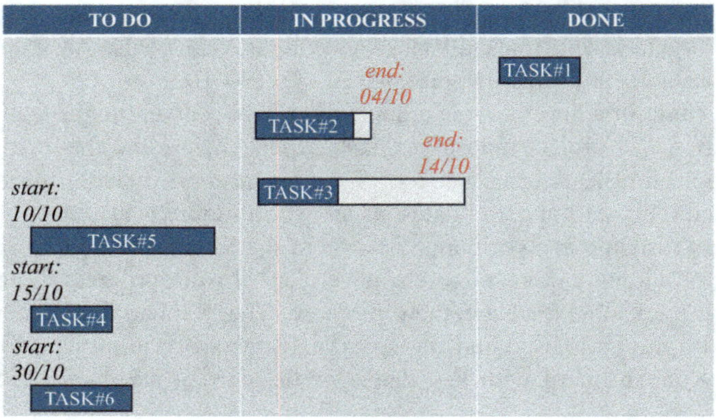

Figure 4.12 An example of a Kanban board enriched with key dates at this stage of the project

4.5 Summary

Time management skills are highly valued both in the labor market and in private life. At university and in the workplace, there are always tasks with fixed deadlines, and activities that must or should be completed within a specific time frame. In the modern world, working to a deadline and achieving results on a specific date has become unavoidable.

The development of time management skills usually starts with formal education, as students are faced with tasks that have set deadlines and must complete daily activities within specific time frames. Studying at the university requires at least basic competencies in time management; however, it is also an ideal time to improve these skills, as a more advanced level is usually required in the labor market. Many university activities, such as attending classes, gradually absorbing course content, preparing for exams, delivering timed presentations, and completing team projects, are directly connected to time management skills. The success of these activities depends largely on the degree of self-organization and the ability to manage time effectively.

There are many time management methods, techniques, and tools that can be implemented. The selection of the most suitable ones should be done on an individual basis, as everyone has different preferences, working styles, and needs. The key is to experiment with various methods to determine what works best for a particular person or situation. Additionally, some universal good practices can be applied to improve time management, as shown in Figure 4.13.

Good practices in time management

- Setting deadlines and reminders for all tasks.
- Creating a daily plan.
- Giving each task a time limit and avoiding multi-tasking.
- Blocking out distractions.
- Establishing routine.

Figure 4.13 Selected good practices in time management

References

[1] Broadbent, J. and Poon, W.L. Self-regulated learning strategies and academic achievement in online higher education learning environments: A systematic review. *The Internet and Higher Education*. 2015;27:1–13, https://doi.org/10.1016/j.iheduc.2015.04.007.

[2] van Eerde, W. Time management. *Encyclopedia Britannica*. 2024;11: https://www.britannica.com/topic/time-management. [Accessed 31 Oct 2024].

[3] Chase, J.-A.D., Topp, R., Smith, C.E., *et al.* Time management strategies for research productivity. *Western Journal of Nursing Research*. 2013;35 (2):155–176, https://doi.org/10.1177/0193945912451163.

[4] Waterworth, S. Time management strategies in nursing practice. *Journal of Advanced Nursing*. 2003;43:432–440. https://doi.org/10.1046/j.1365-2648.2003.02740.x.

[5] MacCann, C., Fogarty, G.J., and Roberts, R.D. Strategies for success in education: Time management is more important for part-time than full-time community college students. *Learning and Individual Differences*. 2012;22 (5):618–623, https://doi.org/10.1016/j.lindif.2011.09.015.

[6] Kang, C.-M., Chiu, H.-T., Hu, Y.-C., Chen, H.-L., Lee, P.-H. and Chang, W.-Y. Comparisons of self-ratings on managerial competencies, research capability, time management, executive power, workload and work stress among nurse administrators. *Journal of Nursing Management*. 2012;20:938–947, https://doi.org/10.1111/j.1365-2834.2012.01383.x.

[7] *Coursera Staff, 7 Essential Time Management Skills*, https://www.coursera.org/articles/time-management-skills [Accessed 10 Nov 2024].

[8] Hanne, K. *9 Key Time Management Skills and How To Improve Them*, https://www.indeed.com/career-advice/career-development/time-management-skills, July 2023 [Accessed 10 Nov 2024].

[9] Pluta A. and Rudawska A. Time management competency of students as prospective employees and its relation to the use of time and procrastination. *HRM [ZZL]*. 2023;155(6):12–34.

[10] World Economic Forum. *The Future of Jobs Report 2023*. Geneva: World Economic Forum; 2016. Retrieved August 16, 2023 from https://www.weforum.org/publications/the-future-of-jobs-report-2023/.

[11] European Commission: Directorate-General for Education, Youth, Sport and Culture, *Key Competences for Lifelong Learning*. Publications Office; 2019. https://doi.org/10.2766/569540.

[12] Vasundhara, S. How to manage your time: Our favorite reads. *Harvard Business Review*. 2023. https://hbr.org/2023/01/how-to-manage-your-time-our-favorite-reads [Accessed 15 Nov 2024].

[13] Cottrell, S. *The Study Skills Handbook*. 3rd ed. London: Palgrave Macmillan; 2008.

[14] Górniak, J., Kocór, M., Prokopowicz, P., and Szczucka, A. Bilans kapitału ludzkiego 2022/2021: Raport z badania pracodawców: powrót do rzeczywistości: drugi rok pandemii oczami polskich firm; 2022. Retrieved June 14, 2023 from https://power.parp.gov.pl/storage/publications/pdf/13-BKL-WCAG_15122022.pdf.

[15] Carson, A.D. Predicting student success from the LASSI for learning online (LLO). *Journal of Educational Computing Research*. 2011;45(4):399–414.

[16] ChanLin, L.J. Learning strategies in web-supported collaborative project. *Innovations in Education and Teaching International*. 2012;49(3):319–331.

[17] Xu, J., Du, J., Wang, C., *et al.* Intrinsic motivation, favorability, time management, and achievement: A cross-lagged panel analysis. *Learning and Motivation*. 2020;72:101677, https://doi.org/10.1016/j.lmot.2020.101677.

[18] Ocak, G. and Boyraz, S. Examination of the relation between academic procrastination and time management skills of undergraduate students in terms of some variable. *Journal of Education and Training Studies*. 2016;4 (5):76–84.

[19] Demırdağ, S. Communication skills and time management as the predictors of student motivation. *International Journal of Psychology and Educational Studies*. 2021;8(1):38–50, https://doi.org/10.17220/ijpes.2021.8.1.222.

[20] Claessens, B.J.C., van Eerde, W., Rutte, C.G. and Roe, R.A. A review of the time management literature. *Personnel Review*. 2007;36(2):255–276, https://doi.org/10.1108/00483480710726136.

[21] F. Cirillo, *The Pomodoro Technique. The Life-Changing Time-Management System*; London, UK: Virgin Books; 2018.

[22] Covey, S.R. *The 7 Habits of Highly Effective People: Restoring the Character Ethic* (Rev. ed.). New York: Free Press; 2004.

[23] Tracy, B., *Eat That Frog!: 21 Great Ways to Stop Procrastinating and Get More Done in Less Time*. San Francisco, CA, Berrett-Koehler Publishers; 2007.

[24] Scroggs, L. *Time Blocking*. https://todoist.com/pl/productivity-methods/time-blocking [Accessed 10 Oct 2024].

[25] Nekvinda, A. *Unlock Success with the Rapid Planning Method (RPM)*. https://hubstaff.com/blog/rapid-planning-method-rpm/ [Accessed 10 Oct 2024].

[26] Groenewegen, C. *How to Use the Rapid Planning Method at Home and Work*. https://www.usemotion.com/blog/rapid-planning-method [Accessed 10 Oct 2024].

[27] Jana, P. and Tiwari M. Lean terms in apparel manufacturing, Jana, P. and Tiwari, M. (eds.). The Textile Institute Book Series, *Lean Tools in Apparel Manufacturing*. Cambridge: Woodhead Publishing; 2021, pp. 17–45, https://doi.org/10.1016/B978-0-12-819426-3.00010-2.

[28] Doran, G.T. There's a S.M.A.R.T. way to write managements' goals and objectives. *Management Review*., 1981;70(11):35–36.

[29] CFI Team, *SMART Goals*. [Accessed 15 Nov 2024] https://corporatefinanceinstitute.com/resources/management/smart-goal/.

[30] Valentine, C. *Use the Alpen Method to Increase Company Efficiency*. Online [Accessed 15 Nov 2024] https://www.timetrackapp.com/en/blog/alpen-method/

[31] Project Management Institute. *A Guide to the Project Management Body of Knowledge (PMBOK® Guide)*, 7th edn. Newtown Square, PA: Project Management Institute; 2021.

Chapter 5
Decision-making

Karol Wojtowicz[1]

Every person makes hundreds of choices every day regarding their personal, social, and professional lives. The ability to make decisions appears to be one of the most important human skills, as it affects the quality of our lives, helps solve problems, and promotes the achievement of goals, as well as enhances our ability to manage difficult situations. In everyday life, at work, or in interpersonal relationships, decision-making is essential to solving problems that we encounter. Fast and accurate decisions can help minimize risks, losses, and make better use of resources. The decisions we make also have a direct impact on our future. Whether it is choosing a career path, deciding on an education, or our personal lives, our choices shape our reality. In addition, people who can make the right decisions often show greater self-confidence because they feel in control of their situation. Decision-making is an important part of self-awareness and building a sense of responsibility. As Peter Drucker points out, decision-making is one of the key skills of an effective manager. The essence of decision-making is not simply to choose between available options, but to actively shape alternatives and solve problems that may not be immediately obvious [1].

Making the right decisions requires a range of additional soft skills, such as analytical and critical thinking. Creativity, problem-solving skills, and sometimes teamwork are also useful. Making decisions requires analyzing the options available, assessing their consequences, and anticipating the outcomes. This develops analytical skills that are valuable not only in professional life but also in everyday life.

The word "decision" comes from the Latin verb "*decidere*" which means "to cut off" or "to decide." In the context of the word "decision," it refers to the process of "cutting off" other options or possibilities, symbolizing the selection of one and the abandonment of the others. This means that decision-making is a process of selection and making a choice that excludes other alternatives. In the literature on organization and management, the concept of decision can be understood as a conscious, non-random choice of one of many possible courses of action [2].

Decision-making is a skill that can be developed so that the choices you make lead to the best possible outcomes. However, even the most competent and

[1]Faculty of Economic Sciences, University of Warmia and Mazury, Olsztyn, Poland

experienced managers and leaders can sometimes make wrong decisions. Decision-making is a complex process influenced by both internal factors (e.g., knowledge, experience, emotions) and external factors (e.g., unpredictable environment, limited access to information). On the other hand, making the right choices is the result of synergy between the decision-maker's skills, access to information, an effective decision-making process, and the environment, as well as a bit of luck. Here are some examples of wrong and right decisions made by the boards and owners of the world's largest corporations.

Coca-Cola

Coca-Cola had the opportunity to acquire the failing Pepsi in the 1930s but declined. Coca-Cola's management did not see the potential in the company and its drink, which turned out to be a strategic mistake, as Pepsi became one of Coca-Cola's main competitors in the following decades.

George Lucas

George Lucas gained a reputation as a talented creator with the success of American Graffiti. However, most studios rejected the next project, considering it too risky. Eventually, Lucas partially financed it himself and founded Industrial Light & Magic (ILM), which became a pioneer in this field of special effects. "Star Wars: A New Hope" (1977) became a huge commercial success, grossing over $775 million worldwide.

Ford

Henry Ford's decision to raise workers' wages to $5 a day and shorten the workday to 8 hours in 1914 is an example of a well-thought-out and effective managerial decision. In the early 20th century, Ford Motor Company introduced an innovative assembly line that significantly increased production efficiency. However, monotonous work and low wages led to high employee turnover, which in turn reduced efficiency and generated additional costs associated with recruiting and training new people. The decision to increase salaries and reduce working hours contributed to reducing employee turnover, increasing productivity, stimulating demand for its own products and improving the company's reputation as a progressive and caring company. Ford's decision also had far-reaching effects, influencing the formation of the American middle class and setting new standards in wages and working conditions in industry.

Kodak

Kodak was founded by George Eastman, who revolutionized photography with his easy-to-use camera and roll film. Kodak became a dominant player in the photography market, with a near-monopoly on camera and supply sales. Ironically, it was Kodak engineer Steven Sasson who invented the first digital camera. However, management ignored the potential of the technology, fearing it would undermine sales of analog film. While competitors such as Sony and Canon invested in the development of digital cameras, Kodak stuck to a strategy based on selling photographic film. The company did not invest in the development of digital technologies until the 1990s, but by then, it was already far behind the competition. Changing consumer behaviors, the development of smartphones with built-in cameras and social media, which revolutionized the way people stored and shared photos, were additional factors in the company's decline.

5.1 Types of decisions

Decision-making situations can involve different areas such as events, forecasts, opportunities, strategies, and needs. However, in all these situations, making a choice implies a commitment to achieving a satisfactory outcome for the people involved. Decisions can be classified into four main categories:

- acceptances: involve yes/no decisions and consist of acceptance or rejection.
- choices: involve choosing among several alternatives.
- creative: involve creative management of resources to achieve an ideal solution.
- evaluation: involves decisions about the value of an action.

Decisions can also be classified in different ways depending on the context and criteria that are considered. The following types of decisions are distinguished in the literature:

- By organizational level:
  - Strategic—relates to the long-term goals of the organization, e.g., choosing a development strategy.
  - Tactical—involves medium-term activities, e.g., allocating resources in a project.
  - Operational—relates to current, short-term activities, e.g., setting up a work plan.
- Due to decision-making criteria:
  - Rational—based on data analysis and logical reasoning.
  - Intuitive—based on experience and intuition.
  - Political—consider the interests of various groups in the organization.

- Due to the number of participants:
  - Individual—undertaken by one person.
  - Group—requires cooperation and consensus in the team.
- Due to the decision-making environment:
  - Made under conditions of certainty—all relevant data is known.
  - Made under conditions of risk—there are some uncertainties, but their probability can be estimated.
  - Made under conditions of uncertainty—insufficient information to predict the consequences.
- Due to time constraints:
  - Short-term—effects visible in a short time, e.g., change in the production schedule.
  - Long-term—consequences are felt after a longer period, e.g., investments in technological development.
- Due to the nature of the problem:
  - Technical—related to technological or organizational processes.
  - Social—relate to interpersonal interactions, e.g., in team relationships.
  - Economic—relates to the allocation of financial resources.

One of the most important ways of classifying decisions is to divide them into individual and organizational decisions. Individual decisions are most often made informally, and their impact concerns primarily our personal lives. These are decisions that we do not delegate. In contrast, organizational decisions are more often made formally. They are intended to support the interests of the organization and, importantly, can be delegated. In seeking to add value to the organization, employees and managers should consider the interests of all stakeholders, and decisions should be made carefully for the benefit of all concerned. Individual decisions are made by an individual and concern his or her own actions, e.g., choice of career, method of performing tasks, and time management. They usually have short and medium-term effects. Organizational decisions concern the whole or part of the organization and are made by the board, management, or decision-making groups, often after consultation or analysis.

When classifying decisions, it is worth considering criteria such as level of importance, time, and complexity. Each of these aspects influences the way decisions are made and their consequences.

The level of importance of a decision affects how you act. If the importance of the choices you make is low, and a bad decision would have little or no impact, you can make decisions intuitively. However, if you are making an important decision, you need to spend more time on it and approach it carefully. The following questions can help you determine the importance of a decision:

- What is the importance of the decision?
- What are the worst consequences of making the wrong decision?

In the case of management decisions, we make a distinction based on the level of importance:

- **Strategic decisions**—which relate to the long-term goals and direction of development of the organization, project, or unit, and have a significant impact on the functioning and success of activities.
- **Tactical decisions**—which relate to medium-term actions and the implementation of adopted strategies. They are not as critical as strategic decisions, but they have an impact on operational effectiveness.
- **Operational decisions**—which relate to current, day-to-day activities and have a relatively small impact overall, but their quality affects the effectiveness of individual processes.

Important decisions become urgent with time. To assess the urgency of a decision, try to answer the following questions:

- Will there be negative consequences if I make the decision later (e.g., next week)?
- Will I have enough time to carry it out?
- Will the decision be limited (due to lack of time or opportunity)?

When classifying decisions according to their degree of urgency, we distinguish between:

- **Crisis (Urgent) Decisions**—requiring immediate action, which may affect their quality as time pressure may limit the accuracy of the analysis.
- **Routine decisions (moderately urgent)**—have a specific deadline, but allow time for analysis and preparation. They are a typical decision-making process of a well-functioning organization or unit.
- **Long-term (non-urgent) decisions**—can be postponed because they do not require immediate implementation, and this puts the decision maker in a comfortable situation because they can thoroughly analyze different options and make the best possible choice.

The degree of complexity of a decision should be a guide to how to proceed. Decisions that we have made many times based on one or two criteria are usually made intuitively. For more complex decisions, a methodical approach is helpful. To assess the complexity of a decision, ask yourself the following questions:

- Have I made similar decisions before?
- Can I draw on experience from other decision-making processes?
- How many factors/variables can affect this decision? [3]

Depending on the level of complexity, decisions are divided into:

- **Simple decisions**—which most often involve issues that are common, understandable, and based on repeatable processes. They require little analysis and can often be made automatically or intuitively.

- **Complex decisions**—which involve many variables and require the analysis of large amounts of data, and therefore may require the cooperation of many people or specialists.
- **Uncertain (risky) decisions**—involve situations where complete information is not available, and the outcome is difficult to predict. They require risk assessment and may be based on predictions or intuition.

A useful tool in the decision-making process, taking into account time and priorities, is the **Eisenhower Matrix**. The author of this concept was the President of the United States, Dwight D. Eisenhower, known for his effective management of time and priorities. His matrix is a great tool for selecting and prioritizing our tasks, using importance and urgency as a reference point. The matrix is divided into four squares, and its main assumption is to assign tasks to the appropriate categories.

The matrix is made up of four quadrants, as depicted in Figure 5.1:

- Important and urgent (DO IT) identifies tasks that require immediate action, often due to crisis situations or deadlines. Examples: crisis problems at work, urgent reports, sudden failures.
- Important but not urgent (PLAN IT)—this includes strategic tasks that will bring long-term benefits but do not require an immediate response. Examples: professional development, project planning, health care.
- Urgent but not important (DELEGATE IT)—these are tasks that require quick action but do not have a major impact on long-term goals. They can be delegated to others. Examples: some emails, meetings of no great value, and minor organizational tasks.
- Not important and not urgent (DELETE IT)—these are all tasks and activities that do not add value and can be removed from the schedule. Examples: excessive use of social media, unproductive conversations, watching TV, mindlessly surfing the web, playing video games, and gambling.

Decisions can also be divided into rational and emotional. Rational factors should be considered when making a decision—"hard data"—but it is worth remembering the

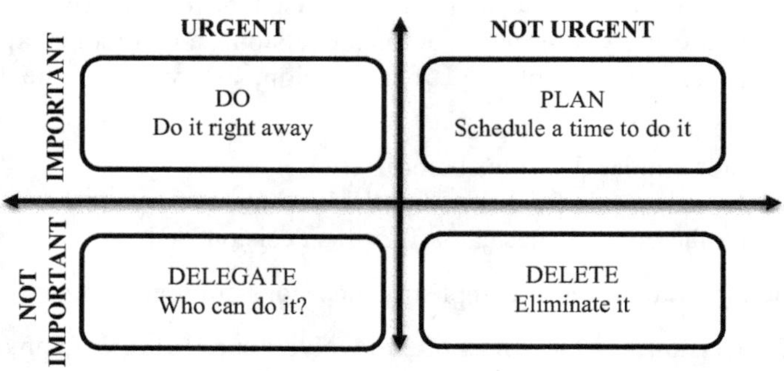

Figure 5.1 The Eisenhower Matrix

emotional factors. They are important too. Rational decisions are based primarily on logic, analysis, and objective data, and the decision-making process is deliberate, thoughtful, and systematic. It involves gathering information, evaluating alternatives, predicting consequences, and choosing the best solution. The main advantages of rational decision-making are that it reduces the risk of error and makes it easier to predict the consequences. On the other hand, the process is usually much more time-consuming, and we do not always have access to complete information. Emotional decisions are mainly based on hunches, intuition, and subjective feelings. They are decisions that are often made quickly, impulsively, and under the influence of emotions (e.g., fear, joy, anger) and result from the individual's experiences, beliefs, and values. Emotional decisions are characterized by the speed of the decision and better alignment with personal values and intuition. Unfortunately, there is a greater risk of making mistakes due to the lack of objective analysis.

In practice, the best decisions often combine rationality and emotion. An example would be the process of hiring an employee—the recruiter considers both the education and experience of a potential candidate (rational approach), but also their soft skills, ability to cooperate, kindness, openness, because these will determine the atmosphere at work (emotional approach).

Another classification (similar to the one above) divides decisions into intuitive and logical. Intuitive decisions are based on intuition, experience, and unconscious processing of information because they are the result of hidden knowledge accumulated over the years (known as heuristics). They are made quickly, spontaneously, and often without conscious analysis. The ability to make intuitive decisions is particularly useful when we make choices under conditions of uncertainty—especially when there is a lack of complete data. It is also effective when decisions are made by an expert in a particular field.

Logical decisions are based on data analysis, evaluation of facts, and logical thinking. The decision-making process is therefore much more methodical and systematic, including gathering information, evaluating alternatives, predicting consequences, and choosing the best option. Decisions based on logic increase the chances that the decision will be correct and limit the influence of emotions—minimizing the risk of mistakes based on momentary moods.

As with rational and emotional decisions, the best solution is to base the decision on both logic and intuition. As Herbert Simon points out, effective decision-making under conditions of limited rationality requires the use of intuition as a supporting tool [4]. When we have limited time and extensive experience in a particular area, we can rely more heavily on intuition. On the other hand, when we have access to the right information and the consequences of decisions are important and long-term, emotions can distort our assessment of the situation, and we should rely more on logic.

5.2 Decision-making process

The decision-making process involves a logically related set of mental operations, arranged in an appropriate order, that enables the assessment of the

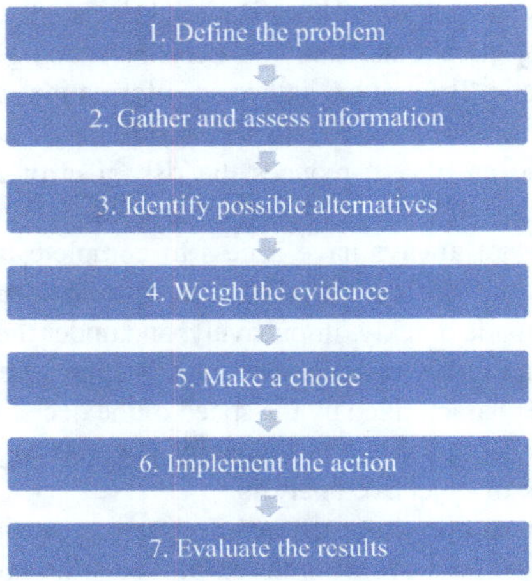

Figure 5.2 Decision-making process

decision-making situation and the selection of the most advantageous variant. The factor that triggers decision-making processes is a problem situation. The need to make a decision arises when the manager's intervention is necessary (or advisable) or when there are many alternatives to this intervention. The manager's intervention is usually linked to the introduction of changes within the organization and/or its environment. That process is put in the right order, which enables the evaluation of the decision situation and the selection of the most favorable option [5].

The decision-making process includes the following steps, summarized by Figure 5.2:

- Defining the problem: clearly define the situation that requires a decision.
- Information gathering: obtaining relevant and objective data.
- Developing alternatives: generating different options and scenarios.
- Evaluate alternatives: analyze the advantages and disadvantages of each option.
- Choose the best solution: making a choice based on the information gathered.
- Implementing the decision: translating the decision into concrete actions.
- Evaluate the results: monitoring the impact of the decision and making adjustments if necessary [6].

5.2.1 Step 1: Define the problem

The first step is to identify the problem or issue that needs to be discussed in the decision-making process. While some problems are obvious and easy to identify, others are complex and involve many factors. If there is no clear vision of the problem, tools such as a cause-and-effect model can be used to help decision-makers identify the real causes of specific problems.

To help you define the decision situation and identify the problem, you can use the following questions:

- Is the decision important, and does it have long-term consequences?
- Is it a problem or an opportunity?
- Are there rules for dealing with the situation?
- To which group does the problem belong?
- Is the problem real or imaginary?
- Is it about people or money?
- What will happen if nothing is done?
- How unusual is the problem?

5.2.2 Step 2: Gather and assess information

To make the right decision, it is necessary to have access to sufficient information about the problem. Information allows decision makers to see different sides of a problem and contributes to a better understanding of it. Gathering information is the foundation of a conscious and thoughtful decision-making process that helps avoid mistakes and make better choices. Tools such as brainstorming and mind mapping allow teams to create visual representations of problems, leading to better decision-making. This stage may also include market research, consultation with experts, or analysis of historical data.

In the search for information and the collection of data, we have to keep in mind that it should be reliable and credible. At this stage, attention should be paid to all factors (external and internal) that may distort or falsify the information. Among the most common errors and distortions of information are: availability bias, bias error, memory error, selection error, first number error, association error, favoring error, and experience error [7].

5.2.3 Step 3: Identify possible alternatives

As you gather information, you are likely to identify several possible courses of action or alternatives. Alternative paths or solutions may also be inspired by previous experience with similar problems, which may not be necessary or exactly the same as the current one. At this stage, you can use your imagination and additional information to construct new solutions. You may find it helpful to use different creative thinking techniques.

The main aim of this step is to create possible ways of solving the problem and to make a decision. Remember, the more options you explore, the better your final decision will be.

5.2.4 Step 4: Weigh the evidence

At this stage in the decision-making process, you need to evaluate the various alternative options, forecasting (imagining) the outcome of each solution, which often results in the successful achievement of the set objectives. Once you are satisfied that you have a good selection of realistic alternatives, it is time to assess the feasibility, risks, and implications of each alternative.

Assess the risks, identify the potential consequences, consider whether your resources are adequate, and consider the financial feasibility of the decision. For management decisions, especially strategic ones, SWOT analysis can be used to identify the strengths, weaknesses, opportunities, and threats of each alternative. Tools such as risk analysis, impact analysis, cost-benefit analysis, NPV, and IRR can also be helpful.

5.2.5 Step 5: Make a choice

After a thorough evaluation of the alternatives, the next step is to make a decision. If one option is clearly better than the others, your choice will be obvious. However, if you still have several competing options, there are many tools available to help you decide between them.

At this stage, the decision is made! This can be an individual or group decision, depending on the context.

5.2.6 Step 6: Implement the action

This step introduces the activity or solution into the process, and the team begins to implement the solution into the project flow. Especially in business and engineering decisions, before implementing the solution, the team should be ready to understand and use it. In this step, further meetings with the team can help them to learn more about the activity, why it is being adopted, and how it will be integrated into the process.

5.2.7 Step 7: Evaluate the results

Once the action has been implemented, the final results should be examined to understand the performance of the solution and the process used to achieve it. The final step is to monitor the impact of the decision taken and analyze whether it is producing the expected results. If the results are not satisfactory, it may be necessary to correct the action or take a new decision. Feedback is also helpful in the ongoing decision-making process, as it can be used in step 2 as part of the information gathered for the next decision-making process.

The decision-making process provides a broader view of the problem and helps decision-makers to explore different perspectives. The time spent on the decision-making process and the team involved will vary depending on the size of the problem and the time available for decision-making. In a crisis situation where time is limited, it may take only a few minutes to go through the seven steps above, whereas for complex problems that require in-depth investigation, the decision-making process may take more time.

5.3 Group decisions—pros and cons

In the process of making decisions, especially those whose consequences will affect a larger group of people, the question arises—should the decision be made alone or as a team? Managers often consider alternatives—on the one hand, making the right decision alone will be a source of success and glory for the decision-maker. On the other hand, there is always the risk of making the wrong decision. Then the negative consequences will also be attributed to a particular person. Individual decisions must consider the risk of failure, in which case the responsibility for the unfavorable outcome must be borne.

Group decisions can be about personal and family matters, but much more often they are about management skills. Management theory identifies situations in which leaders and managers should consider the possibility of team decision-making. Team decision-making is currently one of the most important aspects of management, especially in the context of modern organizations, where the emphasis is on collective action, innovation, and efficiency. Group decision-making allows many perspectives to be taken into account, but at the same time requires effective cooperation and coordination. Furthermore, this process often takes longer than decisions made individually, which can be problematic in situations where quick action is required. [8]

The question then arises, when is it worth making choices with the support of a group? Dawson points to the following five situations [4]:

1. When you think that team decisions will allow for the development of more solutions.
2. When you need an external point of view, expertise.
3. When you want to raise the ethical standards of decisions.
4. When you think that it is necessary to support the group in making the expected decision.
5. When you are not sure whether your idea will be accepted.

Because of the synergy effect, team decisions can be a source of many solutions. Group actions also allow for the development of non-standard solutions and are a form of creative problem-solving. Individual decisions may be associated with a lack of consideration of the wider context, or a lack of appropriate information or skills. In such cases, it is worthwhile consulting external experts. Experience, and especially expertise, will not only be a source of the right decisions, but will also make it possible to identify a problematic situation from a different perspective [7,8].

Some decisions require consideration by a wider group of stakeholders. For example, in situations of potential medical error, setting up a committee to investigate the matter will result in raising the ethical standards of decisions made. Solving problems and making decisions in a larger team provides an opportunity to develop an objective and balanced position—then group actions will influence the raising of ethical standards of decisions. However, it should be remembered that group decisions will not always be wiser than individual decisions. The voice of the group will often be more prudent, but at the same time, it may support less prudent decisions than individual decisions. The opinions of a larger group of people may influence individual team members who, because of the risk of social ostracism, will not be guided by their own judgment but will adopt the position developed by the group (sometimes against their own will).

It is worth making team decisions in situations where involving the whole team in the decision-making process will have a positive effect, both in terms of a favorable decision outcome and in terms of increased enthusiasm and team integration. Many autocratic leaders, convinced of their enormous competence, see no need to waste time involving group members in decision-making processes. Their belief in their own infallibility and excessive self-confidence are often the source of poor decisions. Managers with a more democratic style of leadership [9] are aware that they need to allow their colleagues to make decisions in order to build and develop an effective organization. Involving the team in making decisions that are important to the organization gives people a sense of ownership and makes them feel more connected to the company.

Group decisions also work well in situations where decision-makers are full of fear that their ideas will be rejected. The problem arises when a manager is convinced of the validity of a particular solution or idea but does not fully believe that it will be accepted by the environment. In such a case, allowing decision-making to take place in a team will not only allow the final idea to be refined but will also create a greater chance of it being implemented. Many benefits to group decision-making result from the synergy, diversity, and commitment of the team members [9]. The main benefits of team decision-making are:

- **More information and results**—one of the key benefits of team-based decision-making is access to more data and information, which can lead to better results than the sum of individual contributions (synergy). In addition, decision-makers can support each other emotionally and intellectually, which increases efficiency and team morale.
- **Diversity of perspectives and knowledge**—group members have different experiences, knowledge, and skills, which allows for a more comprehensive approach to the problem; they are able to generate more ideas and solutions than an individual, which promotes creativity and increases the chances of finding innovative solutions.
- **Higher quality of decisions**—the group can collectively identify and analyze more factors influencing the decision, and at the same time, group members check each other and verify their observations, which limits the impact of individual biases and errors.
- **Greater commitment and acceptance of decisions**—group members who have participated in the decision-making process are more likely to accept and support their decisions, and their involvement in the decision-making process is a source of motivation that builds greater commitment to the implementation of tasks.
- **Sharing experience and developing group skills**—team members make decisions together and have the opportunity to share knowledge and experience, learning from each other and increasing their skills at both individual and group levels.
- **Preventing bad decisions and increasing risk resilience**—The risk of making a bad decision is reduced in group decision-making because participants are able to anticipate potential threats more accurately, thanks to different

points of view. It is also important to distribute the responsibility for the outcome of the decision among all team members.

As mentioned above, group decision-making has many advantages. However, it is important to remember that the process also has some disadvantages and limitations. The main difficulties arise from group dynamics, time constraints, and psychological factors. The main disadvantages of group decision-making are:

- **Longer decision-making time**—the specifics of team decisions (meetings, discussions, negotiations, and agreement of positions) tend to take longer than individual decisions. In situations that require quick action, group decision-making can have negative consequences.
- **Over-complicating the decision-making process**—teams can fall into the trap of analyzing too much data or alternatives, leading to decision paralysis, and an over-complicated decision-making process can reduce its effectiveness.
- **The influence of the leader's decision and the risk of groupthink**—in the group decision-making process there is a risk that more dominant, more confident or more senior individuals will dominate the discussion and limit the input of other members, which can result in ineffective influence or in the valuable opinions of less assertive participants being overlooked altogether. There is also a phenomenon, groupthink, where participants seek consensus at the expense of critical thinking. Group members avoid expressing differing opinions so as not to negatively affect the work of the team.
- **Blurred responsibility**—as shown above, collective responsibility for the consequences of a group decision can be an advantage, but on the other hand, diffusion of responsibility can make it difficult to identify the person responsible for a possible failure. It can also weaken the motivation to take risks or limit the responsibility for implementing the decision.
- **Tendency to compromise**—decisions made by the team may be based on the principle of compromise, and therefore may not always be optimal, leading to a reduction in the quality of decisions and a loss of opportunities to find more innovative solutions.
- **Organizational costs**—collective decisions require specific organizational activities related to, among other things, the preparation of meetings and the coordination of work. These activities require time and financial resources, which increase the cost of the decision-making process compared to individual decisions.

Despite the benefits of team decision-making, there are significant limitations that can reduce the effectiveness of the process. Awareness of these limitations is critical to the success of team decision-making. To manage a team effectively and minimize the drawbacks, it is necessary to plan well, facilitate the process by the leader, encourage open communication, and avoid pitfalls such as groupthink or blurring of responsibilities.

Considering the possibility of group decision-making, it is necessary to determine the number of people who will be involved in the process. On the one hand, many participants will provide more information and possible solutions, but on the other hand, it will make it more difficult to reach a consensus and prolong the final decision. The size of the group has a significant impact on the quality of team decisions. The optimal number of participants depends on the context, the nature of the problem, and the dynamics of the group. Both too small and too large groups can cause specific difficulties.

The main advantages of small groups are: effective communication (in small groups, communication is more direct), ease of management (the leader has more control over the decision-making process), shorter decision-making time, and greater responsibility and involvement of the members. However, this is associated with less diversity of opinion and potential ideas, and a relatively high risk of a dominant role for individuals who can easily impose their views on other team members. Teams made up of more people are more diverse in terms of skills, experience, and viewpoints, which increases their innovation potential. Large teams also mean access to greater knowledge resources and a broader information base, which can lead to better analysis of the problem.

In assessing the impact of group size on the quality of team decisions, it is worth identifying the optimal size. According to research by Hackman and Vidmar, groups of 5–7 people are the most effective for decision-making because they provide a balance between diversity and manageability. In the case of strategic decisions, larger groups can be advantageous because they allow different perspectives to be considered, whereas operational decisions are better made in smaller teams where speed of action is critical [10].

In summary, group decision-making plays a key role in management processes and the functioning of organizations. As the literature suggests, the essence of this process is to use the diversity of perspectives and knowledge of the participants to achieve better results than individual decisions. However, the effectiveness of the process depends on skillful team management, conflict resolution, and the use of appropriate support tools. The key challenge is to avoid pitfalls such as groupthink or social laziness, which can weaken the quality of decisions made.

5.4 Barriers to good decision-making

There are several barriers and constraints to making good decisions. One of the most important constraints is the lack of or poor quality of data, which can lead to poor decisions. This is particularly important when decisions need to be made quickly, before all the information is available. On the other hand, too much data can make analysis difficult and prolong the decision-making process, leading to decision paralysis. In the case of group decisions, a barrier to getting the right decisions can be the phenomenon of conformism, where team members avoid expressing dissenting opinions, leading to worse decisions.

It is also worth noting the psychological and social constraints that can negatively affect the decision-making process. Dawson identifies five barriers to good decision-making: acting too quickly, acting too slowly, not knowing how fast to make a decision, unwillingness to admit mistakes, and excessive autocracy [4].

5.4.1 Do not make decisions too quickly

Decisions that are made too quickly are usually made under time pressure, which can have many negative consequences. Rushing to make a decision often overlooks important information, leading to incorrect conclusions and sub-optimal solutions. Lack of time for risk analysis increases the likelihood of unforeseen problems. Impulse decisions are often based on intuition rather than facts and data. Decisions that are made too quickly cause the decision-maker to focus on one scenario instead of looking for alternatives. Quick decisions can therefore limit the ability to analyze different options and choose the best solution.

5.4.2 Do not make decisions too slowly

Taking decisions too slowly can be just as damaging as acting too quickly. Psychologists point out that each of us has a built-in defensive mechanism for avoiding problems. In situations where we must make a choice, we unconsciously try to avoid problems, which leads to three attitudes: (1) It cannot happen to me; (2) I can deal with it later; (3) Let someone else deal with it.

Taking decisions too slowly is particularly dangerous in business, as it means losing opportunities and a competitive advantage. In a dynamic business environment, delaying decisions can mean missing out on opportunities to develop, innovate, or enter new markets. Competition can make a decision faster and gain an advantage, for example, by introducing a product first. Furthermore, delaying decisions can result in additional costs, such as higher raw material prices, logistics and storage problems, costs associated with maintaining inefficient processes, or losses due to inaction. One of the barriers to fast decision-making is the collection of too much information. Of course, optimal decision-making requires gathering the necessary data to accurately analyze the situation, but in many cases, too much data can lead to information noise and, as a result, decision paralysis. Fear of the negative consequences of decisions can also slow down the decision-making process. Fear of failure causes many people to try to avoid making decisions. However, it is important to remember that there is risk involved in every decision, and we cannot completely remove the risk factor from the decision-making process.

5.4.3 How quickly should a decision be taken?

The need to make decisions quickly means that we are forced to break the rules of making good decisions. It pays to rely on the most reliable data that has been collected and to involve the most competent people in the environment who can help us make the right choice. In the case of business decisions, it is worth observing what the competition is doing. Knowledge of the activities of our competitors is very important to decide at the right time. When trying to determine the right time to make a decision, it is worth remembering that even in the case of good ideas, delaying their implementation can cause them to become obsolete. When under pressure, it is also worth asking yourself to what extent the decision is reversible. In some cases, it is worth making a quicker decision, remembering that there is always a way out of a seemingly unfavorable situation.

5.4.4 Be ready to admit your mistake

Not all the decisions we make are successful. Even the best managers make bad decisions, and often, because they are overconfident, they stick with bad decisions. One reason for the reluctance to admit mistakes is the feeling that an admission of failure would be bad for our image. However, the willingness to admit a mistake has a significant impact on the quality of decisions made. People who are able to admit a mistake are quicker to spot irregularities and take corrective action. This minimizes the negative impact of wrong decisions and limits financial, time, and image losses. Admitting a mistake helps to draw conclusions for the future and to avoid similar mistakes. In the case of business decisions, it helps build an organizational culture based on continuous improvement and openness to feedback. The ability to admit a mistake builds trust and authority in the leader—it shows maturity and responsibility, which builds trust in employees. People are more likely to engage in discussion and decision-making when they see that the leader is open to improvement. In an environment where mistakes are accepted as part of the job, people are more likely to experiment and seek new solutions. This encourages creativity and better strategic decisions. The willingness to admit a mistake improves the quality of decisions because it allows for faster correction of mistakes, increases the ability to learn, builds trust, and supports innovation. As a result, organizations and leaders who are not afraid of making mistakes make smarter, more thoughtful decisions.

5.4.5 An excess of autocracy

An extremely important limitation of the quality of decision-making—especially in the organizational context—is the excessive autocracy of managers. Their excessive self-confidence causes them to perceive their decisions as the most effective, thus limiting the decision-making potential of their employees. Management based on strong centralization of power and decision-making by one person or a small group may be good in some cases, for example, in crisis situations requiring immediate response or in specific organizations in the military, medical, or crisis management industries. In the long run, however, it often leads to mistakes due to a lack of consultation, low employee involvement, and difficulties in adapting to change. An autocratic leader often relies solely on his or her own beliefs, which can lead to poor or sub-optimal decisions. Ignoring the opinions of experts and employees leads to a lack of diversity of perspective and increases the risk of underestimating threats or failing to implement the best solutions. Decisions imposed by managers make employees feel excluded and less committed to the organization's goals, leading to lower loyalty, high turnover, and reduced innovation. In autocratic organizations, leaders may ignore or suppress feedback about poor decisions rather than correcting them. Organizations may then find it difficult to adapt quickly to change if the leader does not see the need to change strategy. The lack of free exchange of information and feedback can lead to stagnation and loss of competitiveness. When making a decision, remember that in most situations, a style based on a balance between the leader's authority and openness to the opinions of others works better.

In order to make better decisions, it is useful to be aware of these barriers and apply strategies to minimize their impact, such as critical thinking, risk analysis, or consultation with experts [5].

5.5 Summary

The ability to make decisions is crucial in both our personal and professional lives, as it affects our daily choices, long-term goals, and overall life satisfaction. The ability to make good decisions gives us greater autonomy and independence in our personal lives, allowing us to avoid over-reliance on others and to live life on our own terms. People who make conscious choices are less likely to feel lost or controlled by circumstances. Making conscious choices when interacting with others helps to build healthy relationships and avoid conflict.

> In any moment of decision, the best thing you can do is the right thing, the next best thing is the wrong thing, and the worst thing you can do is nothing.
>
> **Theodore Roosevelt**

Knowing how to make good professional decisions makes you more efficient and productive at work and helps avoid delays. People who make good decisions are promoted faster and achieve career success. Making good decisions quickly is crucial in the dynamic world of business, where situations can change quickly. People who are confident in their decisions are perceived as competent and trustworthy.

In stressful, uncertain, or crisis situations, the ability to make decisions helps to keep calm and act rationally. Lack of decision-making or constant hesitation may lead to the escalation of the problem. The decision-making cycle is complicated, and there are many possible situations, so it is impossible to clearly present a recommended way to act. However, knowledge and skills in the field of making choices can improve the quality of decisions made and determine success or failure. Good decisions can help you be successful in your personal and professional life. People who make considered choices are usually more effective and successful because they can judge risks, pick the right options, and react to changing situations. Developing decision-making skills leads to greater satisfaction, effectiveness, and success in various aspects of life.

References

[1] Drucker P. F. (1967). *The Effective Executive*. Harper & Row, New York.

[2] Klincewicz K. (red.) (2016). Zarządzanie, organizacje i organizowanie – przegląd perspektyw teoretycznych. Wydawnictwo Naukowe Wydziału Zarządzania Uniwersytetu Warszawskiego.

[3] Herzog D., and Brenk L. (2008). 30 Minuten. *Entscheiden*. Gabal Verlag GmbH, Offenbach.

[4] Dawson R. (1998). *Confident Decision Making*. Career Press, Wayne, NJ.

[5] Robbins S. P. (2004). *Decide and Conquer: Make Winning Decisions and Take Control of Your Life*. FT Prentice Hall, Hoboken, NJ.

[6] Kahneman D. (2011). *Thinking, Fast and Slow*. Farrar, Straus and Giroux, New York.

[7] Mintzberg H. (1973). *The Nature of Managerial Work and Simply Managing: What Manager Do – And Can Do Better*. Harper & Row, New York.

[8] Saaty T. L. (1990). *Decision Making for Leaders: The Analytical Hierarchy Process for Decisions in a Complex World*. University of Pittsburgh, Pittsburgh, PA.

[9] Simon H. A. (1987). Making management decisions: The role of intuition and emotion. *The Academy of Management Executive (1987–1989)*, 1(1), 57–64.

[10] Hackman J. R., and Vidmar N. (1970). Effects of size and task type on group performance and member reactions. *Sociometry*, 33(1), 37–54.

Chapter 6
Problem-solving

Edita de Lorenzo[1] and Enrique Costa-Montenegro[2]

This chapter focuses on the soft skill of "Problem Solving," classified by the European Union [1] in the topic S1-communication, collaboration, and creativity skills. It is defined as "developing and implementing solutions to practical, operational, or conceptual problems which arise in various work contexts."

In this chapter, we will explore its relevance for both students and lecturers, whether working individually or as part of a team. We will also discuss its impact on professional development, as it helps us identify the goals we must achieve, assess progress, and verify outcomes. The chapter includes a collection of tools designed to help teachers promote problem-solving skills among students and support them in learning and developing these skills independently. Each tool links to specific activities for skill development. To master this skill, we will center our chapter on the following key: identifying and defining the problem, generating alternative solutions, evaluating and selecting the best alternative, and finally, implementing the selected solution.

6.1 Introduction

A problem is an obstacle that prevents people from achieving their goals. The soft skill of "Problem Solving" is associated with this definition and is the ability to address issues effectively and in a timely manner without unnecessary impediments.

This skill is vital in organizations, as they face daily situations that require efficient solutions. Consequently, developing problem-solving skills is critical for the professional growth of students [2]. It is also important for our daily lives, as individuals face personal challenges that this skill can help them confront [3].

It is relevant to highlight that "Problem Solving" skill integrates and builds on several other soft skills explained in this book [4].

Foundational skills, like time management and interpersonal communication, are essential for achieving success. Nowadays, we must work in teams at any

[1]Departamento Teoría do Sinal e Comunicacións, Universidade de Vigo, atlanTTic, Spain
[2]Departamento de Enxeñaría Telemática, Universidade de Vigo, atlanTTic, Spain

workplace, in any discipline. Sometimes, people from different backgrounds must deal with complex problems, and their ability to work together effectively is key to achieving shared success.

As we will discuss later, decision-making is a key step in problem-solving. To make effective decisions, we must apply skills such as critical thinking, analytical thinking, creativity, and reflective thinking. These four skills work together to generate the best ideas and approaches for solving any problem, following the steps and techniques outlined in the following sections of this chapter. This process is illustrated in Figure 6.1.

First, we will examine in detail what a problem is and provide a possible definition of the skill "Problem Solving." A problem can be understood as any challenging situation that prevents individuals from achieving their goals. Related to this, "Problem Solving" is the ability to address issues effectively and on time, without any impediments. This skill may help resolve both personal or professional problems, whether within a company or group, at work, or in decisions we face in our daily lives, regardless of the problem's size.

This skill can be broken into several steps, the more usual divisions are the following:

- Identifying and defining the problem
- Generating alternative solutions
- Evaluating and selecting the best alternative
- Implementing the selected solution

It is crucial to underline the importance of the first step: identifying the problem. Often, the problem itself is hidden beneath its visible consequences. A common mistake is trying to solve the symptoms caused by the problem, rather than dealing with the cause that generated the problem. To solve this, we must consider the root cause of the problem, trying to find the underlying cause that is generating the situation under analysis.

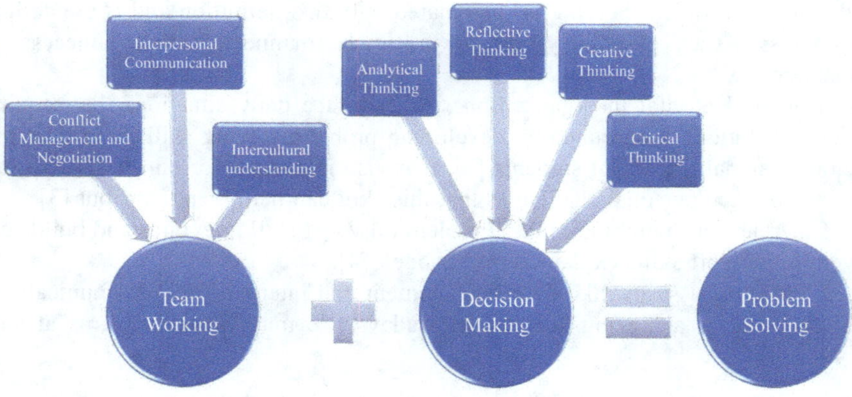

Figure 6.1 Problem-solving skill related to other soft skills

Once the root cause is identified, it is time to start looking for possible solutions. To do this, we can use other skills presented in this book, especially those related to different ways of thinking, creativity, and different tools, like brainstorming, just to mention one example.

After evaluating the possible solutions, selecting the best alternative is essential. However, it is equally important to consider implementing the selected solution, as implementing that solution may not be feasible.

There are many different approaches or methods designed to solve problems, such as the Kepner-Tregoe method, scientific approaches, feasibility analysis, four-phase approaches, and more. Some of these approaches are more focused on solving engineering problems, while others are more related to management, and so on.

Since this book is primarily oriented toward engineering, we decided to base our chapter on a more "engineering approach." Our proposal is based on the Kepner-Tregoe method [5] but also incorporates elements from other methods. We start by asking three key questions:

- What is happening to us now?
- Why is this happening to us?
- What course of action should we take?

These questions guide us through three different phases:

- Situational analysis
- Problem analysis
- Decision analysis

To apply the methodology proposed in this chapter, we suggest the four running examples, which can be found in Activity 6.1. The idea is to work individually or in teams, selecting one of the examples and using it throughout the different activities proposed in the chapter. Alternatively, you can choose to work on a new problem of your own. The problems given here are broad and open-ended. You can add more details or center them around a specific need or topic if that seems more appropriate.

Activity 6.1

Proposed running problem examples:

- Living in the city center or the suburbs
- Next trip: how to plan it and things to take
- Decision between homeschooling or formal education
- Should we expand our company internationally?

The rest of the chapter will be organized as follows: First, we will define situational and problem analysis, e.g., understanding where we are and what problem we are facing. Next, we will explore different techniques for generating as many potential solutions as possible, thus providing a wide range of alternatives. After that, we will learn how to choose the best solutions among the possibilities. Finally, we will finish this chapter with some conclusions.

6.2 Situational and problem analysis

The second section of this chapter discusses "situational analysis" and "problem analysis." As we will see, depending on the problem we are dealing with, we should perform one or both analyses.

Let us start with situational analysis. The first step in solving a problem, especially when working within a company, is to understand the company's current situation. This means that we should start by assessing the current situation.

To do so, we have to understand the situation by asking ourselves and trying to answer these given questions:

- Where are we now?
- What is the current situation?
- Where do we need to be?

The key point here is to find the best way to define the problem we are addressing. We also need to have a detailed description of our current situation, whether we are a company or an individual. Finally, our objective or goal is to be sure that we are approaching the real problem under analysis.

Situational analysis cannot be done just once; it is, instead, an iterative process over time. Some ways to perform this analysis include surveying your customers to gather their opinion about you (as a company, a provider, or about a specific product), examining what the competition is doing (products, campaigns, strategies, etc.) to learn from their performance, and also researching various sources of information to acquire more knowledge about the situation.

For our situational analysis, evidence is a crucial aspect. We should prioritize hard evidence (such as historical data, factual observations, and industry benchmarks) over feelings or soft evidence (personal opinions, reactions, behavior, and gut feelings). It is important to recognize that the final goal of this analysis is to achieve an accurate and precise understanding of the current situation. Several tools can help us in this situational analysis: one of the most well-known being the Strengths, Weaknesses, Opportunities, and Threats (SWOT) tool, as shown in Figure 6.2.

Once we understand the current situation, we can consider the "situational analysis" finished and start with the "problem analysis." The problem analysis is mainly focused on identifying the source or root of the problem to be solved.

The first step is a process of determining what information is relevant and what is not to the actual problem. One tool we can use for problem analysis is the "What

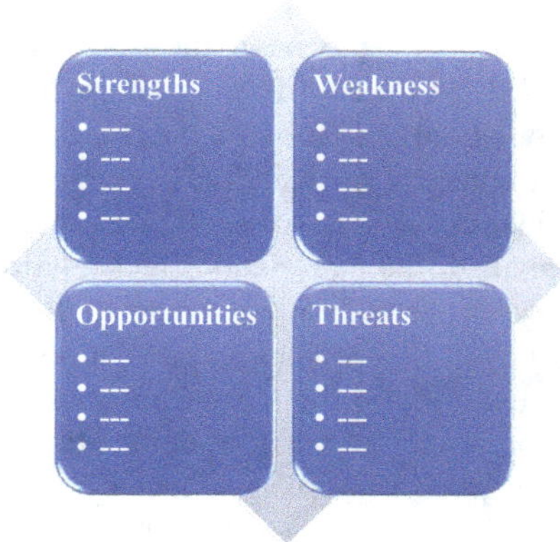

Figure 6.2 SWOT tool description

Table 6.1 Matrix for the tool "What is in/what is out" in problem analysis

Problem parameter	Probable cause	What is in?	What is out?	Conclusion
What				
Who				
Where				
When				
How much				

is in/what is out" technique. We need to identify which factors are potential causes of the problem and which ones can be ruled out, as potential causes, at least in the following factors:

1. Identification of performance gap: we need to discover what is below expectations, such as defects or errors within a process.
2. Problem parameters: Then, we should create a list of probable causes of the performance gap related to "What," "Who," "Where," "When," and "How Much" [5].

For this tool, it is recommended that you create a matrix like the one in Table 6.1 and fill it with all relevant observations. The columns will include all potential causes, what is in, what is out, and the conclusion.

By filling out this matrix, we can describe the problem in detail and reach different conclusions that will help us in this step of problem analysis.

Another popular tool for problem analysis is the "Root Cause Analysis," also known as the "5 Whys." It was created by Sakichi Toyoda, the founder of Toyota, based on a deductive reasoning technique often used by detectives [6]. It is a technique for exploring the cause-and-effect relationships underlying a particular problem by iteratively asking questions, specifically repeating the question "Why?" Each answer forms the basis for the next question. This simple technique of repeatedly asking "Why?" helps us identify the root cause of the problem. We just need to ask a why question about the problem, get the answer to that question, and ask again a why question with that answer. We repeat this procedure at least five times or until we get a final answer. By using this technique, we collect evidence about the problem and analyze the cause-and-effect relationships involved in the given problem. Sometimes, it is helpful to create a flowchart to visualize the method, where we write down the problem and its causes.

Next, we present an example of the Root Cause Analysis method applied to find the solution to a real problem: the deterioration of a monument in Washington, DC [7]. The given example is chosen because of its visibility and because the real cause of the problem would not be found without using a tool like this that uses deep analysis.

***Problem**: One of the monuments in Washington, DC is deteriorating.*

***Why #1**—Why is the monument deteriorating?*
***Answer #1**—Because powerful chemicals are frequently used to clean the monument.*

***Why #2**—Why are powerful chemicals needed?*
***Answer #2**—To clean off the excessive volume of bird droppings on the monument.*

***Why #3**—Why is there an excessive volume of bird droppings on the monument?*
***Answer #3**—Because the large population of spiders in and around the monument are a food source to the local birds.*

***Why #4**—Why is there a large population of spiders in and around the monument?*
***Answer #4**—Because vast swarms of flying insects, on which the spiders feed, are drawn to the monument at dusk.*

***Why #5**—Why are swarms of insects drawn to the monument at dusk?*
***Answer #5**—Because the lighting of the monument in the evening attracts the local insects.*

***Solution**: Change how the Jefferson monument is illuminated at dusk to prevent the arrival of swarming insects.*

We suggest doing an activity using the "Root Cause Analysis" technique, as described previously (see Activity 6.2). You, either individually or as a part of a team, are asked to think about a new problem they would like to address and then find the root cause of that problem after a series of "Why?" questions.

Activity 6.2

Root Cause Analysis Exercises:

- Identify a problem within your group
- Apply the "5 Whys" technique individually to analyze the problem
- Share and discuss your results with your teammates

Another problem-solving technique is to analyze when the problem happens and when it does not. With this technique, we can know the conditions that cause the problem, which will help us find the root of the problem. Here's a brief example of the kinds of questions we might ask to find the reason for a situation, like when there is a lack of Internet connection. We usually ask ourselves these kinds of questions:

- Can you browse websites, or is it only email that's affected?
- Do you have Internet access on your smartphone and on your laptop?
- Are you connected via Wi-Fi or LAN?

Answering these questions helps us find the root of the problem.

The last technique we introduce is the Pareto analysis. This principle tells us that 80% of the problems are often caused by 20% of customers or sources. We are sure that you have already heard about this Pareto analysis, as it can also be applied to other aspects. It can also be used to focus on problem-solving. The idea is to find and then concentrate on solving this 20% of the causes that lead to the majority of issues, allowing for a more targeted solution.

6.3 Generate solutions

Once we have properly identified the current situation, using any of the techniques proposed for the problem analysis, and found the possible cause of the problem, it is time for the next important step: finding potential solutions.

To generate a solution, we can use logic and creativity, which are different ways to approach and solve a problem. Usually, people mainly use only one of these approaches, but the best solutions always come when using both. This is one of the reasons why working in teams usually leads to better solutions; team members bring different perspectives, with some approaching the problem analytically and others creatively. Combining creative thinkers with analytical minds is an excellent way to generate fresh and innovative ideas.

Even though we have not yet seen the chapters dedicated to different ways of thinking (analytical, creative, critical, and reflective thinking), some of the tools discussed in those chapters will also be useful here in our exploration of generating solutions.

We now propose two activities related to logic and creativity (see Activities 6.3 and 6.4). The idea is to use both approaches to solve the proposed exercises. Read the instructions and try to solve them. Share your solutions with your colleagues to compare answers and discuss different approaches.

Activity 6.3

Logic and creativity

EXERCISE 1: The Broken Clock

An analogue clock fell and broke into four pieces
The numbers on all of these pieces add up to 15
What are these pieces?

Note: Image generated by Idiogram.ai using the prompt: an analog clock broken into four pieces

Activity 6.4

Logic and creativity

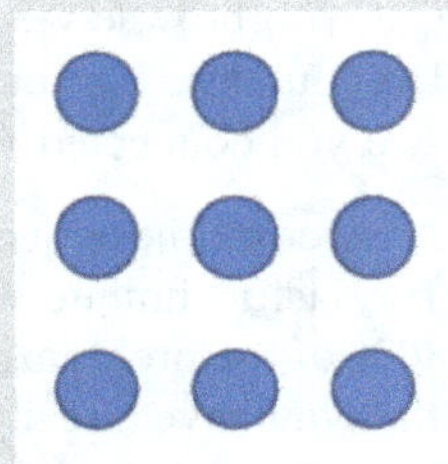

EXERCISE 2: The Line Game
Rule: All lines must be connected and pass through all the dots

Draw 4 connected straight lines through all 9 dots.
Draw 3 connected straight lines through all 9 dots.
Draw 1 straight line through all 9 dots.

Now, we will explore some techniques and methods for generating solutions. One of the most popular methods to generate ideas is Brainstorming. It consists of joining a group of people to provide different ideas to solve a problem. These individuals share their ideas as possible solutions to the given problem. As these ideas are expressed, other participants can build upon them to create new ideas or improve what has already been said. Visualizing is also beneficial; for example, you can write the ideas on sticky notes and place them on a wall.

It is essential to separate the two stages in Brainstorming: the idea generation and the idea evaluation. These stages should not be mixed. In the generation stage, everyone should participate. We should encourage people to express their crazy ideas without judging them. Maybe they are just crazy, but other people in the group could use them to build on them and come up with new, feasible ideas. Once we have many ideas, we move on to the evaluation stage. The authors of the ideas should not feel attached to them or try to defend them against others. Ideas belong to everybody, as they were generated in a group brainstorming session. To stay visual, grouping sticky notes with similar ideas could help us to build on them and arrive at reasonable solutions.

Another way to generate ideas is by using Mind Maps. A Mind Map is a versatile tool consisting of an exploratory method to help understand an issue both in detail and as a whole by identifying critical elements and the relationships

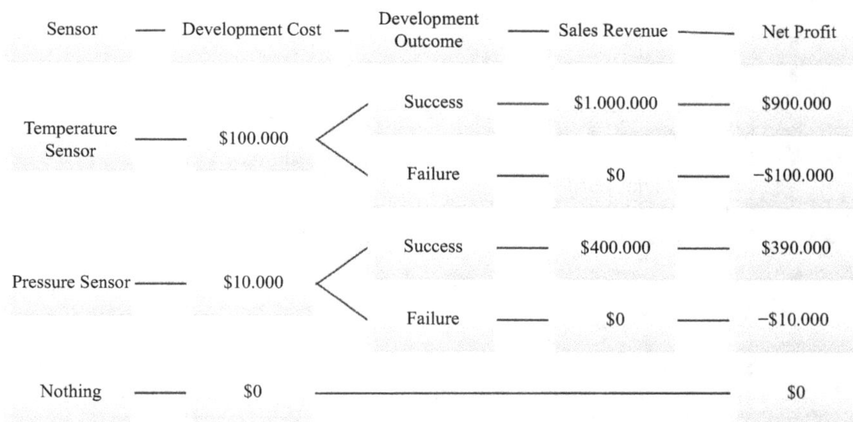

Figure 6.3 Example of a problem solved using different sensors

between them. A mind map is a graphical representation of how our brain thinks, using lines to group related topics and trying to cover all the aspects related to a problem. With a mind map, we try to articulate all the associations related to the problem using words, symbols, and colors. It is also recommended to be visual; it is better to use simple words, symbols, colors, and all relevant associations connected to the problem.

Another visual method for generating solutions is decision trees. With them, we organize the decision process using related questions (nodes or squares) by connecting each alternative shown (represented by lines). To create them more accurately, we can assign probabilities and revenues, calculating the expected values at the end. The decision tree is finished when we have drawn all possible courses of action.

For example, we can see a problem with three possible solutions: using a temperature sensor, using a pressure sensor, or doing nothing, as shown in Figure 6.3. The development cost of the sensors is included, with temperature sensors being ten times more expensive. In both cases, the development outcome is linked to sales revenue, associated with both success and failure. Finally, we can calculate the net profit for each solution, with associated numbers.

6.4 Choose a solution

Now that we have generated different solutions to our problem, we must choose one of the ideas found before. For this task, we can use some of the methodologies seen in the previous chapter, such as "decision-making" skill, or some of the methods proposed below.

When faced with the decision to choose a solution, we have several options. One of the simplest is creating a list of options with pros and cons. In these lists, we place the benefits or positive aspects of a solution on one side and the drawbacks on the other. This can help us quickly identify several factors related to the problem

Table 6.2 List of pros and cons of a business partnership

Pros	Cons
More employees	Less independence
More knowledge	More possibility of conflicts
Less financial burden	Divided profits
Less paperwork	No separation of business
Less tax forms	Individual taxes

Table 6.3 Weighted list of pros and cons

Job on:	Big company	Academia
Salary	3	2
No need to move	1	3
Not tied to a schedule	1	2
Like the job	2	2
Chance of traveling	2	3
"Relaxed" job	1	2
No dress code	1	2
Total	11	16

that will help us select the best solution. The example in Table 6.2 lists the possible pros and cons of a business partnership.

In the example in Table 6.3, we can see a more advanced list of pros and cons. Instead of having a simple list, we assign a weight to each criterion for each possible solution. In this case, we define the maximum weight as 3 and the minimum one as 0, even though it is never used. The problem involves selecting between two job offers: working in a large company or a position in academia. Then, we list all the criteria relevant to the decision, such as salary, need to relocate, schedule, etc. For each job offer, we assign a value to each criterion, as shown in the example in the table. Finally, we sum all the values to get a total. The job with the higher total is considered the best solution, according to the user in this example.

We are going to practice these techniques in the activity 6.5. We suggest creating a list of pros and cons, with or without weighting, related to one of the running problems we proposed in the introduction.

Activity 6.5

List of options with pros and cons

EXERCISE:

Create a list of pros and cons of the problem you selected in the first section of the chapter.

The weighted decision matrix is a more advanced method for choosing a solution to a problem. Given a problem, we create a matrix where we place the different solutions on the x-axis (horizontal). Then, different criteria are used to evaluate the solutions related to the problem and place them on the y-axis (vertical). We assign a weight, as a percentage, of importance to each of these criteria, ensuring that the sum of all these weights is 100%. For each solution and each criterion, we assign a value between 0 and 100, representing how well the given solution meets the criterion. Finally, to calculate the total score for a given solution, we just need to sum all the values for each criterion, multiplied by the weight of the criterion. We can see an example in Table 6.4, related to different applicants that will help us select the best candidate to be hired by the company.

In Table 6.5, we have another example of a weighted decision matrix related to a personal problem: "Which house to buy?" We have four houses, A to D, and six different criteria, as shown in the table. Compared to the previous example, we do not represent the weight as a percentage, but instead as a number indicating its importance, from 0 to 3, in parentheses close to the criteria. For each house and criterion, we assign a value from 0 to 10 depending on our opinion. Now, we have to multiply this value by the weight of the criterion to determine the weighted

Table 6.4 Weighted decision matrix for selection between applicants

Criteria	**Weight**	**Applicant 1**	**Applicant 2**	**Applicant 3**
Experience	25%	90	90	50
Job Knowledge	15%	50	70	50
Achievements	10%	70	90	50
Education	10%	50	90	50
Communication Skills	10%	25	90	50
Teamwork	10%	20	50	50
Adaptability	10%	70	50	40
Motivation	5%	20	20	50
Integrity	5%	80	55	65
Total	100%	58.5	73.75	49.75

Table 6.5 Weighted decision matrix for decision on which house to buy

Criteria	**A**	**B**	**C**	**D**
Square meters (3)	8 → 24	10 → 30	7 → 21	8 → 24
Location (3)	9 → 27	4 → 12	5 → 15	6 → 18
Garage places (2)	5 → 10	8 → 16	5 → 10	0 → 0
Garden (2)	7 → 14	4 → 8	6 → 12	8 → 16
Views (2)	8 → 16	9 → 18	3 → 6	5 → 10
Price (3)	5 → 15	6 → 18	4 → 12	5 → 15
Total	106	102	76	83

importance of that criterion for each house. Finally, we have all these values in a column to determine the total score for each house. The house with a higher score will be the one to buy, according to the values we have assigned to the different criteria.

As the last activity in this chapter, we propose creating a weighted decision matrix related to one of the running problems we proposed in the introduction.

Activity 6.6
Weighted Decision Matrix

EXERCISE
Create a weighted decision matrix of the problem you selected in the first section of the chapter.

Another method for deciding which solution to choose is related to the conditions a given solution "must have" and the ones it would be "nice to have." The "must haves" are characteristics that must be present in a solution; for example, a candidate applying for a job must be an engineer. The "nice to haves" are other characteristics that are not vital to be present in a solution but would be beneficial if present. For example, the candidates for a job may not need to have experience programming in a particular language, but it would be advantageous if they had it. So, from all the possible solutions, we select those that meet all the necessary "must haves." Among these selected solutions, we choose the one with the most "nice to haves" or the best combination of them.

6.5 Conclusions

In this chapter, we have explored how to approach a problem using different methodologies, creating a path to reach a good solution by improving our expertise in the "Problem Solving" skill. First, we have learned how to analyze the situation, both from a situational perspective—understanding where we are—and from a problem perspective—identifying the root causes of the problem, rather than just addressing the apparent ones. Next, we have examined how to generate alternative ideas to solve the problem, using both logic and creativity. Finally, we have concluded by exploring different methods for selecting one of the solutions from the alternatives generated in the previous step.

We do not want to finish this chapter without leaving some advice when dealing with problem-solving situations:

- Never jump straight to a solution: first, conduct situational and problem analysis, then generate various solutions, and finally evaluate and decide on the best one.

- Always work with teams: different people think differently, so collaborating with others can help you see the problem from multiple perspectives and generate diverse solutions.
- Separate the two processes: distinguish between idea generation and solution generation. This way, you can generate a wider range of ideas or solutions and then select the best one when you have more options.
- Use both logic and creativity: when seeking possible solutions, it is important to use both logical thinking and creative approaches. This is another reason why working in groups can help find a broader range of solutions.
- Always consider implementation: the best solution may not be the most feasible if it is challenging to implement or beyond the company's capabilities.

References

[1] *ESCO Classification, Skills & Competences*, European Commission (2024) https://esco.ec.europa.eu/en/classification/skill_main.

[2] Handbook on "Problem-Solving Skills", series: Soft Skills for Good Managers. *Centre for Good Governance* (2017).

[3] J. D. Bransford, and B. S. Stein, *The Ideal Problem Solver*, W. H. Freeman, New York (1993).

[4] T. Richards, *Problem Solving: Proven Strategies to Mastering Critical Thinking, Problem Solving, and Decision Making*, CreateSpace Independent Publishing Platform (2015).

[5] Kepner–Tregoe Method Guide for Problem Solving (2023) https://www.linkedin.com/pulse/problem-solving-kepner-tregoe-method-simplified-guide/.

[6] O. Serrat, "The Five Whys", *Knowledge Solutions* 30 (2009).

[7] *Root Cause Analysis Blog, 5 Whys and the Monumental Mystery* (2020) https://www.sologic.com/en-gb/resources/blog/english-uk/5-whys-and-the-monumental-mystery.

Further reading

Courses on Problem Solving by Linkedin https://www.linkedin.com/learning/search?keywords=problem%20solving%20techniques.

Problem Solving and Decision Making – Excellence in Financial Management: https://exinfm.com/training/M2C1/index.html.

Chapter 7
Learning strategies

Fátima M. García Doval[1]

7.1 Introduction

So far, we have analyzed what soft skills are and their importance in acquiring and developing the professional competencies for a fully qualified expert needs. But, due to their nature, we have to bear in mind that this process of acquisition is permanent and, in fact, never ends. Soft skills put into action several of the executive functions of the brain. Executive functions are the set of cognitive processes behind our controlled behavior. Although there is no consensus on how to divide or name these functions, they include attention, self-control, and working memory, among others. Most authors agree that memory, inhibitory control, working memory, reasoning, emotion regulation, etc., play a significant role when it comes to creative thinking, problem solving, conflict management, communication, and, in general, in any personal and professional skill that is not restricted to a specific domain.

This said, we must bear in mind that these executive functions and, as a result, the soft skills they sustain, are strongly content-dependent. This means that once an ability is cultivated in a given set of tasks, the person becomes skilled in those tasks but not necessarily in others not closely related [1]. Numerous experiments have shown [2] that memory very much depends on the prior knowledge of the individual and that the transferability of a soft skill is limited. A person might become highly skilled in memorizing Pi digits or DNA sequences, but this ability does not naturally transfer to other domains, such as chemical structures or play scripts. Many soft skills depend on a deep understanding of concepts or the mastery of certain abilities. To come up with creative, innovative solutions, a person must draw on previously acquired knowledge, that is, on formerly learned material, a material that is due to become partially obsolete sooner rather than later.

The natural consequence of the above-mentioned is that any professional needs to spend their entire life learning new up-to-date information, procedures, etc., to stay relevant in their field. This is known as lifelong learning, and it is one of the major challenges in continuing professional development (CPD) and in-service training.

[1]Departamento de Pedagoxía e Didáctica, Universidade de Santiago de Compostela, Spain

Because of the enormous amount of time a professional is going to spend devoted to learning, the most interesting, cost-efficient approach is to optimize this process to make the most of it and to allow the person to be completely autonomous, self-guided in fulfilling their learning needs in any area, taking advantage of formal, non-formal and informal opportunities of education. In other words, we want professionals to be able to deploy a wide range of quality learning strategies to cater to any need that may arise.

Learning is inevitable. When we talk about learning strategies, we are talking about maximizing results and, somehow, controlling what is to be learned. This chapter is useful for both learners—basically anybody—and teachers. All activities and recommendations can be put to use for oneself or in the process of instruction.

In this chapter, we will focus on what learning strategies are and why deliberately teach them. Then, we will face some of the myths and misconceptions about learning that are widespread among the public. We will focus on the idea of learning as a trace of thinking, that is, on the importance of purposeful thinking for learning. Memory is the foundation of learning, and it is inherently linked to forgetting. We will present the understated importance of forgetting in any learning process. And finally, we will analyze the role of motivation and success and how these two concepts relate to each other to foster learning.

7.2 Learning strategies: what they are and why teach them

We must begin this journey by understanding what learning strategies are, as they tend to be identified with so-called study techniques. The former is based on scientific knowledge, whereas the latter appears to be packed with myths and biased perceptions. For this reason, we will not refer to study techniques as the general public understands them, and to prevent confusion, we will avoid the term completely.

7.2.1 What are learning strategies?

To understand what learning strategies are, it is fundamental to understand the nature of learning. There is a plethora of definitions for it, but all of them refer to the acquisition of knowledge, skills, and values that have a perceivable effect on a person's behavior in a wide sense.

> "Acquiring knowledge and skills and having them readily available from memory so you can make sense of future problems and opportunities." [3]
>
> "Learning is a process that occurs within nebulous environments of shifting core elements – not entirely under the control of the individual. Learning (defined as actionable knowledge) can reside outside of ourselves (within an organization or a database), is focused on connecting specialized information sets, and the connections that enable us to learn more are more important than our current state of knowing." [4]

It has been suggested that the term learning defies precise definition because it is put to multiple uses. Learning is used to refer to (1) the acquisition and mastery of what is already known about something, (2) the extension and clarification of the meaning of one's experience, or (3) an organized, intentional process of testing ideas relevant to problems. In other words, it is used to describe a product, a process, or a function. [5]

The mere interaction with a variety of contexts generates learning as a result of experience. In that sense, learning is, as we have already mentioned, inevitable, and it is what defines our behavior, the decisions we make, the actions we take, how we perform them, etc. But, although learning is natural, when we talk about learning, we imply influence or control over that learning process, maximizing its results, or improving the chances of success when succeeding is hard. In other words, most of the time when we talk about learning, we refer to deploying learning strategies.

How we learn is well studied and determined by cognitive psychology, whereas how to combine that core knowledge with the findings in other areas—such as sociology, anthropology, or philosophy—is the realm of pedagogy. Although there is currently a great hype about neuroscience and its impact on education and learning, it is a level of analysis that does not shed light on them. At least, no more than chemistry or quantum physics. How neurons or clusters of neurons form, for example, engrams—a network of interconnected neurons that sustain the physical substrate of memories—or how a certain stimulus is capable of triggering a neural response here or there is of no practical use in implementing a professional development plan—or any other study plan whatsoever, the same way that deep understanding of quantum physics is of no use in understanding cancer, even though it has an impact in its detection and treatment.

7.2.2 Metacognition

As we have just mentioned, neuroscience cannot be directly transcribed into learning strategies [6]. It can only confirm or explain, at a physiological level, what cognitive psychology already knows. And, among its certainties, it tells us about the importance of knowing what we know and also knowing how to learn more to develop our skills, widen our knowledge, and shape our attitudes and values.

That is what we call metacognition:

- What we know about what we know, that is, assessing our prior knowledge.
- What we know about how to learn more and how much we are progressing in our knowledge.

Humans are particularly clumsy when it comes to this due to the biased nature of our thinking [6]. With regard to what we know, we often confuse how knowledge feels and the awareness of ignorance. When we know a little, we are not aware of our ignorance, and we tend to be affected by the Dunning-Kruger effect [7]. Due to this popular cognitive bias, people with low competence in one particular task or aspect tend to overestimate their abilities in the field. It is important to stress that it is not a question of intelligence, as the list of renowned intellectuals falling spectacularly into the Dunning-Kruger effect is endless. Highly intelligent people may be subject to this Dunning-Kruger effect mainly because they are overconfident in their abilities to solve any given matter. When it comes to this effect, the focus is always on the task or area of knowledge itself, and the level of expertise we have about it.

On the other hand, when we know a lot, we are painfully aware of the myriad of details, particular cases, exceptions, and specific aspects we still ignore. When this happens, we fall into the Impostor syndrome [8]. Behind this name lies a behavior pattern in which an individual, who is undoubtedly competent in a subject, experiences a persistent lack of confidence, self-doubt, and fear of being exposed as a fraud. In other words, experts tend to overestimate the level of competence of others on their subject matter while dramatically underjudging their own. Both cognitive biases illustrate effectively how poor is, in general, our ability to assess our knowledge and competence.

Sadly, if confronted with the task of acquiring new knowledge and competence, that is, whenever we need to learn more, we are, by no means, any better. We often confuse familiarity and learning. We feel we are learning a lot when, in fact, we are just revisiting material we have previously seen. That way, we choose learning strategies based on revising, re-reading, copying, underlining, or highlighting parts of the text, etc. Such futile strategies are often linked to traditional study techniques. They may have an impact in the short term (some degree of effectiveness if the test or assessment takes place in the near future), but they fail in generating a robust knowledge that stands the test of time.

Instead, learners should resort to activities that require recalling and putting memory to the test: writing with their own words (without the original material at hand), drawing diagrams and mind maps by heart, inventing relevant questions about the material and answering them, etc. Those strategies that mean elaboration, recalling, and testing against the forgetting curve are the ones that most effectively tell our brain that certain information is valuable and might well be needed with a specific degree of frequency in the future. So, it is a good idea to save it in long-term memory.

This is key because when activating the neurological systems that clean up our memory by throwing away the irrelevant information and consolidating the important information, something that happens primarily during our sleep, the brain will fix and make ready-to-use that knowledge that has proven itself necessary to perform tasks or attend to demands. We will come back to this idea later in this chapter.

So, apart from learning strategies based on recalling (remembering by heart, without cues or indicators), sleep also plays a key role in effective learning. Without a good sleep, all the efforts invested in remembering and testing our knowledge might well be futile [9].

7.3 Myths and misconceptions about learning

There are plenty of myths and misconceptions about learning, and what is most worrying is that many of them are prevalent in texts on education aimed at the general public, and also in professional texts and teacher training manuals.

Some of them have no clear origin and seem to derive from a biased, unscientific analysis of the learning process, the same way the mere observation of the sun "revolving around the Earth" made humanity come up with the idea of the Earth being the center of the universe.

Some others are the consequence of a clumsy, not methodical or uninformed (amateur) interpretation of advanced scientific research, concluding too simply to be correct, and directly applying them to educational contexts.

This is mostly the case with neuromyths, such as that which states that we only use 10% of our brain or that there are right and left-brained persons, depending on the predominant hemisphere in a given task. That is why an interest in neuroscience by itself, from a non-specialized point of view, is no protective factor against neuromyths whatsoever. In fact, exposure to popular science about how the brain works appears to increase the likelihood of believing in neuromyths [10], rather than dispelling them.

Although apparently contradictory, it all makes perfect sense. What most strongly shapes what we see and how we interpret it is a combination of our prior knowledge, beliefs, and expectations. This bundle accounts for what is known as "confirmation bias." This thinking pattern gives the utmost importance to the clues and pieces of evidence that tend to confirm our point of view while ignoring or, at least, minimizing those that challenge our previous ideas. As we just adore being right, there is a variable discomfort in being wrong—technically called "cognitive dissonance"—that we try to avoid at all costs.

All strategies based on recalling the learning material or any other thing we need to learn feel uncomfortable because it makes it clear that we do not know as much as we would like to think. On the contrary, strategies based on repetition feel reassuring and pleasant because familiarity makes us feel competent and skillful, with an apparent high degree of mastery of the matter.

As we will analyze later, to improve our learning, we need to resort to a combination of:

- reflective practice,
- deep thinking,
- sense of achievement (self-efficacy and expectations of success).

All of them need to overcome a sense of underachievement. That is natural to any learning process. To get this right, the most powerful tool is managing expectations correctly, and to do so, it is important to avoid those based on biased knowledge about the learning process.

7.3.1 Popular myths and the truth behind them

Some popular myths and misconceptions [6] that stand in the way of effective learning strategies are:

- mere repetition is an appropriate strategy for learning;
- our learning capabilities depend solely or primarily on our intelligence or any other innate potential;

- intelligence or executive functions are determined by genetics, and there is nothing to do about it;
- the more time invested in studying, the better. The key factor is time versus the efficacy of the techniques deployed;
- we are good at assessing what we know and what to do to learn more (metacognition);
- passing exams with high grades is a more accurate indicator of successful learning as it implies a good understanding of the subject and a durable knowledge.

But the facts behind these myths are quite a different story. Repetition *per se* is not our best learning choice. It improves familiarity with what we need to learn, but this sense of recognition and command is illusory.

When we learn something, we build strong neuron networks called engrams. Later on, whenever we need to recall that learned material, we need to activate that physical brain structure that is where knowledge resides. Not being able to do so might be due to two different errors:

- a wrong or faulty engram, which occurs when the physical imprint of knowledge was not formed properly or vanished due to its weakness;
- a failure in the identification of the appropriate route to that engram; in other words, we have that piece of knowledge securely stored in our brain, but we cannot find the way to reach it. We know that we know, we have it on the tip of our tongue, but we are unable to reach it.

Both errors can occur at the same time and are more probable when learning is superficial or weak, that is the reason why it is so important to build a strong corpus of knowledge with minimum gaps and multiple interconnections with other pieces of information. The more pathways leading to a certain knowledge, the easier it will be to reach it when necessary, even using alternative routes when needed. Recognizing something is quite straightforward, as the clue to reach the information necessary is given to us, so we know from the start which engrams to activate and the secure route to reach them.

7.3.2 Nature versus nurture

As for intelligence, it is true that it has a genetic base, but it is also influenced by nurture and the circumstances in which the individual grows. But what is most important is that the intelligence itself is no more than the power of the engine [11]. It is important, of course, but what is not inherently determinant is how that power itself is put to work, what we do with it, and how we employ it. That is, the learning strategies used.

There is a wide range of learning opportunities for any given intelligence, provided that we put into practice all the activities and tasks that have been backed up by an enormous amount of quality research evidence.

The same applies to all executive functions: working memory, attention, etc. Although there is an innate component, nurture plays an important role, and what is determinant is the resources a person is able to put into action to make the most of them.

For some people, learning is natural, and they deploy appropriate strategies with no prior guidance. Others may require more guidance, but this does not mean they are unable to perform well when it comes to learning. They are a bit more dependent on what they do and how they optimize their natural abilities. It is also important to state that everybody takes advantage of the use of effective and efficient learning strategies, even those who consider themselves "natural born learners." The obvious conclusion is that study time is not the key factor. The amount of time spent on study activities is a poor predictor of the quantity and quality of the knowledge acquired.

Due to the counterintuitive nature of many aspects involved in learning, we are severely limited in our perceptions about our own learning. It is necessary to make a deliberate, explicit, guided effort, rooted in scientific knowledge (primarily, learning and cognitive psychology), to adequately assess what we really know and how to broaden that knowledge.

7.3.3 A word about tests

The same essential reasons are behind the difficulty in assessing the knowledge of a person using an exam or any other sort of test. Designing and implementing an exam is a very delicate, complicated question that involves not only content knowledge but also pedagogical knowledge and a solid background in didactics.

One of the most delicate aspects of teaching is, in fact, that of assessing and providing formative feedback.

We identify as formative feedback the relevant information provided to learners about their progress and results, about how they perform in comparison to expectations, about the difficulties we observe or the misconceptions they are developing... All this valuable information should be handed over during the formative period, when the students/learners still have time to reorient their work and efforts. In other words, formative feedback is not a qualification but some good pieces of pedagogical advice given when the learner still has time to do something about it.

Besides, a considerable number of tests are designed in such a way that mere familiarity is enough to pass them, or the questions they include do not show the misconceptions or conceptual errors the student has developed about the subject, because they can be answered by heart and without elegant elaboration.

For all this, passing an exam might not be a good indicator of our actual command of any given matter.

7.4 Learning as a trace of thinking

Many of the myths and misconceptions explained in the last section have their roots in the altered nature of our perceptions about learning itself. Understanding that, as we already mentioned, learning is, basically, creating imprints in our brain, is not enough. We must bear in mind that these imprints (engrams) are knitted in dense neural networks. The denser the network, the deeper and richer the knowledge will be. On the one hand, as every piece of knowledge is linked to a myriad of others, reaching it and, thus, recalling it to put it into use, will be easier. On the other hand, this intricate network will facilitate the creation of novel associations that have the potential to root meaningful creativity. The term meaningful creativity refers to insights aimed at solving problems and challenges instead of just random ideas.

Memory works as a whole [12], but depends on different brain structures. While we will not discuss this interesting matter in depth—since it is not essential for our purposes—the key idea to understand is that when we refer to memory in a general sense, we are actually combining the roles of two distinct types of memory:

- episodic, and
- semantic.

7.4.1 Episodic and semantic memory

Our episodic memory files trace of what has happened to us. It is, then, the housing of our memories in the most popular sense. As for semantic memory, it stores all the concepts, ideas, and meanings we recognize and understand.

In the first case, episodic memory takes advantage of strong, vivid experiences. Emotions play an important role in this type of learning. Some personal events are associated with physiological responses packed with reward or stress neurotransmitters, such as dopamine, oxytocin, or adrenaline, to mention just a few well-known examples.

Dopamine is a neuromodulatory molecule that works as a neurotransmitter. It plays a major role in reward-motivated behavior, although it also has an impact on other brain pathways. In popular culture, it is normally presented as the hormone of pleasure, but the most prevalent scientific opinion is that dopamine signals salience, conferring prominence to any given event and linking it to motivation.

Oxytocin is also a hormone and a neurotransmitter with a primary role in the molecular pathway of social bonding. It is, therefore, present in love, reproduction, and childbirth, as well as in social activities, boosting cohesion. It is the hormone that tells apart "us" from "them."

Adrenaline is a hormone that functions as a neurotransmitter in the regulation of visceral functions. It plays a key role in essential role in the fight-or-flight response. It does so by increasing blood flow to muscles and heart output and blood sugar level in a way that gives the body all the fuel and power that will eventually be needed. It also has some other outputs, such as pupil dilation, restricting all non-vital functions, and improving perception performance.

Chances are that these events are relevant, in one sense or another, to our survival. So, probably, it is interesting to file them to use the derived knowledge later on in our lives, providing a faster, more adjusted response.

Strong feelings and emotions make experiences more vivid and memorable as an adaptive response to the perils of life, but, as a consequence, they get in the way of semantic memory. Our brain is too busy fixing the key details of an important moment that might save our lives in the future, and it consumes all the resources in our reasoning. Even our perception is altered. If the emotion is overwhelming, we focus with such concentration that we switch to tunnel vision, ignoring all peripheral details and accessory information. It is not the time and moment for reflections and meaning. All the relevant aspects of the episode will be imprinted in our autobiography, and this is the primary mode in which we construct it.

But, most probably, we will not be able to immerse ourselves in deep reasoning, and we will resort to automated patterns of response. Basically, this is the reason why vivid and exciting learning experiences might, in fact, interfere with the learning process. While busy fixing episodic memories, it is difficult to devote brain resources to reflect on meanings, that is, on building semantic memory.

To summarize it, episodic memory takes great advantage of strong feelings and emotions, but semantic memory is quite another story. A very different one, in fact. When it comes to learning about concepts, facts, principles, etc., the activity that matters the most is "reflection" [9]. It is often said that "learning is the trace of thinking." The more and deeper we think about something, the more connections we will build in our semantic network linking it to other previous pieces of knowledge.

7.4.2 What thinking leaves behind

When we refer to active learning, then, we are not implying the physical activity of moving around. What has to be active is our brain. Our hands and the rest of our bodies may or may not be active. That is mostly irrelevant when it comes to semantic learning with one important exception: when the learning material is so novel that we lack the abstraction tools that would make it possible to manipulate it solely with our intellectual resources.

For the purposes of this chapter, we will consider thinking as a behavior in itself. And, like all behaviors, it needs meaningful repetition to command it. As the popular saying declares wisely, "practice makes perfect."

When approaching a new area or domain, it is often beneficial to visualize the problem by drawing it, outlining a diagram, or simulating it using any available objects. This transitory help will be gradually retrieved, but what needs to remain is the purposeful act of thinking about the learning material.

Some useful ways of thinking are, among others:

- recognizing the concept among a varied set,
- finding examples for yourself,
- linking this new knowledge with that previous one the learner already possesses,
- identifying it in your own past experience.

The question is whether this practice can be performed as an empty transversal ability. Thinking is not an ability; thinking about something is. Although it is true that the more we think about different matters, the more chances we have of performing well in a new domain, it is also important to remember that thinking is very much content-dependent. Our capability of thinking about a subject is determined, to a great degree, by our prior knowledge, both in terms of amount and depth.

7.5 Memory and forgetting

The thinking activity that has proved itself more efficient and effective is recalling, that is, trying to remember by heart the learning material. As we think about whatever we are learning and try to retrieve the acquired knowledge from our memories, we are sending a powerful message to our brain, stating its importance and relevance.

Our brains invest a lot in preserving those pieces of knowledge that might be useful or needed in the future. The rest of the information we come across during a regular day—which is a considerable amount—is disposed of in a process that we identify with the term "forgetting."

Although this process runs non-stop, it is during the different phases of sleep that most of the major clearance takes place. It is then that our brains get rid of all the bits and pieces of reflections, perceptions, and meanings left around during wakefulness.

It is also when what is considered important is appropriately filed for future occasions.

A robust, well-structured knowledge needs quality sleep time between learning activities, and bad sleep habits or sleep deprivation lead to fragmented, less useful knowledge that will be difficult to access in the future. In severe cases, it ends up in memory problems, amnesia, and the lack of effective learning (that is, the impossibility or extreme difficulty in acquiring new knowledge due to brain damage).

7.5.1 Distributed mixed practice

To make the most of our learning efforts, apart from making sure we guarantee our brain the quality rest it needs to neatly fix new knowledge, there are other tips we should bear in mind [13,14]:

- distribute practice [15],
- mixed (also called varied) practice [16].

The first one refers to the benefits of spacing learning sessions. A practical example would be that, instead of devoting 3 h to a learning material concentrated in 1 day, we split that available time into three sessions of 1 h on separate days. In doing so, the first obvious consequence is that we guarantee the above-mentioned sleep time. But, also, we force our brain into recalling the learning material again and again over a period of time, and, as we already know, this is a powerful signal that marks any piece of information or ability as relevant and, thus, fixes knowledge in our semantic network.

The spacing between learning activities (study or practice sessions) should be variable. In the beginning, learning activities should be concentrated on consecutive or alternative days. Then, progressively, the interval between sessions should be widened following a sort of inverted forgetting curve (see Figure 7.1).

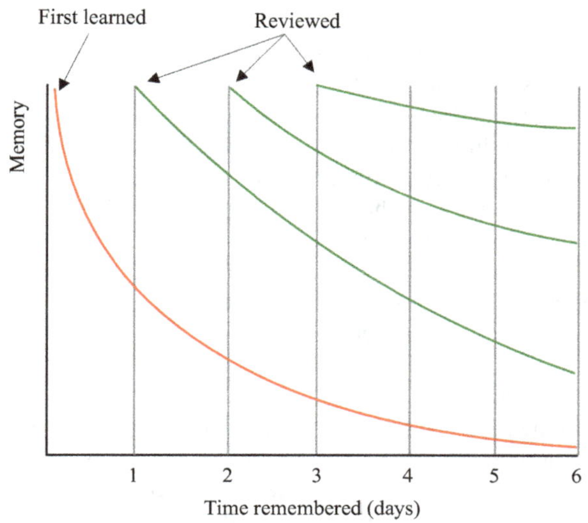

Figure 7.1 Representation of the forgetting curve without repetition (red) and with spaced repetition (green)

In the beginning, learning material should be revised often to progressively space study sessions. The appropriate moment is that in which recalling the material is still possible (at least, in most part) but hard.

As for mixed practice, the idea behind it is exactly the same. Let us suppose that we have to learn three different subjects, or three different types of activities within the same domain. Instead of dedicating one hour to each subject in uninterrupted blocks, once we have a minimum command of each subject, it is a better approach to mix all the subjects or activities. This way, our brain strives to switch from one to the other, and it is never sure which neural network to activate or which response pattern to put into action.

This thrilling state, as that generated by a partially forgotten material or the mere testing against the forgetting curve, is what has been called "desirable difficulties" [17]. The difficulties that boost learning must be challenging but not frustrating or out of the learner's reach. Desirable difficulties are, in other words, manageable challenges. They put our brains in a state of alert that makes us prone to fixing knowledge, that is, learning.

7.6 Motivation and success

Finally, it is essential to consider the importance of motivation and success in the learning process and how it affects the efficacy of learning activities.

It is a common belief that motivation leads to success and, in fact, many people claim that the sole origin of their difficulties when it comes to learning—or even the cause of an eventual failure—is, indeed, such a motivation shortage. This might be apparent, but certainly, it is a bit more complicated than that. Although motivation affects success, it is success that mostly affects motivation.

When we have to deal with a task, we unconsciously make predictions about our chances of succeeding in it based on the perceived difficulty and, mostly, on our previous experiences performing similar tasks (not necessarily identical). This judgment is what we call the perception of self-efficacy, and it shapes the expectations about the outcome of the tasks we face. When we have previous successful experiences, we have a positive self-efficacy in our abilities to complete that given task. The opposite works the same way round. A personal record of failure instills the perceived idea that we are clumsy or unable when it comes to that activity.

The question is that nobody likes failing. Any given task we perceive that is "not for us" is immediately avoided and, when avoidance is not an option, poorly attempted. It is important to make clear that this might not be an explicit decision. But, by not devoting genuine effort in repeated attempts, we have the perfect alibi in the probable event of a failure. This "not really trying" approach relieves part of the uncomfortable (or even painful) feelings associated with failure.

On the other hand, when we have a history of successes at our back, any opportunity to perform that task is welcomed and, even if we fail by accident or due to the difficulty of the assignment, it is highly probable that we will retry again and again, generating multiple opportunities for success. This positive credit is not endless, of course. If failure persists, our perception of self-efficacy will change, and things will gradually be different in the future.

The same way a history of failure induces a perception of negative self-efficacy about a task—which is the case of a considerable amount of students about learning in one or more subjects—a history of success might turn around negative self-perceptions into a positive one that will boost the options of performing considerably better in future occasions by trying harder, longer and with a positive attitude.

So, motivation owes more to success than success owes to motivation. It is possible to revert a vicious circle in which negative past experiences (mainly failure) develop into negative self-perception about the task, eventually boosting evasive approaches finally resulting into a most probable failure; into a virtuous circle in which successful experiences progressively generate a perception of positive self-efficacy that generates a genuine interest and perseverance in the task that, chances are, will result in success (see Figure 7.2).

In this sense, it is important to assume that errors are an inherent part of the learning process and should never be felt or perceived as failure. But, most importantly, it is key to understand that errors do not result in learning except "learning to avoid." It is success that generates effective learning. That is why all learning strategies deployed and teaching activities proposed should be carefully tailored to be challenging but with high chances of a positive result. They have to be challenging because any routine activity does not activate the cascade of neurophysiological and psychological events leading to a sense of reward. Nobody

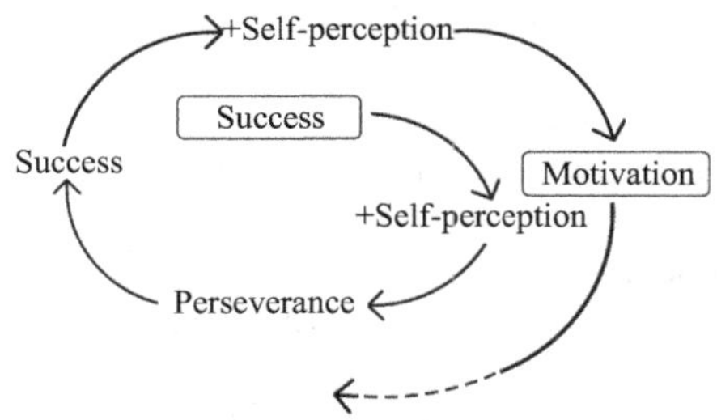

Figure 7.2 Virtuous circle of success and motivation

finds interesting and stimulating routine activities. Practice makes perfect, it is true, but mainly if it is reflective practice and if it includes desirable difficulties. And those learning activities have to be success-oriented because we need to maintain failure at bay to improve self-efficacy and, eventually, motivation.

7.7 Conclusions: a useful plan

To outline all of the aspects and issues addressed in this chapter, it is useful to set a learning plan that might be used as a whole or adapted to someone's needs and restrictions.

- Draw a diagram or a mind map that includes all the key factors of the material to learn without referencing the original fonts (studying notes, manuals, teaching materials, etc.) or any other external source (printed and online complementary materials). When the diagram or mind map is ready, then go to the original source to complete the gaps you already know you left behind (what you know you do not remember). Then go through the original material to find any other aspects or pieces of information your diagram is missing (what you did not remember at all).
- Reconstruct the information by (orally or in writing) performing a coherent discourse. Try to be as exhaustive as possible. Use your mind map as a reference (now complete) to fill in the gaps you might have, only after a serious effort to remember for yourself. If necessary, go back to the original material, but only if the clues in the mind map are not enough after a deep, purposeful thought.
- Write down questions for you to answer, and/or feed an Artificial Intelligence (AI) with the original material (if it is not confidential or includes personal data) and ask it for questions. Use only questions, not the answers, as you will provide them yourself as part of your learning process. Once you have completed the task, ask the AI for its answers, compare them with yours, and see how you can combine both to arrive at the best possible answers. Refer to the original material to ensure that nothing important or relevant is missing.

- If you have the opportunity, explain what you are trying to learn to another person. They do not need to be aware of the matter and can be completely unfamiliar with the subject. Explaining something to someone who is completely new to the information and knowledge is one of the most powerful ways of consolidating our memory, as it will make us clearly aware of the gaps and lacks in our understanding.
- All along these steps, try to link the learning materials with what you already know.
  - How does this relate to any prior knowledge?
  - Can I find examples of this other than the ones in the learning material?
  - Can I make valid analogies to explain this piece of knowledge to other individuals?
  - In what obvious and not-so-obvious situations may I put this knowledge to use?
  - How does this learning material change or polish my previous knowledge on this (or any other subject)?
  - What would I like to learn now, taking this knowledge as a starting point?
- Try to take each step on different days. Ideally, 1 or 2 days between the first steps and a week or more between the last ones. This will allow time for sleep between study sessions and will also highlight the effects of the forgetting curve.
- It is always a good idea to mix other learning material in the in-between days and even on the same day, as the effort of changing the focus tells our brain to fix and retain that learning strongly and durably.
- When the learning process involves a skill or ability, rather than declarative knowledge (facts, concepts, principles, etc.), the same general scheme applies, but instead of recalling, we must put the knowledge into practice:
  - First, practice the skill or ability until you get, reasonably, a hang of it. When a decent amount of attempts is correct, given the stage you are at (beginner, intermediate, advanced, etc.), stop practicing for the day. When you reach that desired level, practice just a little bit more to get confident, but do not keep on persisting over and over again. At that stage, learning makes no difference, it is a waste of time and effort.
  - Start mixing the practice of different skills early. Mix programming in Python with C++, for example, or different types of physics problems. When all the practice focuses on the same topic, you already know what set of strategies to deploy to solve it. But when you face mixed practice, you have to select, among a wider range of options, what resources are useful in that specific task. This makes the practice more memorable and helps to strengthen a resourceful network of skills, abilities, and knowledge. This is one of the bottom lines for creativity.
- Always set yourself learning goals that contain a bit of a challenge but are realistic and achievable. Completing tasks is the best way to experience a sense of success. This feeling builds up on our perception of self-efficacy, that is, our capability of completing a certain task. The higher our self-efficacy, the more

we will persist in it until we achieve success. This virtuous circle is one of the most efficient paths to motivation.

References

[1] Ripp, I., Emch, M., Wu, Q. *et al.*, Adaptive working memory training does not produce transfer effects in cognition and neuroimaging, *Translational Psychiatry* 2022;12:512, https://doi.org/10.1038/s41398-022-02272-7.

[2] Sala, G, and Gobet, F., Does far transfer exist? Negative evidence from chess, music, and working memory training, *Current Directions in Psychological Science*, 2017;26(6):515–520, https://doi.org/10.1177/0963721417712760.

[3] Brown, P. C., Roediger III, H. L., and McDaniel, M. A., *Make It Stick: The Science of Successful Learning*, 2014, Belknap Press: An Imprint of Harvard University Press, Cambridge, MA.

[4] Siemens, G., Connectivism: A learning theory for the digital age. *International Journal of Instructional Technology and Distance Learning*, 2004; 2:1–9.

[5] McCaughan Smith, R., *Learning How to Learn: Applied Theory for Adults*, 1982, Pearson Prentice-Hall, Upper Saddle River, NJ.

[6] Ruíz Martín, H., *Edumitos. Ideas sobre el Aprendizaje sin Respaldo Científico*, 2023, International Science Teaching Foundation, London.

[7] Kruger, J., and Dunning, D., Unskilled and unaware of it: How difficulties in recognizing one's own incompetence lead to inflated self-assessments. *Journal of Personality and Social Psychology*, 2000;77(6):1121–1134, https://doi.org/10.1037//0022-3514.77.6.1121.

[8] Feenstra, S., Begeny, C. T., Ryan, M. K., Rink F. A., Stoker, J. I., and Jordan, J., Contextualizing the impostor "Syndrome". *Frontiers in Psychology* 2020;11:575024, https://doi.org/10.3389/fpsyg.2020.575024.

[9] Willingham, D., *Why Don't Students Like School?: A Cognitive Scientist Answers Questions About How the Mind Works and What It Means for the Classroom*, 2021, Jossey-Bass Inc., San Francisco, CA.

[10] Ferrero, M., Garaizar, P., and Vadillo, M. A., Neuromyths in education: Prevalence among Spanish teachers and an exploration of cross-cultural variation. *Frontiers in Human Neuroscience*, 2016;10:496. https://doi.org/10.3389/fnhum.2016.00496.

[11] Boring, E. G., Intelligence as the tests test it. *New Republic* 1923;36:35–37.

[12] Sigman, M., *The Secret Life of the Mind: How Your Brain Thinks, Feels, and Decides*, 2017, Little Brown & Co., New York.

[13] Ebbinghaus, H., *Memory: A Contribution to Experimental Psychology*, 1913, Teachers College Press, New York. https://doi.org/10.1037/10011-000.

[14] Carlson, R. A., Sullivan, M. A., and Schneider, W., Practice and working memory effects in building procedural skill, *Journal of Experimental Psychology: Learning, Memory, and Cognition*, 1989;15:517–526.

[15] Baddeley, A. D., and Longman, D. J. A., The influence of length and frequency of training sessions on the rate of learning to type, *Ergonomics* 1978; 21(8):627–635.

[16] Battig, W. F., Intratask interference as a source of facilitation in transfer and retention. In J. F. Voss (ed.), *Topics in Learning and Performance* (pp. 131–159), 1972, Academic Press, New York.

[17] Bjork, R. A., Institutional impediments to effective training. In D. Druckman and R. A. Bjork (eds.), *Learning, Remembering, Believing: Enhancing Human Performance* (pp. 295–306), 1994, National Academy Press, Washington, DC.

Chapter 8
Planning

Francisco Cima[1] and Ana María Canto Esquivel[2]

Planning is indispensable for team performance and project management, serving as a cornerstone of the methodological competencies recognized in the European Union´s Skills Panorama. Various studies highlight the importance of enhancing planning skills in engineering to effectively navigate collaborative projects in both academic and professional settings. In recent years, positive experiences have been documented in diverse educational programs, highlighting the crucial role of engineering education in fostering methodological competencies among future engineering professionals, including planning skills. This chapter discusses theories and strategies that foster planning skills in educational environments. Examples are provided to illustrate how the successful implementation of these strategies has contributed to the development of teamwork skills among participants. This chapter aims to serve as a valuable resource for educators and students seeking to enhance teamwork effectiveness by developing planning skills.

8.1 The importance of planning

The literature on team effectiveness and project management emphasizes the importance of planning. Numerous studies on the first perspective indicate that a critical driver of success and the viability of teamwork is the degree to which a team develops an effective initial plan of action to move toward its goals [1]. From the project management perspective, planning is a phase of the lifecycle that focuses on establishing the project's deliverables and requirements, which are essential for effective execution [2]. Building on these perspectives, we first discuss how planning contributes to team and project success to underline the meaning and importance of individual planning skills.

8.1.1 Planning and team effectiveness

One of the most widely accepted theoretical models for studying teams and work groups is the Input-Mediator-Output-Input (IMOI) framework of team

[1]Departamento de Ingeniería Industrial, Tecnológico Nacional de México – Instituto Tecnológico de Mérida, México
[2]Department of Engineering Management and Systems Engineering, Old Dominion University, USA

effectiveness. This framework describes the input factors, such as team composition and individual characteristics, mediating processes, such as strategy formulation and coordination, and the resulting outputs of teamwork, such as the quality of the outcomes and team members' satisfaction. The IMOI framework proposes that these outputs serve as feedback to the team, acting as new inputs in an iterative process as teams go through several performance episodes [1].

Planning is a crucial process for team effectiveness that occurs in the early stages of team development, as group members identify and allocate the necessary resources for task execution. Effective planning comprises two interconnected elements: (1) gathering information from group members and stakeholders regarding the team's mission, and (2) evaluating this information to formulate the appropriate strategies for mission accomplishment [1]. Gathering information may involve sharing/exchanging information, seeking information, and communication processes that enable the team to collect data to support their initial plan or actions. Subsequently, the team needs to critically assess the information retrieved to develop plans that describe their goals and the best strategies to achieve them [1]. Thus, planning is a crucial transition mechanism that helps identify the main tasks, goals, and resources, as well as how to translate them into desirable outcomes and performance.

Furthermore, the temporal view of teams suggests that they are not static entities but rather dynamic systems that operate through a series of related episodes and phases that require several team processes [3]. Over time, teams may encounter new challenges, change priorities, or acquire additional resources. In these circumstances, as teams and work groups most often operate over multiple performance episodes [4], they need to re-evaluate new information and circumstances to close performance gaps, ensure goal alignment, and adapt strategies to overcome potential problems. Planning becomes a recurring process that allows teams to gather information about emerging opportunities, reassess risks, and adjust their strategies and courses of action.

In sum, planning is critical for sustained team effectiveness as it facilitates the alignment of team efforts and resources with objectives by evaluating information and developing strategies. Moreover, teams can ensure their actions remain focused on desirable outcomes by regularly reviewing their plans.

8.1.2 Planning and project management

Project management refers to the "application of knowledge, skills, tools, and techniques to project activities to meet the project requirements" [5, p. 6]. It involves the five key processes: initiating, planning, executing, monitoring and controlling, and closing [5]. Planning defines a project's direction since identifying requirements and constraints while balancing these with the stakeholders' needs, concerns, and expectations is at the core of project management. The planning phase of project management involves a comprehensive preparation that covers different facets, including scope, objectives, resources, schedule, and sequences of actions [5,6].

The main outputs of the planning process are the project management plan and the project documents, which will become the guide for the subsequent processes

[5]. These outputs delineate the project's critical aspects, including time management, cost management, quality management, procurement management, communication management, and risk management. Thus, the level of detail and flexibility of the planning process's outputs significantly impact the outcome of a project because they establish the groundwork for future steps and choices [7]. As the project is carried out, it is possible that additional planning will be needed to gather, analyze, and understand new information and how changing conditions may affect the project [5]. Therefore, planning in project management can be seen as an iterative process that has implications for the overall operation and results.

8.1.3 Planning skills

Grounded on the dominant framework of teamwork effectiveness, teamwork skills represent an individual's set of abilities to contribute to collaborative work effectively [8]. Planning skills encompass a set of abilities that are critical for gathering relevant information about the group's tasks and resources, and for developing strategies to achieve goals [9]. Due to their relevance for processes occurring at the early stages of task development, planning skills are strongly linked to other competencies necessary to effectively transition toward the execution of task work, such as mission analysis and strategy formulation.

The behavioral view of teamwork, grounded in the taxonomy of team processes, ties planning to mission analysis and strategy formulation, with both being integral parts of the transition phase. Mission analysis involves the identification of the team's tasks and challenges and the evaluation of conditions and resources for mission accomplishment [10]. Mission analysis skills are necessary for defining major goals and tasks required for articulating the team plan and for understanding one's role in goal accomplishment [8]. Planning skills relate to mission analysis skills as both refer to an individual's capacity to identify the primary elements of the task, the essential requirements, and the potential challenges. Strategic formulation skills are also relevant for planning, as these competencies focus on developing actionable approaches to achieve team outcomes. From the perspective of team processes, strategy formulation primarily supports the evaluation and establishment of directions to guide goal accomplishment [8]. Thus, an individual's planning abilities also involve strategy formulation skills, including the capacity to assess alternative courses of action in response to changing scenarios.

What are planning skills?

Planning skills represent a set of abilities critical for gathering relevant information regarding the group's tasks and resources and developing strategies toward the achievement of goals [9]. These skills relate to critical competencies to effectively transition toward the execution of task work, such as mission analysis and strategy formulation.

An alternative view of planning skills is described in the taxonomy of knowledge, skills, and abilities (KSA) most relevant for teamwork, developed by Stevens and Campion [11]. This taxonomy places planning and coordination competencies within the domain of self-management skills. Accordingly, self-management skills, including planning, play a crucial role in enabling teams to allocate individual tasks and roles effectively from the initial stages, ensuring a balanced distribution of workload among group members during task work [11]. Planning contributes to coordination by identifying key aspects of the task work, such as the main activities, timelines, and responsibilities, which enable individuals to establish interdependencies that lead to coordinated task execution.

In summary, planning skills relate to various abilities essential for effective collaboration and task accomplishment. These skills include the capacity to gather and analyze relevant information, set the major team goals, and anticipate challenges and requirements. Effective planners can contribute to defining strategies to transition toward task execution and ensuring that the team´s efforts align with its goals.

8.2 Planning in the EU skills agenda

Planning skills are a fundamental aspect of professional development. They hold particular relevance in the context of the EU's Skills Panorama due to their intrinsic connection with the effective use of resources and adaptability. Proficiency in planning relates to several core competencies described in the digital, lifelong learning, and environmental sustainability frameworks. These frameworks are integral to the EU Policy Agenda for the upcoming years, underscoring the significance of developing strong planning capabilities among future professionals.

8.2.1 Planning for lifelong learning

The LifeComp framework describes the key competencies that every person should continuously develop through education and lifelong learning [12]. While planning is not an explicit competence category, it can enable individuals to develop and implement their goals and plans, assess risks, adapt strategies to changing circumstances, and collaborate in civic life. For example, individuals with solid planning skills can set clear and achievable learning goals focused on specific knowledge or abilities. Using their planning competencies, a lifelong learner can reflect on their current knowledge base, identify areas for improvement, and prepare strategies to track progress through continuous development.

Lifelong learners can also apply their planning skills to assess the potential constraints they may encounter as they engage in continuous learning. Individuals can elaborate contingency plans and strategies to mitigate or overcome problems such as time constraints or lack of financial resources. Furthermore, the ability to evaluate risks and constraints is linked to another important aspect of lifelong learning: adaptability [12]. Adaptation enables learners to adjust their strategies and approaches to navigate the changing circumstances, needs, and requirements. As lifelong learning is a dynamic journey, learners must be able to rapidly adapt their strategies in response to new development opportunities, emerging technologies, or skill demands.

Planning skills are also relevant in collaboration and participation in civic life. Effective planning may enable individuals to identify common goals appropriately, align individual efforts toward collective needs, and actively participate in resource allocation and strategy development with their community. Individuals' planning skills may contribute to developing a collective plan to approach societal issues or to drive positive change.

Planning aids lifelong learning competencies outlined in the LifeComp framework as it enables individuals to develop action plans, manage their own learning time efficiently, and participate in the planning process of collective initiatives to navigate the dynamic social, economic, and technological world.

8.2.2 Planning for a digitalized world

The DigComp framework identifies the competencies individuals need to succeed in today's increasingly digitalized world [13]. Planning skills are crucial for effectively managing and optimizing the digital tools that are prevalent in modern work environments. Effective planning helps streamline workflows and guide resource allocation, including the necessary digital tools and technologies to achieve organizational goals. In today´s digitalized world, individuals must be able to gather relevant information and assess how new technologies can enhance both personal and organizational performance.

In an organizational context, individuals with strong planning abilities can contribute to identifying and evaluating relevant digital tools from the early stages of project development. Because technology is constantly evolving, individuals with solid planning skills frequently revisit and adjust their strategies. Effective planning, therefore, enables adaptation to technological advances and fosters innovation.

The DigComp framework describes some competence areas where planning skills become relevant, including information and data literacy and digital content creation. Since planning involves the capacity to gather and analyze relevant information for achieving objectives, individuals with strong planning skills can locate, retrieve, organize, and analyze relevant digital information to support strategy development in a rapidly changing technological environment. Likewise, digital content creation benefits from individuals who are well-equipped with planning skills, allowing them to develop and integrate digital content strategies aligned with project goals and needs.

Developing planning skills in the context of the DigComp framework enhances an individual´s ability to identify digital technologies relevant to their work and adapt to technological changes in an increasingly digitalized world. Planning skills play a critical role in the context of the EU Skills Agenda by supporting the digital competencies needed to thrive in the digital economy.

8.2.3 Planning for sustainability

The GreenComp framework provides definitions and guidance for sustainability competencies [14]. Within this framework, planning is crucial for integrating sustainability goals into mission analysis and strategy formulation. By considering these goals in their plans, individuals can align their actions with the vision of promoting a sustainable world.

The concept of planning for sustainability involves integrating the environmental, social, and economic impacts of proposed actions into decision-making to develop strategies that ensure sustainable outcomes [15]. Moreover, planning for sustainability requires employing a holistic, long-term approach with continuous assessment and improvement [16,17]. This approach has applications in several fields, including urban planning, corporate strategy, education, and engineering [15]. For instance, engineering designs might incorporate renewable energy sources and efficient transportation systems in their projects. Other applications could involve developing strategies to reduce waste and promote a sustainable culture across organizations.

8.3 Enhancing planning skills

The ability to navigate today's working environments through effective communication, multidisciplinary collaboration, adaptability, and continuous learning has become increasingly crucial for engineering professionals as they face progressively complex challenges [9]. However, employers frequently express concern about deficiencies in so-called "soft skills" when hiring new engineering graduates, highlighting the need for engineering education to go beyond technical mastery and equip future professionals with the essential skills for success in their professional life [18,19].

Teamwork and collaboration are at the core of contemporary engineering practice. Engineers no longer work in isolation and increasingly need to convey their knowledge and skills to colleagues and teammates. Today's engineering challenges require the coordinated efforts of diverse teams to address complex, multifaceted problems. The ability to work effectively with others, understand and manage group dynamics, leverage collective strengths toward goal accomplishment, and share knowledge is thus a cornerstone of professional engineering competence.

Under the EU's Skills Panorama, engineering programs represent a natural scenario to address the teamwork skills gap and ensure their graduates not only master technical tasks but also thrive in team-based environments. Although both accreditation agencies and employers expect engineering graduates to develop professional competencies such as planning skills during their academic journey, not many programs provide impactful opportunities for their development or improvement [19]. By integrating evidence-based, team-based active learning approaches throughout the curriculum, engineering programs could contribute to enhancing critical soft skills that their students need to navigate the challenges of the 21st century.

Fortunately, a range of innovative teaching and learning strategies has been successful in promoting the acquisition of teamwork skills in engineering education. This section will examine specific methodologies and case studies to highlight the value of learning from the best practices across various institutions. By understanding the pedagogical approaches, assessment techniques, and institutional support structures that enable students to develop abilities like planning, engineering educators can better prepare their graduates for success.

8.3.1 How can planning skills be learned?

As the development of soft skills in engineering becomes more prominent, educators are constantly looking for strategies to prepare their students effectively. Although there is no one-size-fits-all approach to teaching and assessing these skills, many examples of successful and promising interventions can inform further strategies targeting teamwork skills, including planning. Some of the most common experiential learning methods for teaching teamwork skills range from short activities—such as design thinking exercises and game-based learning—to course-long approaches, like project- and problem-based learning [9,20]. These methods are often preferred over traditional strategies, such as lectures or conferences, because they directly allow individuals to experience the dynamics of a team. These methods are summarized in Table 8.1.

Design-thinking teamwork is used to promote group collaboration to solve complex problems driven by customer needs [21]. Collaboration is crucial in

Table 8.1 Common teaching methods for teamwork skills in academic settings

Method	**Description**
Design-thinking teamwork	Design thinking is an iterative approach to problem-solving that emphasizes customer-centricity and, often, interdisciplinarity. In design thinking collaboration, teams work together to develop customer-centric solutions within project deadlines and specific requirements.
Game-based learning	Game-based learning is an approach commonly used in educational and organizational contexts that incorporates gamification elements such as rewards and a sense of mission to accomplish team training objectives (e.g., coordination).
Team-based learning	A structured, active learning approach where students first study new conceptual material independently and then work in groups during class to apply these concepts through various activities. This method focuses on fundamental concepts and brief applications to teach technical content and develop professional skills.
Problem-based learning	Problem-based learning is a constructivist approach that uses open-ended problems as the core of the learning process, requiring students to research and acquire new knowledge to solve these challenges. This method involves collaborative problem-solving in small teams, with instructors acting as facilitators, and is effective in developing problem-solving and teamwork competencies.
Project-based learning	A strategy that requires students to work collaboratively on projects that address real-world, complex problems. Student teams often work over extended periods (e.g., one semester) to deliver tangible outcomes while instructors guide their efforts and learning.

design thinking because it often involves multiple disciplines and requires individuals with different mindsets and backgrounds to combine their knowledge and skills to address ill-defined problems. Besides developing cognitive skills such as creativity and analytical thinking, design-thinking collaboration provides opportunities to learn interpersonal (e.g., communication) and methodological skills (e.g., time management and planning). For instance, participants can exercise their planning skills through the identification and evaluation of consumer requirements, as well as the project constraints, to develop suitable design solutions.

Game-based learning is a pedagogical approach that integrates gamification elements (e.g., rewards and a sense of mission) into non-game settings to support the learning process [9]. With the continuous technological advances in game simulations, game-based learning has become a common method for providing team-based experiences through artificially constructed environments in educational and organizational contexts [22]. Although its implementation can be costly, this approach can help engage digital-native learners more effectively than traditional strategies for teaching soft skills. Some relevant outcomes related to teamwork achieved in past implementations of game-based learning approaches include improved communication, coordination, and backup behaviors [22]. Studies suggest that implementing effective game-based learning should incorporate serious games within real-world scenarios and include personalization and collaboration [23] to attract and engage participants in the activity while they apply critical skills. Although game-based learning is implemented in short-term exercises, it also provides opportunities for developing planning competencies as teams work on their strategies.

Team-based learning is a group-based active learning approach used to teach new concepts and their applications [24]. It relies on the inverted classroom pedagogical strategy, in which students first revise new conceptual material independently and then use classroom time working in groups on the applications of such concepts through active learning activities [25]. Team-based learning differs from other group-based strategies, such as problem-based learning and project-based learning, because it focuses on fundamental concepts and relatively less extensive applications of newly acquired knowledge [24]. Hence, this approach can be used in lower-level courses over shorter periods to facilitate both learning technical content and developing professional skills, such as problem-solving and interpersonal teamwork skills [24,25].

Problem-based learning is an approach that uses problems at the core of the learning process to enhance students' understanding of new knowledge [26]. By engaging with a problem, students undergo a process of research to determine what they need to know to solve the problem [27]. In other words, the problem delineates what content knowledge is needed. Problem-based strategies involve open-ended problems that students need to solve by following a set of predefined steps [24]. Other defining characteristics of problem-based learning include collaborative problem-solving and the flexible role of the tutor [26]. Similar to other inverted classroom strategies, instructors guide small student teams during the problem-solving process, while traditional methods, such as lectures, are limited. Problem-based learning is the natural strategy for teaching problem-solving competencies

[9]. However, it could also be used to develop teamwork competencies during group work by guiding teams in defining their schedules, roles, and planning their work and communication strategies [28]

Project-based learning is an educational strategy that reflects the principles of social constructivism by emphasizing collaboration as a key component in knowledge generation. Project-based learning involves students working on real-world, complex projects over extended periods, culminating in tangible outcomes [29]. This approach mirrors authentic engineering challenges and helps prepare students for their future careers while simultaneously developing critical professional competencies. Implementations of project-based learning across various engineering disciplines have significantly improved students' performance, technical skills, and the development of problem-solving, collaboration, and leadership abilities [9,30]. Project-based learning environments are student-centered [29]. In this context, instructors assume diverse roles beyond traditional teaching, guiding students through the learning process. This dynamic shift fosters a more authentic learning experience, allowing learners to engage with projects as they would in professional settings [31]. Project-based learning offers several opportunities to teach teamwork skills, as it naturally integrates collaborative work into the learning process.

Among the different strategies used by educators, project-based learning emerges as the primary mechanism for developing teamwork-related skills in engineering. Structuring project-driven courses provides an excellent opportunity for individuals to experience first-hand collaborative work from start to end, as classes are designed around teams. Furthermore, the benefits of team-based implementations on teamwork skills, such as planning skills, can be strengthened by providing support from the instructor, guided hands-on activities that facilitate effective group transition processes, and supplementary tools that stimulate project planning and management.

An important consideration in the design of strategies to foster planning skills relates to the role of the instructors and what artifacts are implemented to guide the learning process. Findings from past interventions have shown that instructors must play a proactive role in facilitating the project work and assessing the skills learned; it is not enough to simply require participants to complete a set of tasks in teams [19]. Instructors can also incorporate additional elements to simulate real-world teams to make the experience more authentic. For example, past implementations of team-based learning have relied on web-based tools, project timelines, flow chart forms, and team charters that enable accountability and support project-based experiences. In the following sections, we summarize some promising examples of how to enhance planning skills in engineering.

8.3.2 A couple of exemplary cases

8.3.2.1 Teaching teamwork in engineering and computer science

An approach to enhance team collaboration skills in engineering and computer science using team projects supported by freely available web-based collaboration tools showed promising results in improving participants' abilities, including planning skills [19]. This approach considers eight key factors that contribute to successful team-based project implementations:

1) Adoption of agile principles. Teams are encouraged to follow the agile principles and take full responsibility for organizing and managing their work. This is supported by a CASE tool, which simplifies the role of working code while integrating essential work products and deliverables.
2) Guided definition of project objectives. Instructors use a set of assignments to guide participants through developing project objectives. These assignments include an analysis of the problem, a discussion of solutions and ideas, and a team activity to establish the final project objectives under change control.
3) Team formation strategies. A web-based tool is used to form the project teams. Participants are allowed to recruit members and join the team of their preference, which is expected to create cohesive teams.
4) Adoption of project asset libraries (PAL). Teams create their own PAL to store their work products. Each team designates a PAL administrator, who is responsible for approving the storage of team-approved products. In addition, each member maintains a personal product library to store individual contributions, which are peer-reviewed as needed.
5) Web-based communication platforms. In addition to external communication methods (e.g., email and phone calls), freely accessible tools facilitate group communication through messages.
6) Artifacts for project planning, execution, and tracking. Teams must develop detailed plans and schedules, which include information about activities, products, and responsibilities related to the project work. Furthermore, teams evaluate their progress against the plan and take necessary actions as needed.
7) Enabling accountability. Team members assume responsibility for completing specific tasks and peer reviewing the contributions of other members. Individual products are stored, and reviews are saved as comments.
8) Avoiding lengthy technical reports. Formal presentations and lengthy technical reports are avoided. Instead, the CASE tool automatically generates reports based on the working and final products.

The combination of artifacts and tools to facilitate group collaboration encourages participation and provides valuable insights for instructors about team progress and the level of collaboration within each team [19].

8.3.2.2 Developing critical collaboration skills in engineering students

Another example involved a project-based program that integrated web-based collaborative tools and instructional scaffolding tasks and artifacts to develop virtual collaboration skills among engineers [32]. The scaffolds targeted key cognitive and behavioral team processes to guide teams through the different phases of their projects. These included:

- A team charter artifact, a project plan artifact, and a team identity-building task to support mission analysis, goal formulation, and planning.
- A communication strategy artifact and a project plan artifact to support strategy formulation.
- Freely accessible tools to track work document progress, update task status, and conduct web conferencing for progress monitoring. Additionally, the teams completed a mid-project evaluation activity.
- A project plan evaluation and adjustment task to enhance team coordination.

Using a quasi-experimental design, the researchers examined the impact of the tools, tasks, and artifacts on developing engineering students' teamwork competencies. The results indicated that using the scaffolds positively influenced participants' self-managing skills, including goal setting and performance management competencies, as well as planning and task coordination abilities [32].

8.3.3 Challenges and opportunities

It is clear that engineers require not only technical knowledge within their field of expertise but also the skills to work effectively in teams [20]. Providing team-based experiences that allow students to develop these skills is not easy, as it requires structure and proactive guidance from faculty [19]. In addition to the lessons learned from past implementations of active pedagogical methods to incorporate teamwork skills discussed in the literature, there are also challenges and opportunities to consider. The most common challenges include:

- Assuming a proactive role in facilitating team-based experiences. Despite the growing interest in developing teamwork skills in engineering using active learning approaches (e.g., project-based learning), many programs are often implemented with minimal guidance on critical aspects of group collaboration, including team development, conflict management, and leadership [20]. Thus, instructors must adopt a proactive approach to assist in developing teamwork competencies [19].
- Time and effort are needed to guide effective teamwork. A significant challenge for instructors using active learning strategies is the time and effort needed to guide teams and assess teamwork skills development [19]. Learning from best practices and using empirically tested tools and artifacts might help optimize instructors' resources.
- A standard criterion for framing and assessing teamwork skills. Several conceptualizations and categorizations of teamwork skills complicate the establishment of an integrated framework to guide researchers and practitioners [33]. This issue also presents challenges in assessing teamwork skills development, as the methods employed vary between contexts. An opportunity exists regarding the use of theoretically driven and validated assessment techniques to measure skills development effectively [20,34].

Providing team-based learning experiences for engineering students is essential but challenging. It demands considerable time and effort because instructors must take a proactive role in facilitating group collaboration and the development of teamwork skills. It also requires a well-structured program supported by tools and artifacts to guide participants through the process of collaboration. To address the challenges, instructors can draw on evidence-based approaches when designing and implementing programs aimed at developing planning skills in engineering. Furthermore, there is an opportunity to establish a standardized framework for defining and assessing teamwork-related competencies.

8.4 Conclusions

Planning is essential for success in today's fast-paced, digitalized, and sustainability-focused world. Planning is a critical process that helps individuals and teams to align resources, develop strategies, and adapt to changing circumstances. It is essential for engineers, who often work on complex and interdisciplinary projects that require strong teamwork and project management skills. Engineering education is an ideal environment for cultivating the abilities outlined in the EU's Skills Agenda among future engineers. By teaching students to gather and evaluate relevant information, establish team goals, and contribute to strategy development, engineering programs can foster the essential soft skills for success. One effective approach to achieving this is by incorporating evidence-based, active learning methods that provide students with opportunities to practice and enhance their planning competencies. A structured focus on these skills not only prepares future engineers to excel in their careers but also equips them to address complex global challenges, driving innovation and sustainability in an increasingly interconnected world.

References

[1] D. R. Ilgen, J. R. Hollenbeck, M. Johnson, and D. Jundt, "Teams in organizations: From input-process-output models to IMOI models," *Annual Review of Psychology*, vol. 56, pp. 517–543, 2005, https://doi.org/10.1146/annurev.psych.56.091103.070250.

[2] P. Serrador, "The importance of the planning phase to project success. *Conference paper presented at PMI Global Congress*," *Project Management Institute, North America, Vancouver, British Columbia, Canada, Newtown Square*, pp. 1–10, 2012.

[3] M. A. Marks, J. E. Mathieu, and S. J. Zaccaro, "A temporally based framework and taxonomy of team processes," *Academy of Management Review*, vol. 26, no. 3, pp. 356–376, 2001, https://doi.org/10.5465/amr.2001.4845785.

[4] J. E. Mathieu, P. T. Gallagher, M. A. Domingo, and E. A. Klock, "Embracing complexity: Reviewing the past decade of team effectiveness research," *Annual Review of Organizational Psychology and Organizational Behavior*, vol. 6, pp. 17–46, 2019, https://doi.org/10.1146/annurev-orgpsych-012218-015106.

[5] Project Management Institute, *A Guide to the Project management body of knowledge (PMBOK® guide)*, Fourth Edition ed. (no. 1). USA: Project Management Institute, 2008, p. 507.

[6] K. H. Rose and G. Indelicato, "Book Review: A guide to the project management body of knowledge (PMBOK® guide), Fourth Edition," *Project Management Journal*, vol. 40, no. 2, pp. 104–104, 2009, https://doi.org/10.1002/pmj.20125.

[7] Y. G. Ayele, "The significance of planning and scheduling on the success of projects," *Economics & Management Information*, vol. 2, no. 1, pp. 1–10, 2023, https://doi.org/10.58195/emi.v2i1.66.

[8] O. Varela and E. Mead, "Teamwork skill assessment: Development of a measure for academia," *Journal of Education for Business*, vol. 93, no. 4, pp. 172–182, 2018, https://doi.org/10.1080/08832323.2018.1433124.

[9] M. Caeiro-Rodríguez, M. Manso-Vázquez, F. A. Mikic-Fonte, *et al.*, "Teaching soft skills in engineering education: An European perspective," *IEEE Access*, vol. 9, pp. 29222–29242, 2021, https://doi.org/10.1109/ACCESS.2021.3059516.

[10] J. A. LePine, R. F. Piccolo, C. L. Jackson, J. E. Mathieu, and J. R. Saul, "A meta-analysis of teamwork processes: Tests of a multidimensional model and relationships with team effectiveness criteria," *Personnel Psychology*, vol. 61, no. 2, pp. 273–307, 2008, https://doi.org/10.1111/j.1744-6570.2008.00114.x.

[11] D. L. Miller, "Reexamining teamwork KSAs and team performance," *Small Group Research*, vol. 32, no. 6, pp. 745–766, 2001, https://doi.org/10.1177/104649640103200604.

[12] A. Sala, Y. Punie, V. Garkov, and M. Cabrera, "LifeComp: The European framework for personal, social and learning to learn key competence," *Joint Research Centre (Seville site)*, 2020. [Online]. Available: https://EconPapers.repec.org/RePEc:ipt:iptwpa:jrc120911

[13] R. Vuorikari Rina, S. Kluzer, and Y. Punie, "*DigComp 2.2: The Digital Competence Framework for Citizens - With new examples of knowledge, skills and attitudes,*" *Joint Research Centre (Seville site)*, 2022. [Online]. Available: https://EconPapers.repec.org/RePEc:ipt:iptwpa:jrc128415

[14] G. Bianchi, U. Pisiotis, and M. Cabrera Giraldez, "GreenComp The European sustainability competence framework," *Joint Research Centre (Seville site)*, 2022. [Online]. Available: https://EconPapers.repec.org/RePEc:ipt:iptwpa:jrc128040

[15] P. R. Berke and M. M. Conroy, "Are we planning for sustainable development?," *Journal of the American Planning Association*, vol. 66, no. 1, pp. 21–33, 2000, https://doi.org/10.1080/01944360008976081.

[16] K. Johnson, C. Hays, H. Center, and C. Daley, "Building capacity and sustainable prevention innovations: a sustainability planning model," *Evaluation and Program Planning*, vol. 27, no. 2, pp. 135–149, 2004, https://doi.org/10.1016/j.evalprogplan.2004.01.002.

[17] G. F. G. Teixeira and O. Canciglieri Junior, "How to make strategic planning for corporate sustainability?," *Journal of Cleaner Production*, vol. 230, pp. 1421–1431, 2019, https://doi.org/10.1016/j.jclepro.2019.05.063.

[18] M. F. Ercan and R. Khan, "Teamwork as a fundamental skill for engineering graduates," in *2017 IEEE 6th International Conference on Teaching, Assessment, and Learning for Engineering (TALE)*, 12–14 Dec. 2017, pp. 24–28, doi:10.1109/TALE.2017.8252298.

[19] R. Lingard and S. Barkataki, "Teaching teamwork in engineering and computer science," in *2011 Frontiers in Education Conference (FIE)*, 12–15 Oct. 2011, pp. F1C-1–F1C-5, https://doi.org/10.1109/FIE.2011.6143000.

[20] L. J. Shuman, M. Besterfield-Sacre, and J. McGourty, "The ABET "Professional Skills" — Can they be taught? Can they be assessed?," *Journal of Engineering Education*, vol. 94, no. 1, pp. 41–55, 2005, https://doi.org/10.1002/j.2168-9830.2005.tb00828.x.

[21] E. Vardakosta, G. Priniotakis, M. Papoutsidakis, M. Sigala, A. Tsikritsis, and D. Nikolopoulos, "Design thinking as a co-creation methodology in higher education. A perspective on the development of teamwork and skill cultivation," *European Journal of Educational Research*, vol. 12, no. 2, pp. 1029–1044, 2023, https://doi.org/10.12973/eu-jer.12.2.1029.

[22] S. L. Marlow, E. Salas, L. B. Landon, and B. Presnell, "Eliciting teamwork with game attributes: A systematic review and research agenda," *Computers in Human Behavior*, vol. 55, pp. 413–423, 2016, https://doi.org/10.1016/j.chb.2015.09.028.

[23] D. Zhao, C. H. Muntean, A. E. Chis, G. Rozinaj, and G. M. Muntean, "Game-based learning: Enhancing student experience, knowledge gain, and usability in higher education programming courses," *IEEE Transactions on Education*, vol. 65, no. 4, pp. 502–513, 2022, https://doi.org/10.1109/TE.2021.3136914.

[24] R. M. O. Connell, "Adapting team-based learning for application in the basic electric circuit theory sequence," *IEEE Transactions on Education*, vol. 58, no. 2, pp. 90–97, 2015, https://doi.org/10.1109/TE.2014.2329650.

[25] L. K. Michaelsen, A. B. Knight, and L. D. Fink, *Team-based learning: A transformative use of small groups in college teaching*. Taylor & Francis, 2023.

[26] P. Ellingsen, T. Tonholm, F. R. Johansen, and G. Andersson, "Learning from problem-based projects in cross-disciplinary student teams," *Education Sciences*, vol. 11, no. 6, p. 259, 2021, https://doi.org/10.3390/educsci11060259.

[27] D. R. Woods, R. M. Felder, A. Rugarcia, and J. E. Stice, "The future of engineering education: Part 3. Developing critical skills," *Chemical Engineering Education*, vol. 34, no. 2, pp. 108–117, 2000.

[28] A. Carvalho, "The impact of PBL on transferable skills development in management education," *Innovations in Education and Teaching International*, vol. 53, no. 1, pp. 35–47, 2016, https://doi.org/10.1080/14703297.2015.1020327.

[29] N. F. Jumaat, Z. Tasir, N. D. A. Halim, and Z. M. Ashari, "Project-based learning from constructivism point of view," *Advanced Science Letters*, vol. 23, no. 8, pp. 7904–7906, 2017.

[30] V. Hayashi, J. Dutra, F. Almeida, *et al.*, "Implementation of PjBL with remote lab enhances the professional skills of engineering students," *IEEE Transactions on Education*, vol. 66, no. 4, pp. 369–378, 2023, https://doi.org/10.1109/TE.2023.3243532.

[31] O. Ayala, K. Gutierrez, F. Cima, *et al.*, "Experiences during the implementation of two different project-based learning assignments in a fluid mechanics course," *Paper presented at 2022 ASEE Annual Conference & Exposition*, Minneapolis, MN, 2022, doi:10.18260/1-2-41801.

[32] P. Pazos, N. Magpili, Z. Zhou, and L. J. Rodriguez, "Developing critical collaboration skills in engineering students: Results from an empirical study," in *ASEE Annual Conference and Exposition*, New Orleans, Louisiana, 2016, doi:10.18260/p.26750.

[33] V. Rousseau, C. Aubé, and A. Savoie, "Teamwork behaviors: A review and an integration of frameworks," *Small Group Research*, vol. 37, no. 5, pp. 540–570, 2006, https://doi.org/10.1177/1046496406293125.

[34] F. Cima, P. Pazos, M. Lee, *et al.*, "Developing teamwork skills in undergraduate engineering students: A comparison between disciplinary and cross-disciplinary projects," *IEEE Transactions on Education*, vol. 68, pp. 86–94, 2024, https://doi.org/10.1109/TE.2024.3456098.

Chapter 9
Digital skills

Fátima M García Doval[1] *and Enrique Costa-Montenegro*[2]

9.1 Introduction

What we commonly know as “digital skills” has raised deep interest in various institutions worldwide because of their crucial role in driving the digital transformation of different areas of expertise and society in general. These digital skills, along with their corresponding knowledge and attitudes, constitute what is known as “digital competence.” Recognizing its importance, the European Union has developed – and recently updated – the Digital Competence Framework for Citizens (called DigComp 2.2). Additionally, given the relevance of digital competences in education, a specific framework for educational sector (known by the name of DigComEdu) has been created for students, teachers, and educational organizations. Both frameworks include detailed guidelines for assessing and progressively developing digital expertise in different professional fields. Along this chapter, we will examine the key elements of this digital competence (including knowledge, skills and abilities, as well as ethical issues and attitudes toward digital technologies and their use). We will explore the theoretical foundations of this approach and outline effective pedagogical strategies for developing digital competence in both formal and non-formal higher education contexts. Practical tips and examples will also be provided to illustrate these concepts.

9.1.1 Necessity of digital competences

In May 2018, the Council of the European Union updated the Recommendation on Key competences for Lifelong Learning (2006/962/EC) [1]. These recommendations initially identified eight essential competences for citizens, including those necessary for personal fulfillment, a healthy and sustainable lifestyle, employability, active citizenship, social inclusion, and digital competence. Although some competences were revised in the 2018 update, digital competence remained relevant and consistent, highlighting its central role in today’s world.

[1]Departamento de Pedagoxía e Didáctica, Universidade de Santiago de Compostela, Spain
[2]Departamento de Enxeñaría Telemática, Universidade de Vigo, atlanTTic, Spain

So, digital competence was included in this group of eight essential competences, along with literacy; multilingual; science, technology, engineering, and mathematics (STEM); personal, social and learning to learn; citizenship; entrepreneurship; and cultural awareness and expression competence. For each of these eight competences, essential knowledge, skills, and attitudes were defined.

Digital competence is described as involving "the confident, critical and responsible use of, and engagement with, digital technologies for learning, at work, and for participation in society. It includes information and data literacy, communication and collaboration, media literacy, digital content creation (including programming), safety (including digital well-being and competences related to cybersecurity), intellectual property related questions, problem solving and critical thinking."

As shown in Figure 9.1, in 2023, 56% of European citizens aged 16–74 had at least basic overall digital skills [2]. According to "Europe's Digital Decade: Digital Targets for 2030" [3], the objective is to have, at least, 80% of all adults with basic digital skills by 2030. Referring to Figure 9.1, in 2023, we were 24% points below the target, with significant differences between European countries. The highest proportions were found in the Netherlands (83%) and Finland (82%), while the lowest were in Romania (28%) and Bulgaria (36%). Additionally, in most countries, men reported slightly higher levels of digital skills than women.

This information was obtained using the digital skills indicator (DSI), a comprehensive composite indicator that covers the five key areas of the DigComp framework: information and data literacy skills, communication and collaboration skills, digital content creation skills, safety skills, and problem-solving skills. The DSI has played a crucial role in providing reliable and comprehensive data on the digital skills landscape across Europe.

As we can see, a significant portion of the European population still needs to acquire at least basic digital skills, meaning there is still a long way to reach the target of 80% by 2030. It might be assumed that those lacking basic digital skills are primarily elderly individuals, but this is a misconception. Many young people also lack basic digital skills, even though the common belief that younger generations are inherently digital natives. The fast and massive spread of this idea "digital natives," originally proposed by Marc Prenski with a different perspective, has often been misunderstood. The oversimplification and rapid dissemination of this concept have contributed to many ineffective approaches to developing digital competence among young people [4].

9.1.2 Knowledge and skills in digital competences

According to the Council of the European Union recommendation on key competences for lifelong learning [1], all competences integrate the three types of learning content: knowledge, skills, and attitudes. In our case, digital competence, the basic knowledge is related to "understanding how digital technologies can support

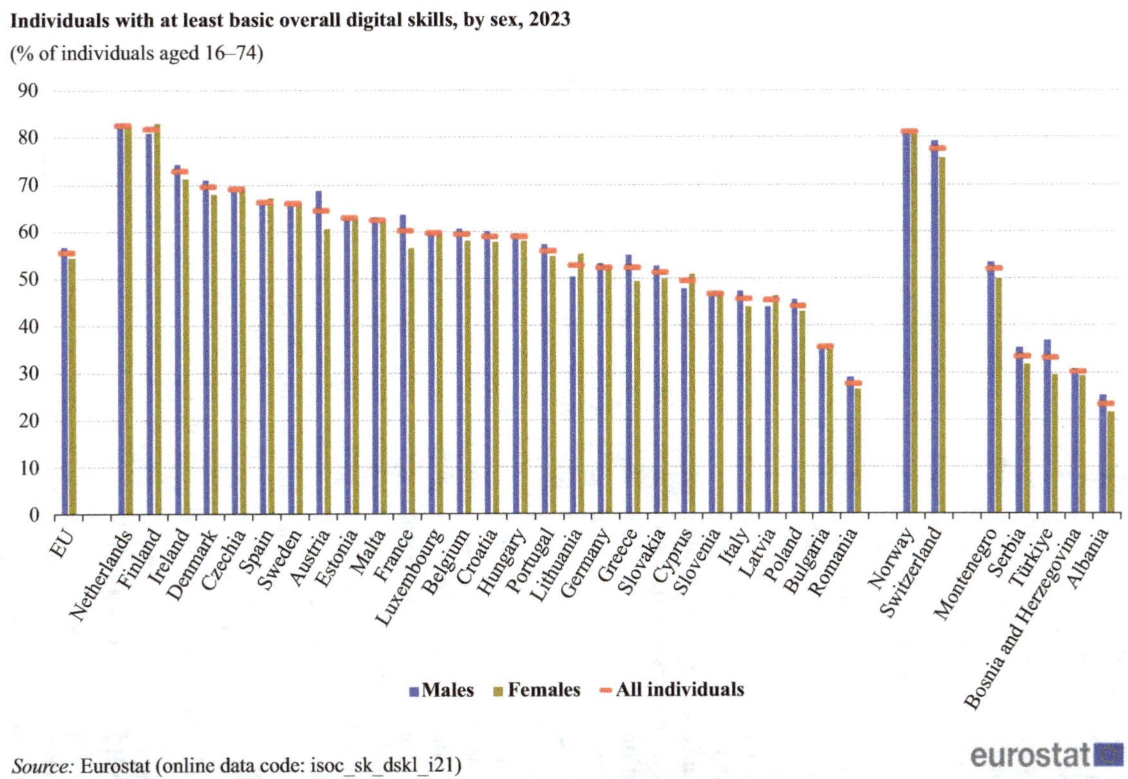

Figure 9.1 *Individuals with at least basic overall digital skills, by sex, 2023 (% of individuals aged 16–74 years). Source: Eurostat (isoc_sk_dskl_i21).*

communication, creativity and innovation, and be aware of their opportunities, limitations, effects and risks." This also includes "knowing the basic function and use of different devices, software, and networks" without forgetting "a critical approach to the validity, reliability and impact of information and data made available by digital means." Regarding skills, digital competences emphasizes the ability to "use of digital technologies to support their active citizenship and social inclusion, collaboration with others, etc.," as well as "the ability to use, access, filter, evaluate, create, program and share digital content" and "be able to manage and protect information, content, data, and digital identities, as well as recognize and effectively engage with software, devices, artificial intelligence or robots" are remarked. Finally, the identified attitudes promote "a reflective and critical, yet curious, open-minded and forward-looking attitude to the evolution of digital technologies." Additionally, it "also requires an ethical, safe and responsible approach to the use of these tools."

It is important to note that digital competence is a specific domain that, while relevant to a vast majority of human activities, is displayed within a particular context. Therefore, the knowledge, skills, and attitudes needed to develop it require a focused and intentional learning process. Transferability from other areas of competence is not as immediate and easy as we often assume. For example, critical thinking is not restricted to the digital world. However, the sheer volume of misinformation, disinformation, and fake news online makes it significantly more challenging to exercise this ability in digital contexts compared to traditional ones. The same applies to other ethical areas, where potential risks and dangers might be underestimated or even completely ignored.

Another important example is data protection. The capabilities of digital computing in processing, analyzing, and cross-referring vast amounts of data are difficult for regular citizens to comprehend. So, although most individuals are pretty cautious about their personal image and information in real, analogue world, they tend to be far more careless online.

Behavior is contextual, and digital environments often lead individuals to minimize the consequences of certain decisions. Therefore, developing digital competence involves more than transmitting valuable concepts, training specific skills, or promoting core ethical attitudes. It is also, and maybe more important, understanding the particularities and determinants of a global, fast-changing context that is, in essence, overwhelming on a human scale.

9.2 Selected digital skills

As it exceeds the scope of this chapter, we will not analyze all the skills outlined in the European Digital Framework in depth. Doing so would be unnecessary, given the existence of comprehensive manuals that already fulfill this task excellently. Instead, to provide a helpful starting point, we have selected a series of digital skills from the DigComp 2.2 document, which we consider fundamental to approaching digital competence in today's world.

9.2.1 Accessing information: how to browse, search, and filter it

The internet has revolutionized how we access and consume information, transforming it into a vast and universal digital library. With the rise of the internet, thanks to its digital repositories, primarily in the form of web pages, a staggering amount of knowledge is now at our fingertips in just a moment. According to Statista [5], as of July 2024, there are 5.45 billion internet users worldwide, representing approximately 67.1% of the global population of nearly 8 billion.

However, what software do we use to find information on the internet? The most common and widely used tool is the web browser. And what online tools helps us locate specific information? Normally, search engines. It is important to note that these two concepts are not always clear, as they are often used interchangeably. Understanding the distinction between them can significantly enhance digital literacy.

A web browser and a search engine are two distinct tools that serve different purposes. A web browser is a software application that allows users to access and view websites on the internet. Web browsers can be used on different devices, like computers, laptops, smartphones, tablets, and smart TVs. Popular examples of web browsers include Mozilla Firefox*, Google Chrome†, and Safari‡, to cite some of the more popular ones.

However, a search engine is a specialized online tool that searches for and retrieves information from the internet based on user queries. Search engines such as Google§, Bing‖, and Yahoo¶, among the most popular examples, use complex algorithms to index and rank web pages, providing users with relevant results according to the set queries.

The key difference lies in their functionality: a browser is used to access and view websites, allowing us to navigate and display them, whereas a search engine is used to find websites with specific information or content across the vast expanse of the internet.

So, to access the information available on the internet, we need a web browser to view it. If we do not know the exact location (we do not know the specific webpage), a search engine may help us search for and find the desired information.

Usually, both web browsers and search engines are free to use, but there is often a hidden cost. We must be aware that the online content available to us at no monetary charge is often funded through advertising or by selling user data. For instance, when using search engines, the first results displayed after a search are usually marked as "sponsored." This means that these results may not necessarily be the most relevant to our queries but are instead the ones advertisers have paid to

*https://www.mozilla.org/en-US/firefox
†https://www.google.com/intl/en_en/chrome/
‡https://www.apple.com/safari/
§https://www.google.com/
‖https://www.bing.com/
¶https://search.yahoo.com

appear first. The goal is to maximize the chances that the users click on them, thereby promoting the sales of the advertised products.

Additionally, all queries made by a user are stored to create what is called a user profile. Not only are the search records tracked, but also the web pages we visit. Cookies (a small data file in our web browser) can track our browsing behavior, impacting both our privacy and security. Using this information, the owners of search engines can learn about our preferences, interests, and tastes. This information is often sold by the companies owning the search engines to advertisers who highly appreciate knowing which personalized publicity to offer in order to maximize product sales.

Another important aspect to understand is the personalization of search engine results. Search engines use artificial intelligence (AI) algorithms to generate responses tailored to individual users based on their user profile. However, these results are also influenced by various factors, such as context (e.g., geographical location), the device being used (e.g., computer or a smartphone), region local regulations (which can dictate what type of content can be shown), and the behavior of other users (e.g., trending topics at that moment). All of these factors influence the response to our queries. This can be observed when comparing the results for two people making the same query or when using the private browsing features of a web browser.

Finally, we must recognize that not all the information on the internet is accessible. Some of it is private, while other may require a fee or registration for access.

Given all the above, one crucial skill we must develop when accessing information on the internet is recognizing that the results of our queries are usually influenced by a series of opaque factors, especially our user profile as interpreted by unknown algorithms. Moreover, this profiling process is typically carried out by complex, non-explicative AI systems. As a result, even the enterprises behind search engines may not fully understand why certain decisions are made or which relevant factors (and their relative weights) influence the tailored results. What is particularly concerning is that most users are unaware that the results they trust might be mediated by the hidden agendas of search engine companies or by some random algorithmic biases.

The quintessential version of this AI powered search is the use of generative AI to provide specific answers. Tools like Perplexity** or Copilot†† are examples of this technology, capable of providing elaborate answers to complex questions, sometimes even referencing scientific papers. Setting aside the possibility that some of this research evidence might be completely made up, the risks are clear. When using a traditional search engine, the selection of which results to explore is on the human side. This highlights the importance of human judgment and responsibility in the search process. Sure, this selection might be influenced by availability or a perceived sense of relevance, but there is still some margin for the

**https://www.perplexity.ai/
††https://copilot.microsoft.com/

user to contrast and verify information. In contrast, when we ask a generative AI system to provide an answer to a complex question, the AI itself selects the relevant sources and, within them, the important pieces of information. The human role is only limited to the very end of the process, when a coherent answer has been generated. We will come back to this challenge further in this chapter.

If we want to obtain more neutral search results (without influences), we may choose some of the following options:

- Use the private (incognito) mode of our web browser: In this mode, the browser does not use information from existing cookies, and the search is not linked to a specific account, so the results should not be adjusted to our profile. Additionally, any cookies generated during the session in private mode are not stored, so they do not affect future browsing.
- Use private browsers: For even more objective results, we can use a browser that operates exclusively in private mode. These browsers keep your privacy by automatically blocking tracking mechanisms and do not save any browsing history or website data. An example of this technology is Firefox Focus.[‡‡]
- Use private search engines: These search engines are designed for users who value privacy and do not want to be tracked online. They do not monitor our activities or store our search history. A well-known example is DuckDuckGo.[§§]

Another useful related skill is improving our searches efficiency by utilizing advanced features available in most search engines:

- Use quotes: When searching for an exact word or phrase, enclosing it in quotes ("...") ensures that the results include the exact phrase in the same order.
- Exclude words: By using a minus sign (−) or the operator "NOT" before a word, you can exclude results that contain that specific term.
- Combine multiple terms: Using the "AND" operator between words, the search engine will return results containing all of them. (Note: Modern search engines like Google often imply AND by default.)
- Site-specific search: Using the "site:" operator followed by a website domain restricts the results to that particular website (e.g., "site:wikipedia.org [search term]").
- Filetype search: Using the "filetype:" operator followed by a file extension (e.g., "filetype: pdf") filters the results to show only files of that specific type.

Last but not least, it is essential to mention that this tailored selection, designed to fit our needs and preferences, can be manipulated or biased by different groups of interest, lobbies, and other influential entities. This is not a new phenomenon. The same mechanism has long existed in traditional media, where newspapers or magazines choose certain stories to feature prominently on the front covers. However, within the digital world, such biases are often less obvious to general public, making them harder to identify and scrutinize.

[‡‡]https://www.mozilla.org/en-US/firefox/browsers/mobile/focus/

[§§]https://html.duckduckgo.com/html/

9.2.2 Evaluating information

As discussed in the previous section, the results of a search engine query are influenced by a series of factors, and it is crucial to understand this process. More importantly, we need to realize that not all the information on the internet is accurate, and we must fight against misinformation, disinformation, and fake news. Therefore, we need to critically analyze, compare, and evaluate the credibility and reliability of the information we find on the internet. Learning to distinguish trustworthy sources from unreliable ones is a key skill in navigating the digital landscape effectively.

According to the European Commission [6], "The spread of both disinformation and misinformation can have a range of harmful consequences, such as threatening our democracies, polarizing debates, and putting the health, security and environment of EU citizens at risk."

For this reason, we, as concerned citizens, must:

- Be skeptical about the information we read online.
- Understand that current AI technologies can generate or alter images, videos, and audio.
- Check several sources to contrast any information to assess its reliability.
- Fact-checking before dissemination. We must ensure the accuracy of any information, especially if it is from an untrusted source like social media. This responsible approach helps prevent the spread of misinformation.

By following these simple steps, we can help and protect ourselves – and others – against manipulation in the form of misinformation, disinformation, and fake news. These terms are often used interchangeably, but understanding their distinct definitions is crucial to recognizing the nature of these corrupted pieces of information and the intentions behind them [7]:

- Misinformation: False information that is disseminated, regardless of any intent to mislead. Although there is no harmful intent, the effects can still be harmful.
- Disinformation: Misinformation that is deliberately disseminated to mislead. It is spread with the intention to deceive or secure economic or political gain. It may cause public harm.
- Fake news: False information, often of a sensational nature, that mimics news media content.

Even though there is a component of intention behind distinguishing them, both misinformation and disinformation can lead to harmful consequences. The other term "fake news," often used when facing fact-checking discussions, refers to information that is deliberately manipulated to deceive the public. As we know, this is an increasingly visible global phenomenon. It is not just a problem in one country, but rather represents a challenge that affects us all. Around 85% of Europeans find disinformation a problem in their own country, and 83% see it a threat to democracy. However, we also play a role in the spread of fake news, as

60% of news items shared on social media were never read first by the user who shared them.

Most of the time, there is a well-engineered strategy behind the rapid, unconscious spread of false information. At other times, lies just run like prairie fires. In both cases, the way we emotionally process information is at the root of the issue. Those infuriating new pieces tend to be shared rapidly, boosted by our confirmation bias, that is, our natural tendency to believe anything confirming our existing beliefs and ideas. Taking a moment to think critically is always necessary. However, when the information we are about to trust and share is outrageous and appeals to our deepest values or strongly reinforces our vision of the world, seeking counter-information becomes not just important, but urgent. We must actively seek out information that challenges our beliefs to ensure we are not contributing to the spread of misinformation.

But how can we determine whether a given piece of contents is disinformation or misinformation? Even if the content aligns with our beliefs and values (particularly in such case), we need to approach it critically and learn how to recognize misinformation patterns. Here are some helpful guidelines for identifying when a piece of news may be fake, false, or otherwise fabricated:

- Check the content: We must look for biased information, verify whether the facts and figures presented are accurate, and establish whether the information has a known source or comes from a credible source.
- Check the source: We must verify whether we are familiar with the source and, if not, investigate it is reliability. Ideally, we should also find who is behind it or funding it.
- Check the author: Identify if the piece has an author, if it comes from a journalist or an unknown individual, or if it is attributed to an anonymous source. In the latter cases, there is a higher likelihood that the information is fake.
- Check the data: Make sure the information is based on data from trustworthy entities or comes from renowned experts in the topic.
- Check the picture: With AI, we know that media can be easily manipulated, but that is not the only way a picture can mislead us. So, check if the image is original or run a reverse image search to see if it comes from another source such as a different event in another country or at a different time (ensuring it matches the time and place).
- Think twice before sharing: if you are unsure if the information is fake news, it is better not to share it, as it can deceive others. Moreover, most fake news appeals to our deepest feelings, like fear or anger, triggering a hasty reaction that usually leads to sharing without careful fact-checking. Break this tendency by holding back. After a while, the small details that reveal the piece as false or fake will likely become clear. Most doubts about the trustworthiness of a piece of news will disappear by simply leaving it on the back burner.
- Fight against fake news: If you know a piece of information is fake news, inform the person who shared it with you. Maybe this person will think twice before sharing it again, helping to stop the spread of misinformation. But, most

importantly, never share a link to false information, even to report it as fake news.

As it is clear, it is impossible to verify every single piece of information we receive daily. That is why societies have promoted entities in charge of fact-checking, which is the process of verifying if a piece of news is accurate by questioning the statements in that information. For example, we have the European Digital Media Observatory (EDMO)[‖‖]. On its website, we can find information about recent fact-checks and a list of fact-checkers across different European countries.[¶¶]

As mentioned before, much of the disinformation and fake news we receive owes its legitimate appearance to advances in AI. This is why we must promote the ethical use of this technology. The rapid development of AI, especially generative AI since the release of ChatGPT in November 2022, has welcomed a new era of innovation and efficiency, completely transforming fields like education, industry, and research. We now have systems that use this technology to create long texts, images, and even videos using a simple prompt: a command containing just a few words describing what the user wants. However, these intelligent technologies present various ethical challenges, such as privacy, safety, security, fairness, non-discrimination, and sustainability. Next, we will address some of these issues.

AI algorithms are typically trained using vast amounts of data, and their performance is closely tied to both the fairness of the algorithm and the quality of the data. It is often said that AI algorithms are only as good as the data they are trained on. This underscores the critical role of unbiased data in AI training, a key factor that influences algorithm's performance.

Consider this: if the data used to train an algorithm are biased, AI will learn from and operate based on these biased data, producing results according to the information it has received, thus perpetuating or even amplifying this bias.

But what kind of bias are we talking about? Mainly, social inequalities. Some examples could be the use of different amounts of data for different genders, giving better results to one gender above the other, or using unbalanced data from racial or ethnic groups that could lead, for example, to biased facial recognition, misidentifying certain groups more frequently, and so on.

So, the solution lies in introducing fairness into our AI systems. Furthermore, this can be achieved by implementing, among others things, some of the following ideas:

- Enhance the diversity and representativeness of the data used to train an AI system. This ensure that it caters to a wide range of users, including gender, ethnic groups, religion, culture, ideology, perspectives, and diversity in capabilities. The potential benefits of this approach are vast, offering a more inclusive and fair AI landscape.

[‖‖]https://edmo.eu/
[¶¶]https://edmo.eu/about-us/associated-institutions/

- Use techniques to detect and mitigate bias like fairness metrics and bias audits. Regularly employing these methods can help us identify biased situations and address them effectively.
- Regulations and standards play a crucial role in the development of AI. By adhering to these norms throughout the process, we can detect and mitigate bias more effectively, providing a sense of control and reassurance in the AI landscape.

One key issue with AI systems is their capability to process vast amounts of information, including personal data, which can create privacy issues. Examples of this include the risk of surveillance or misuse of personal information. Data protection regulations require the owners of these systems to clearly explain the intended use of the data we provide, and if it can be used to retrain the system, creating a possible leakage of information.

Another critical issue with AI systems is the lack of transparency. Many of the algorithms used, such as those based on deep learning neural networks, are considered "black boxes." We know the input data and the output results, but we do not know what happens in between, the processes that lead to a specific decision, conclusion, or recommendation. Therefore, explainability must be added to the design of AI systems; that is, the system must be able to provide explanations for its decisions based on the input data. This allows users to better understand them and trust in their decisions, as they can be scrutinized for potential biases.

Although in the beginning, AI-generated images, videos, and audio were obviously fake and even disturbing, the technology in this field has improved rapidly. "Deepfakes," false audiovisual pieces created to mimic real ones, are now almost impossible to distinguish from authentic, legitimate content. Some "deepfakes" may be innocent jokes never intended to deceive anybody, while others might have an artistic purpose. However, some are created with a misleading intent, or even to harm, undermining the reputation of public figures or anonymous citizens. We must be aware of this risk. If something seems too good or too bad to be true, it probably is.

Finally, it is essential to talk about what is now called "echo chambers" or "filter bubbles." As algorithms and machine learning systems (systems which are capable of learning from experience) tend to suit our ideology, values, and principles, we are more likely to see news, results, and options that align primarily (or exclusively) with our personal preferences. This can lead to a false consensus effect, where we overestimate the prevalence of our own thoughts and beliefs within society. This biased effect poses a challenge for pluralistic societies, counting against diversity and progressively reducing the variety of viewpoints that contribute to healthy democracies.

9.2.3 Protecting our devices

Much has changed in the world with the arrival of "information technologies" and the way we interact with each other. Mobile devices have become widely used by the global population and are used continuously, sometimes excessively. This creates a series of problems. One of these is the need to understand the risks and threats in digital environments in order to protect both our devices and our digital content. To address this, we must implement safety and security measures and learn to care for our information and its privacy online.

The first way to protect our devices is by implementing strong authentication methods to prevent unauthorized access. The access methods we use can vary depending on the device. We will focus on two main types: mobile devices (mainly smartphones and tablets) and computers (either laptops or desktops).

Regarding mobile authentication methods, we have two main categories:

- knowledge-based: something we know such as a password, PIN, or pattern.
- biometric-based: using or scanning something that is part of us such as our fingerprint or face.

None of these methods is perfect, but both provide sufficient security as long as we follow other complementary precautions. For example, we need to ensure that no one is watching us while we enter our PIN or pattern.

Regarding computers, the most typical method is the use of passwords. Another option is the use of USB keys. However, we must keep in mind that not all passwords are secure. Many lists on the internet contain commonly used passwords, so when selecting a password, we must check that it is not in one of those lists.

It is also advisable to use a strong password that is at least 11 characters long, includes capital letters, symbols, and numbers, and does not contain actual words. As a security measure, we suggest using different strong passwords for each online service, just modifying them slightly by adding two letters depending on the site we visit. There are different methods to create strong and memorable passwords, and we suggest the following one:

- Take the first letter of each word in a sentence that is easy to remember for you. For example, the beginning of the sentence "All those moments will be lost in time, like tears in rain" from the movie Blade Runner.
    - All those moments will be lost in time → Atmwblit
- Swaps letters for numbers, typical ones are: A → 4, E → 3, I → 1, and O → 0
    - Atmwblit → 4tmwbl1t
- Enter some symbol in a known position, like the middle
    - 4tmwbl1t → 4tmw-bl1t
- Modify the current password depending on the site we want to access, using, for example, the first two letters of the site:
    - Gmail: Gm4tmw-bl1t
    - TikTok: Ti4tmw-bl1t

As we can see on the news, sometimes there are security breaches in companies, and our information, such as our passwords, may be compromised. If we have used different passwords for different sites, we can mitigate the negative effects of an account being compromised or hacked, as the stolen password will not work on other sites. There are websites that allow us to check whether our personal data has been compromised by data breaches like Have I Been Pwned?*** By entering our email address, we can see if it was involved in a data breach.

Another way to protect our devices is using encryption techniques, but what does this mean? Encrypting means saving your content in a modified format using an encryption key. Without that key, an outsider who gains access to our device will be unable to see its content such as our photos and documents.

Most mobile devices and computers already offer encryption by default, and the decryption key is usually the PIN, password, or pattern we use to unlock them. Without that key, an outsider who gains access to your mobile device or computer cannot view its content, either by connecting the mobile device to another computer or by removing your computer's hard drive.

We can make our devices extra secure by installing and configuring protection software. This type of software is designed to protect our devices (desktops, laptops, and mobile devices) and all the information and data they host from unauthorized access, intrusions, viruses, and other threats.

At the user level, the most important types of protection software are:

- Antivirus: This comprehensive program is designed to prevent, detect, and remove computer viruses and other types of malware, providing a robust shield for our devices. These products can also include protection from malicious URLs, spam, and phishing, offering a complete security solution. Most current operating systems come with it already installed, like Play Protect in Android or Microsoft Defender in Windows.
- Firewall: A network security program or hardware that monitors and controls incoming and outgoing network traffic based on predetermined security rules. It creates a wall or barrier between the network we are connected to and an untrusted network such as the internet.
- Virtual private network (VPN): A VPN is a powerful tool that allows our devices to access private content in a different network in a secure way, giving us the freedom to work or browse from remote locations with confidence. This can be used to remotely access our work or home network, minimizing the associated risks. Also, many companies offer access to VPNs, allowing us to access the internet without revealing our connection location, IP address, or identity.

Apart from the ways to protect our devices mentioned above, there are also some good practices we should consider to effectively protect them such as being cautious with untrusted Wi-Fi connections, always keeping our device updated, backing up our data, restricting app permissions, updating to the latest versions of our software, using multi-factor authentication, being aware of phishing scams, etc.

9.2.4 Protecting our personal data and privacy

As we have seen in the previous section, there is a lot of information on our personal devices, so it is necessary to use the appropriate security mechanisms to

***https://haveibeenpwned.com/

protect it. However, sometimes, we are unaware of the sheer amount of information we make available to everyone on the internet. Once it is uploaded, we lose any control over it whatsoever, and we must realize that sometimes this information may be used against us. This is the main reason why we must protect our personal data and our privacy in digital environments.

Some applicable definitions on this matter, according to the European Commission [8], are:

- Personal data: any information that relates to an identified or identifiable living individual. Different pieces of information, when collected together, can lead to the identification of a particular person, also therefore constitute personal data.
- Personal data that has been rendered anonymous in such a way that the individual is no longer identifiable is not considered personal data. For data to be truly anonymized, the anonymization must be irreversible.
- Examples of personal data include a name and surname; a home address; an email address; an identification card number; location data (e.g., the location data function on a mobile phone); an Internet Protocol address; a cookie ID; the advertising identifier of your phone; data held by a hospital or doctor such as a symbol that uniquely identifies a person.
- Data privacy: the ability of an individual to determine when, how, and to what extent their personal data is shared with or communicated to others. As individuals, we are entitled to have our personal data kept private.

On the internet, websites, applications, and social media platforms often collect and store personal data from their users to provide services. But sometimes, they may exceed the scope of necessary data collection and usage, leaving users with less privacy than they realize. Additionally, inadequate security measures around this data can lead to a data breach, compromising personal information.

The lack of data privacy can result in personal information being misused in various ways such as being used by criminals for blackmail, fraud, or harassment, or being sold by companies without our consent. It can also lead to profiling based on our online activities. These risks underscore the critical importance of safeguarding our online privacy. Just like the right to free speech, privacy is a fundamental human right essential to a free society.

As mentioned in the previous section, the best way to secure our online accounts is by using a strong password. If the website supports it, it is also highly recommended to use two-factor authentication. But securing alone is not enough; we must also actively manage our privacy settings. Most devices and websites we interact with have their own privacy policies that allow us to adjust the default settings. This empowers us to take control of our data privacy, especially on social networks.

We suggest the following tips containing relevant information to manage your privacy settings:

- Review your current privacy settings on the different websites. Since these settings may differ from one side to another, take the time to adjust the options that are adequate for you.

- Check who can see the information you share on social networks such as posts, pictures, shares, likes, dislikes, and comments. Typically, most platforms allow us to choose between a public and private profile. It is recommended that this information is visible only to known friends.
- Review what others can see on your profiles, avoiding share personal details that could reveal your location, interests, or social circle, making it easier for someone to identify or locate you.
- Create a friend list to limit who can view your posts and exchanges on social networks. Be cautious adding people "as friends" you have not met in real life. You cannot be sure of who is behind an online account.
- Check if you are sharing location data, like the current position or photo metadata including where the pictures were taken. This information can be used to track your movements.
- Think carefully about the information in your profile. Consider who you share it with and if you really need to share all this information. Taking a moment to review this can help protect your privacy.

Sometimes, it is difficult to locate the privacy settings on the different websites. To make this easier, you can visit, Stay Safe Online,[†††] which provides shortcuts to directly access the privacy settings for major websites.

Another important way to protect our data is by using data backups and knowing how to recover them in case of need. Data backup is the process of creating a copy of your data to safeguard it in case of loss. This includes storing copies of files from your different devices and making them accessible in case of data corruption or deletion.

Fortunately, most operating systems already come with built-in tools to help with these tasks. Additionally, different programs allow you to back up your data, and in many cases, these backups can be made automatically and stored online for added security.

However, what we should do if we have many data to back up or, for privacy reasons, prefer not to share our data with online companies? In these cases, we must make our own off-line backups, and for this, we suggest following the golden standard of backups, the 3-2-1 rule:

- 3 copies of data: You should have the primary copy of your data, along with at least two additional copies. This ensures you have multiple options for recovery in case of failure.
- 2 different media: Your primary data and backup copies should be stored on at least two different types of storage media, like two different hard drives. The key is that the backups should not be on the same device (e.g., do not use two hard drives within the same computer).
- 1 copy offsite: Avoid keeping all backups in the same place, as they could be completely wiped out in the event of a natural disaster or emergency, like a

[†††]https://staysafeonline.org/resources/manage-your-privacy-settings/

fire. It is recommended to store one or more backup copies in a different location such as one at home and another at work.

By following the 3-2-1 principle, you ensure that your data is safely backed up and can be recovered when needed.

9.2.5 Protecting our health and well-being

We understand that protecting health and well-being in digital environments involves taking proactive actions to avoid health risks and threats to both our physical and psychological well-being. For these actions, we need to establish clear routines to protect ourselves from potential dangers in these environments and use adequate digital tools that promote social well-being and inclusion.

However, the objective is not only to protect our own physical, emotional, and mental health but also to ensure the health and well-being of others while we navigate and interact within digital spaces.

When discussing protection, we often focus on the "physical and real" world, forgetting the fact that protection is also as important in digital realm. So, how can we protect ourselves in the digital world? We will explore this in more detail during this section, but we can start by considering ways to protect ourselves from harmful behaviors, like cyberbullying, or how to mitigate the negative impacts of social media.

The first step in protecting yourself is understanding whether you are safe in digital environments. You must be aware of your personal situation, which is known as digital health risks awareness. Several risks associated with using digital environments have already been discussed in this chapter:

- Accepting friend requests from unknown users can compromise your safety and expose you to potential threats.
- Posting personal photos frequently increases your visibility and can elevate the risk of your information being misused.
- Using public Wi-Fi is risky, as it can make your data vulnerable to breaches and cyberattacks.
- Checking in on social media can expose your location to strangers, putting your privacy at risk.
- Reusing passwords across multiple accounts makes them more vulnerable to hacking.

Additionally, there are other risks that you should consider. One important question to ask yourself is: Have you searched yourself on the internet? The answer to this question can help you realize how much of your personal information is exposed and how it might contribute to potential risks in digital environments.

Another problem that is very difficult to solve, especially in the case of children and teenagers, is controlling screen time. Screens are an integral part of our daily lives right now, whether at schools, home, work, for entertainment, or while traveling. Nevertheless, it is important to find a balance between screen usage and our overall health and well-being. This balance can be very difficult to strike, as many applications are specifically designed to keep us engaged. For instance, infinite scrolling on platforms like TikTok or by continuously earning rewards in online games are examples of how these technologies keep us hooked. Many techniques such as clickbait, gamification are designed to manipulate our behavior, weakening our ability to control how much time we spend online. These strategies are tailored and implemented based on psychological principles aimed at modifying human behavior, encouraging excessive screen time, consumerism, and more. These methods are highly effective because they are carefully designed to exploit human psychology, making it even harder to resist the urge to engage with digital content for extended periods.

So, the amount of time we spend in front of a screen can be excessive, and it is important to recognize the impact this can have on our physical and mental health. Physically, excessive screen time can contribute to issues like eye strain and sleep deprivation. Mentally, some studies suggest a link between too much screen time and anxiety, depression, and disruptions to sleep patterns. While there is no universal agreement on the extent of these effects, it is essential to consider the opportunity cost. We only have a limited amount of time each day, and the minutes and hours spend in front of a screen could be better spent satisfying other activities that would improve our health and overall well-being.

How can you fight against this? The most important effective strategy we can do is to set boundaries:

- Create tech-free zones: Establish places, such as the dining table or bedroom, where the use of technology is prohibited. This increases face-to-face interaction with family or friends and helps promote better sleep hygiene by removing screens before bedtime.
- Create a schedule: Set clear limits on screen time. It can be defined through specific time blocks or an amount of time per day. There are several applications to use, like Google Family, that can inform you of the amount of time you have been using a device or even restrict its use.
- Take breaks: Stop from time to time from your device to stretch, move around, or focus on something other than a screen. This helps relieve eye strain, activates your body, relax your eyes, and can reduce mental fatigue. Short breaks every 30–60 min can significantly improve your focus and well-being.

We must be vigilant to detect known signs of problematic digital use such as loss of control, withdrawal symptoms, and dysfunctional mood regulation. We must also realize that problematic digital use can be the root of some serious problems, even causing both psychological and physical harm.

To conclude this section, we suggest some actions that we should all follow to prioritize our self-care:

- Recognize the potential negative impacts of excessive screen time and constant connectivity.
- Excessive time on social media platforms might lead to feelings of loneliness, anxiety, and depression.

- Set boundaries around how much time we spend on digital devices and locations where we use them.
- Establish a work-life balance, avoiding access to work-related matters outside of work hours, and keeping personal lives separate.
- Avoid comparing yourself to others on social media, as it can lead to decreased self-esteem and feelings of inadequacy. Not all that we see on social media is real or accurate.
- Be aware of your online presence, try to disclose as little information as possible, and avoid engaging with harmful content.

9.3 Conclusions

As we have already seen, digital competence is a key factor in our increasingly digitized world. To achieve the adequate level of competence, it is necessary to acquire not only the required skills but also specific knowledge, and what is often bypassed, the right set of attitudes and values needed to use digital resources from a secure and ethical point of view.

There are many aspects within this competence, all of them of the utmost importance. In this discussion, we have just focused on a group of them:

- Accessing information: Learning how to browse, search, and filter information effectively to improve the objectivity and relevance of results and data retrieved from various digital sources, while minimizing incidental biases.
- Evaluating information: Detecting and combating intentional biases and different forms of manipulation, as well as controlling for the harmful effects of disinformation, misinformation and fake news.
- Protecting our devices: Safeguarding against unauthorized access, fraudulent data retrieval, and other cybersecurity threats.
- Protecting our personal data and privacy: Managing our data, proactively, to avoid sharing more than is necessary or advisable, mainly due to an excess of confidence or digital ingenuity.
- Protecting our health and well-being: Prioritizing our health to lead a fulfilling and satisfying digital life, acting as responsible citizens while taking care of ourselves and others.

It is important to consistently apply these competencies, particularly given the blurred boundaries between personal and professional life in the digital realm. These competencies are not restricted to professional life and, like the other soft skills discussed in this book, contribute to all aspects of our personal lives.

References

[1] European Commission: Directorate-General for Education, Youth, Sport and Culture, Key Competences for lifelong learning, Publications Office, 2019, https://data.europa.eu/doi/10.2766/569540 [Accessed 04 Nov 2024].

[2] Eurostat, Statistics Explained, Towards Digital Decade Targets for Europe, 2023, https://ec.europa.eu/eurostat/statistics-explained/index.php?title=Towards_Digital_Decade_targets_for_Europe#Digitally_skilled_citizens_and_professionals [Accessed 04 Nov 2024].

[3] European Commission: Europe's Digital Decade: Digital Targets for 2030, 2022, https://commission.europa.eu/strategy-and-policy/priorities-2019-2024/europe-fit-digital-age/europes-digital-decade-digital-targets-2030_en [Accessed 04 Nov 2024].

[4] Bennett, S., Maton, K., and Kervin, L., The 'digital natives' debate: A critical review of the evidence, 2008, *British Journal of Educational Technology*, 39: 775–786. https://doi.org/10.1111/j.1467-8535.2007.00793.x.

[5] Statista, Number of Internet and Social Media Users Worldwide as of July 2024, 2024, https://www.statista.com/statistics/617136/digital-population-worldwide/ [Accessed 04 Nov 2024].

[6] European Commission: Tackling Online Disinformation, 2024, https://digital-strategy.ec.europa.eu/en/policies/online-disinformation [Accessed 04 Nov 2024].

[7] Lewandowsky, S., Cook, J., Ecker, U. K. H., *et al. The Debunking Handbook 2020*, 2020, https://sks.to/db2020. DOI:10.17910/b7.1182 [Accessed 04 Nov 2024].

[8] European Commission: What Is Personal Data?, 2024, https://commission.europa.eu/law/law-topic/data-protection/reform/what-personal-data_en [Accessed 04 Nov 2024].

Part III

Social skills

Chapter 10
Teamwork

Manuel Caeiro[1] and Iñigo Cuiñas[2]

10.1 Introduction

The European Commission WG-MHE [1] framework on soft skills and the Occupational Information Network (O*NET) [2] define the social soft skill "teamwork" as "the skills required to work well as part of a team". While this definition seems to be obvious, after a closer analysis reveals several underlying concepts. The first one, it refers to "skills", meaning that there are many elements taking place when working as a team. In the second one, the skill is described as necessary to "work well", which suggests that it is also possible to work poorly in a team. And, finally, the term "team" is repeated to emphasize the various ways people can be grouped, and not all of these groupings qualify as teams. The definition of teamwork gives us some tips into the elements to be trained for acquiring good performance when dealing with teams: having a clear understanding of what a team is, being conscious that a team is not always working well as it is, and knowing the abilities related to a good team worker. These elements will be the core of this chapter, providing additional theoretical support and some practical activities to improve how we work within a team.

A team can be defined as a set of interdependent people who, voluntarily, work together to achieve a common objective. In other words, a team is a group of individuals, united by a shared purpose. The individuals comprising a team ideally should have common goals, common objectives, and a similar mindset.

There are some differences between a team and a group, as summarized in Table 10.1.

A group is not necessarily a team: a team is a step forward; it is something more than a group. A group can have individuals with varied interests, attitudes, as well as thought processes. It is not necessary that the group members have a common objective or a common goal to achieve. Thus, a set of people is not a team, and a team is not the sum of its members.

[1]Departamento de Enxeñaría Telemática, Universidade de Vigo, atlanTTic, Spain
[2]Departamento de Teoría do Sinal e Comunicacións, Universidade de Vigo, atlanTTic, Spain

Table 10.1 A group and a team

In a GROUP, its members ...	In a TEAM, its members ...
are focused on their individuality	work in a cooperative environment
work on the same subject, but without any coordination	have a sense of belonging
respond individually	all respond for the final result, not independently
can have a particular way of working	work in a cohesive way and with close collaboration

This means that a team has specific characteristics. Here are some qualities of a successful team [3]:

- Team members communicate effectively with each other. They openly share their thoughts, opinions, and ideas with members of their team and take into consideration what others have to say.
- Team members focus on common goals and results. They set and agree on team goals based on outcomes and results, rather than just on the amount of work completed.
- Everyone contributes their fair share. Each member feels a sense of belonging to the team, is committed to their work, and really cares about the success of the project.
- Team members offer each other support. Teams tends to be more productive when they received support from the organization, and access to the required resources.
- Team members bring diverse perspectives. Diversity ensures that all required skills are covered by somebody on the team, allowing each individual to play a role based on their strengths. A variety of personalities, age groups, cultures, and experiences can also bring creativity and introduce a wide range of ideas.
- Teams are organized. Though everyone should be responsible for organizing their own workload, management should ensure that everything is running to plan and each member of the team is getting their tasks completed efficiently. Holding regular meetings can help everyone aligned and ensure deadlines are met.
- Teams have fun. It should not be all work and no play! This can lead to burnout and lack of productivity, so it is important to inject a bit of enjoyment into working life.

In summary, a team is not simply the sum of individual contributions; its members work in a coordinated way to execute a project. They act in harmony, performing different functions and activities, and responsibilities are shared by all its members. Under these assumptions, this chapter is divided into sections focused on explaining different aspects of teamwork: the characteristics of a team, the stages of team development, the emotions in the relationships among team members, the different roles adopted by team members. and, finally, the enemies of effective team work. These topics will provide the reader with a foundational understanding of teamwork.

> **A group or a team?**
>
> A collection of photos can be presented to the students to help them reflect on the differences between a group and a team.
>
> Photos should reproduce various situations in which several people are participating in different activities. Examples may include players during a football match, people attending a church service, a symphonic orchestra performing a concert, students attending a lesson, villagers working together to build a wooden barn, healthcare workers in an operating theatre during surgery, participants at a demonstration, or technicians changing the wheels to a Formula One car during a race.
>
> Students should answer the question "group or team?" and provide arguments to support their choice.
>
> The key difference is that in a team, everyone works to achieve a common goal, while in a group, each member is usually focused on their individual goals.

10.2 Characteristics of a team

The characteristics of a team can be summarized by the five C's, as defined by [4]: complementarity, coordination, communication, confidence, and commitment. Let us comment these concepts:

- Complementarity: Each member is a specialist in a specific area and applies their knowledge and abilities to the objective of the team. Diverse and heterogeneous capacities and attitudes lead to more innovative and creative solutions.
- Coordination: It is necessary to understand both the common and individual objectives, establish a schedule, and implement a monitoring system that takes into account the needs of the different members. A decision-making and consensus-building system is also crucial. Additionally, it is important to set goals, organize work, and manage relationships. Finally, efforts should be oriented toward both the task and the maintenance of the team.
- Communication: It must be constant, open, and bidirectional. Everyone must be able to express their views freely, and no member should be excluded.
- Confidence: Each member must be able to trust in the competence of others and prioritize the team's success over personal recognition.
- Commitment: Each member is engaged, not just involved, in being part of the team and contributing their best efforts.

Other works identify different sets of 5 C's, but the concepts are in general coincident [5]. Some examples include common purpose, connection, communication, collaboration, and celebration [6], or communication, cooperation, contribution, compromise, and commitment [7]. A new "Five C" framework has been proposed by [8] focusing on the core aspects of team building: context, composition, competencies, change, and collaborative leadership.

Cooperative ball

The objective of this activity is to demonstrate how a coordinated team can perform better when they are well organized and follow some established rules.

Participants are divided into groups of four or five. Each team receives a ball. The challenge is to pass the ball through all participants' hands as quickly as possible. There is just one restriction: no one can pass the ball from his left to his right hand, or vice versa, without another person touching the ball in between. Once they complete the task in one minute, additional rounds can be proposed with progressively less time. How many rounds are your team able to complete?

How do you coordinate? How do you feel? Have you acted as a team?

10.3 Team building

Following the model proposed by Tuckman [9], during the life span of a team it is possible to identify five phases: forming, storming, norming, performing, and adjourning. These stages are typically experienced by any group of people who begin to work as a team until they evolve into a cohesive team. All stages are important, as teams need to go through each of them. It is valuable to understand that these stages exist, and to be aware of the implications of each, as this knowledge allows teams to be better supported and oriented toward next stages.

The following subsections will focus on each of these phases, as summarized in Figure 10.1.

10.3.1 Forming

The first stage when a group begins to work as a team is known as forming. At this point, members are usually highly motivated, but they do not know how to proceed. They are organizing themselves and defining procedures and standards, whether implicitly or explicitly. It is recommended to do this explicitly and involve all the members. A good practice is to share personal stories, motivations, and goals among team members to begin building strong team commitment. An effective exercise for team members is to write about their hobbies, future goals, aspirations, and so on in order to get a better knowledge of all the team members and feel more confident.

Some of the characteristics of this first phase of team building are:

- Team processes are proposed and accepted, performance standards and meetings are established, and information is exchanged.

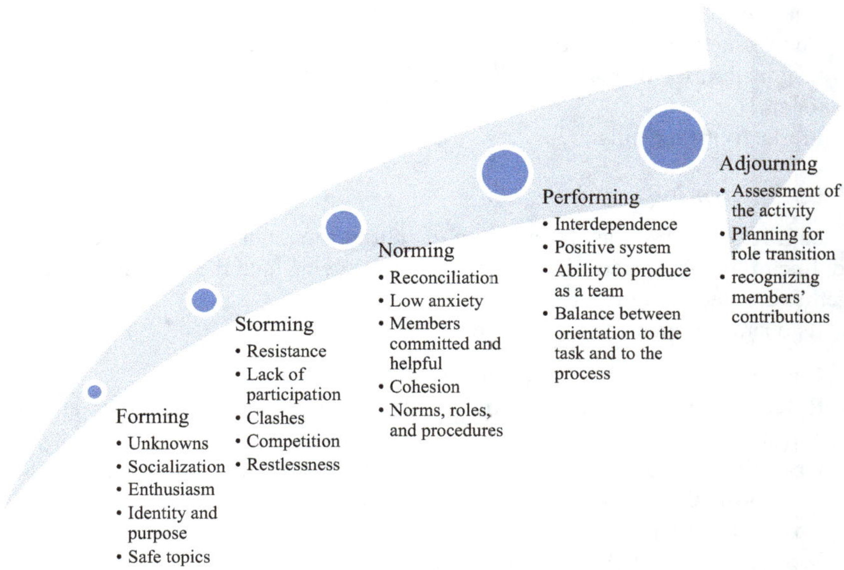

Figure 10.1 The five phases of team building

- Team awareness is generated, and goals are established.
- This is a period of some restlessness and insecurity.
- It is time for exploration, when each team member evaluates others: measuring their skills and aptitudes. The focus is on understanding and learning about the motivations and expectations of team members, rather than on the work itself.
- The members are usually motivated, but roles and responsibilities are not yet fully clear.
- The members are considered strangers.
- The performance is commonly low.

10.3.2 Storming

The second stage is characterized by conflict. During the forming stage, we tried to know each other and establish some procedures and rules. Now, we begin to work as a team, we discover problems. These problems need to be solved, and each team member of the team can have a different view about how to approach them. This is a normal issue. The key is to discuss the problems and possible solutions all together, giving the opportunity to participate all members and trying to achieve a consensus.

The main characteristics of this second phase of team building are:

- Problems begin to appear; doubts arise, and many members do not feel integrated into the group. Impatience due to the lack of progress.
- Each member struggles to influence, lead, and adopt roles. Rivalries appear among the members.

- There is defensiveness, with confrontations, tension, and hostility.
- The team fights to establish its objectives, set expectations, define roles and responsibilities of team members, and establish agreements on how to work together.
- Productivity remains low.

10.3.3 Norming

The conflict stage may last for a longer or shorter time, but after it, the team begins to define protocols and standards. The team is growing, and the benefits of working together are becoming clearer.

Its characteristics can be listed as follows:

- Clashes are replaced by agreements and consensus.
- Roles and responsibilities are clearly defined, and an internal structure appears.
- Operating standards are established (whether explicit or not), including decision-making processes, tools and methods of work and interaction.
- Communication becomes more open, and confidence increases. Team members adjust their behavior toward each other. As a member of the team, you want to solve problems and aspire to harmony.
- The team starts to cooperate instead of competing. Sense of belonging, union, commitment, and the development of a team identity, arise during this stage.
- Motivation increases among team members.

10.3.4 Performing

At this point, the team achieves a stage where it is able to be very productive without conflicts or problems. Performance is optimal.

We can identify this phase by the following conditions:

- The team knows what it does, why it does it, and for what purpose. Team members create and use structured procedures to communicate, resolve conflicts, distribute resources, and interact with others.
- Conflict is channeled productively, and creative solutions are reached.
- Energy is focused on work, and everyone is involved in achieving their objectives.
- This is the most productive phase: the team exchanges knowledge, carries out projects, solves problems, takes risks, makes group decisions, and achieves the designated goals.

10.3.5 Adjourning

Finally, we need assess the work and to decide what to do with the team:

- Have the roles of the team been effectively completed?
- Has the goal of the project been achieved?
- Can the team be permanent?

- Perhaps some members of the team will move to other projects, with other mates, and return to the first stage. Other may remain in the team, which grows with new members and tasks. Or perhaps some may lead other teams or other projects.

Where are you as a team?

This is a reflection activity. The goal is to identify the current stage of team building.

Taking into account the stages of team construction (forming, storming, norming, performing, and adjourning), ask participants to think about on which stage they are as a team and provide some evidence to support their position. Join people in teams according to their previous relationships. First, each participant works individually, and then shares their ideas with the rest of the team members.

- Does everyone agree on the team´s development stage?
- What evidence is proposed?

In case of new teams, members could be asked about a previous positive team experience, moving them to reflect on the factors that contributed to that success or the reasons why they remember that specific experience.

10.4 The emotions within a team

Another important aspect of teamwork is the management of emotions. At this stage, it is important to recognize the existence of different emotions: happiness, anger, sadness, fear, and disgust. Emotions are significant, and we need to manage emotions while working in teams. Thus, it is crucial to understand that emotions can be contagious. Therefore, when you working in a team, it is positive to create a positive atmosphere doing some activities that are enjoyable for everyone, after all, emotions are contagious!

It is said that "30% of team's results can be altered depending on their collective emotional state" [10], which reinforces the importance of understanding these concepts. Working in a team, having a good management of empathy, self-esteem and assertiveness can resolve many difficulties. These three emotions should be trained for assuring good team performance.

- Empathy can be defined as the ability to recognize and understand the emotions of others. "Putting yourself in someone else's shoes" is the descriptive concept of empathy. When you place yourself in the position of the other, you are able to understand their feelings and, probably, the reasons that move them to act as they do.
- Assertiveness is the ability to express feelings, ideas and opinions in a free, clear, and simple way. This concept can be introduced confronting it, against passiveness and aggressiveness. When we work in a team, our behavior should

neither be passive nor aggressive. A passive person accepts everything without complain, while an aggressive person always imposes their position. The key is to be assertive, to be able to share your ideas and defend them without imposing them on, while also being understanding and sensitive to other's point of view. A good example of an assertive sentence might be "I understand your point, but under these circumstances I think ...".

- Self-esteem is the capacity of a person to value, respect, and accept themselves. Having positive self-esteem greatly contributes to being assertive and empathic, and in helps individuals interact freely and confidently with the rest of the team.

Empathy: are you feeling as a king?

This is an ice-breaking activity. The goal is to realize that how you interact with others has consequences on how they perceive themselves.

The lecturer deals each of the participants a card from the deck (having previously made sure to have some figures – king, queen, knight – and several number cards).

Each participant, without seeing the card, places it on his forehead so that the others can see what card it is, but they cannot. In this arrangement, all participants move freely around the room, greeting each other every time they pass someone. The greeting should be more or less ceremonious depending on the category of the card of the person holds: reverence for the king, a cordial greeting for high numbers, contempt for low numbers, etc.

After a while of greetings, the lecturer asks the participants to arrange themselves based on which card they think they have, based on how important they felt during the process.

Are you feeling important or well considered?

The final reflection is about how we can make others feel, whether good or bad, as a result of how we treat them.

10.5 Roles within a team

Belbin's team roles [11] are an effective way to assess the relative strengths and weaknesses of a team. The balance and compatibility between the roles of the members of the team are key to success. Thus, studying the roles developed by each member supports the analysis of the performance of teams. For example, certain teams that are expected to achieve excellent results (due to the high capacity of their members) may experience a disappointing performance, which can be often be related to the dominant roles within teams. Then, understanding the different roles helps the team to identify ways to improve its performance.

How to make assertive phrases?

We divide the participants into groups of three after they have understood assertiveness as the way to defend one's own rights without violating the rights of others.

We will present a series of hypothetical scenarios:

- A friend asks you to take care of her child while she goes to the movie.
- The boss of the company gives you supplementary work for today.
- The neighbor asks you to do her shopping at the same time you are doing yours.
- Your parents try to convince you to go visit them.
- Someone asks you to drive them home.
- Someone asks you to make a fake phone call.
- You are playing football, and someone tells you: you do not hit one!
- You are carrying a new backpack, and your friend tells you: what a ridiculous backpack!
- Someone in your family asks you to lend them money.
- A friend asks you to borrow something that you do not want to share.
- You meet with your friends and wear a very nice t-shirt that leaves your belly button in the air. One tells you that you look ridiculous.
- A colleague asks you to share unpleasant news with your group of friends.
- A friend is trying to convince you to go out tonight.

The facilitator will read each scenarios aloud one by one. After each scenario, there will be 30 s for each participant to write their assertive response to a situation with which they disagree. Each group of three has 1.5 min to read each other's answers and decide which one seems most appropriate. This will be done for all the scenarios. Afterward, there will be a sharing with the large group.

Are you able to create sentences that express your true opinion respecting the others' interests?

The final reflection will focus on how to be trained in assertive behavior.

Belbin defines a team role as "a tendency to behave, contribute, and interrelate with others in a particular way", and classifies them into three orientations: mental, action, and social roles. Let us take a closer look at each of these to understand them better.

- Mental-oriented roles:
  - The *plant* is creative, intelligent, and introverted. In general, they do not communicate effectively with other roles. They are good solving complex problems but do not distinguish what is truly important.

 - The *monitor evaluator* is intelligent, introverted, insightful, and objective. They can evaluate large amounts of information and make judgments, which are rarely wrong. However, their behavior is skeptical, they do not inspire others. Besides, they may appear serious or uninteresting.
 - The *specialist* is smart and highly knowledgeable in a specific area. However, their interests are often narrow and focused primarily on their profession. This specialization allows them to bring deep expertise to the team, but at times, they may concentrate too much on details that are not essential to the team's goals.
- Action-oriented roles:
 - The *shaper* is competitive, oriented to results, entrepreneurial, dominant, and outgoing. They do not avoid conflicts, as they prefer action over ideas and works well under pressure. This role injects vitality to the group, but there should be no more than one shaper per team to maintain efficiency.
 - The *implementer* is organized, disciplined, and highly efficient. They are more concerned with the interests of the project than their own. However, they may be slow to recognize new possibilities and usually compete to raise their status.
 - The *completer finisher* is introverted, anxious, and concerned with ensuring that nothing fails. They are very good detecting errors and handling details. On the positive side, they meet set deadlines. On the negative side, they do not like to delegate, and their intolerance for informality can create tension among team members.
- Social-oriented roles
 - The *coordinator* is dominant and extroverted. They promote team decision-making, and are focused on achieving objectives. A natural planner, they are good at detecting talent within the team.
 - The *team worker* is outgoing and cooperative. They avoid confrontation and are sensitive to the emotions of the rest of the team. Thus, people with this profile are usually popular, as they listen attentively to others. However, as they are very diplomatic, they rarely make decisions.
 - The *resource investigator* is very sociable, communicative, and pleasant. They are always on search for new opportunities, with a strong entrepreneur mindset. Although they are good sellers, they are less effective at completing tasks.

Role knowledge provides insight into the complexity of building efficient teams: simply putting together several people and expecting them to work as a cohesive team is not enough! Belbin's tool helps describe an individual's "preferred" team roles and is designed to indicate how they would ideally operate in a team environment. Strength in one team role is often at the expense of what might be considered a weakness in another context.

It is interesting to recognize that each role has its own weaknesses, which are tolerated in the context of teamwork. Plant may be unorthodox or forgetful. Resource investigators may forget to follow up on initiatives. Monitor evaluators may be too critical and slow. Coordinators may delegate excessively, leaving little work for themselves. Implementers may find it difficult to give up their plans in favor of positive changes. Completer finishers may be accused of taking their perfectionism to the extreme. Team workers may be undecided when it is necessary to make unpopular decisions. Shapers are at risk of becoming aggressive or bad-tempered in their desire to get things done. Specialists may tend to focus almost exclusively on their chosen area of expertise.

A team should ideally have a healthy balance of all nine team roles. Strong teams normally have a strong coordinator, a plant, a monitor evaluator, and one or more implementers, team workers, resource investigators or completer finishers. A shaper should serve as an alternative to a coordinator, rather than having both roles. In practice, the ideal is rarely achieved: thus, it can be beneficial for a team to identify which of the team roles are either overrepresented or absent, and to understand individual's secondary roles.

As we stated at the beginning of this chapter, a team is not simply a group of people assigned to specific tasks, but a collective of individuals where each person plays a role that is understood by the other members. Thus, the members of a team should negotiate the distribution of roles among themselves and then perform more those that feel more natural to them more effectively

Team roles tend to develop and mature over time and may change with experience and conscious effort. If a role is absent from the team, then it is often filled by someone who has not recognized this role as a dominant one. The team should share their team roles to increase understanding and enable that mutual expectations are met.

Anyway, Belbin's team roles are used to classify the behavior of people in a work environment, not their personalities. They indicate our particular tendencies to behave, contribute, and interact socially.

Belbin's team roles?

This is a reflection activity. The goal is to identify your Belbin's team roles, or the team role you adopted in a natural way.

Taking into account the Belbin's team role definitions, each participant should try to identify their own roles. Then, ask for some volunteers to share their findings.

If participants have a good knowledge of each other, you can ask them to identify the roles of their partners. In this case, divide the participants into groups of four or five, and ask each person to identify the roles of the persons sitting to their left and right. After 3 min, they should share and discuss their observations.

Have you learned something new about yourself? And ... what about your colleagues? What about the role other people have assigned to you?

10.6 Enemies of teamwork

Even Belbin's team roles seem to cover all possible personalities within a team; in fact they are focused on those that allow the team to be productive and to work effectively. However, there are some additional profiles that could be identified as enemies of teamwork. These individuals are well known for their specific way of interacting with the rest of the members (or perhaps their way of non-interacting!). The main roles for such persons are *hitchhikers* and *couch potatoes* [12], and they deserve some paragraphs explaining their characteristics and the way the team can deal with them.

10.6.1 Hitchhikers

We can easily identify a hitchhiker. He is always busy when trying to set up meetings, even at the beginning of the semester, when everybody has almost free schedules. And, when the meeting is finally set, he misses it. He difficultly finishes his part of the tasks on time, and often those tasks are incorrect. He never answers phone calls, messages, or e-mails, claiming that "he is too busy" all the time. Apparently, he appears to have good writing skills, but he commonly makes grammar mistakes, template inaccuracies, or misses deadline. He is always complaining about the tasks, the lecturers, the teaching materials, the proposed activities, the short deadline . . . to the point that the rest of the team feels sorry for him and is moved to help him.

Most groups have the same attitude toward hitchhikers: the rest of the members absorb the tasks and activities that are supposed to be done by this person, in order to meet the project deadlines. Ultimately, the whole team ends up doing the work of this member. Hitchhikers rely on others to act self-sacrificially [12]. As a result, they reinforce this attitude and apply it along their lives: at university, in the workplace, and through his life. They finally believe it is acceptable to take credit for the work of others.

Instead of tolerating this situation, the team should mirror the poor performance by reflecting the behavior of the hitchhiker and never accepting blame, criticism, or accusations from them. It is important to establish red lines for behavior that will not be tolerated, and to communicate these limits effectively. For example, being unable to set up meetings is not acceptable; the hitchhiker must understand this, and perhaps the professor or the team leader should be made aware as well. In case of no improvements, the name of this person should be removed from the final work. If the individual is being abusive, the team could show the lecturer their poor contributions to the task. If there is no responses to messages, add the lecturer in copy of all communication; perhaps this passive behavior will change or, at least, the lecturer will be informed about the situation.

Anyway, only the hitchhiker can handle their own problems. The task for the rest of the team is to complete the project on time, not to rescue this person from their situation. You cannot allow such individuals to use your time and efforts. However, identifying this person is the first step to keep the health of the team, even if the individual is not part of it but the lecturer themselves.

10.6.2 Couch potatoes

This type of people is irritating. They are nice, but they never put in the effort to do a good job. They prefer to watch TV or surf the social media instead of doing the work. They are lazy people, even they have good social skills and manage to be sympathetic with the rest of the team. They are less manipulative than hitch-hikers, but for sure they will explore the limits of the team for taking advantages of them.

How to deal with a couch potato? Again, setting firm expectations and clearly explaining the limits are effective approaches. Generally, couch potatoes are benign and may be guided toward reasonably good work.

Working with a couch potato (and with a hitchhiker) can be an experience that helps you to grow as a person and learn the importance of firmness.

10.7 Concluding remarks

Nowadays, working in teams is one of the most demanded soft skills. It is difficult to find an institution, a company, or an organization that promotes individual work as the only approach. Thus, gaining some training in teamwork and understanding its characteristics and stages have become a must have in our soft skills toolkit.

In addition to the theoretical content and exercises proposed along this chapter, organization is crucial. A good tip is to set a weekly work agenda for the teamwork progress and to identify at least three roles: a spokesperson, a coordinator, and a secretary, to ensure continuity in the task.

It is also important to highlight that teamwork does not always mean being together; in fact, it often means the opposite. Typically, 80% or 90% of the time, team members should work independently, with the remaining 10% or 20% spent working together, sharing individual results, discussing them, planning tasks, and making decisions. In case team members need to joint meetings for specific collaborations, they can schedule ad hoc sessions, but most of the time, they should work individually.

References

[1] European Commission, Directorate-General for Education and Culture, “Developing future skills in higher education”, ET2020 – Peer Learning Activity (PLA), Brussels, 25–26 February 2016.

[2] The Occupational Information Network (O*NET). Available from https://www.onetonline.org/ [Accessed 23 Feb 2024].

[3] Amber Acosta (ed.), *Small Group Communication*, 2019, Maricopa Open Digital Press.

[4] Thomas J. Peters, and Robert H. Waterman, *In Search of Excellence: Lessons from America's Best-Run Companies*, 1982, Harper Collins, New York.

[5] Jill Nemiro, Lori Bradley, Michael M. Beyerlein, and Susan Beyerlein, *The Handbook of High Performance Virtual Teams. A Toolkit for Collaborating Across Boundaries*, 2008, Wiley, New York.

[6] Adam Hickman, *The 5 Tactics of Teamwork: A Blueprint for Team Management*, 2020, Gallup, Washington, DC.

[7] Sudeep Basnet, "5 C's of TEAMWORK", 2014, LinkedIn. Available from https://www.linkedin.com/pulse/20140708024726-71964275-5-c-s-of-teamwork/ [Accessed 17 Apr 2024].

[8] W. Gibb Dyer, Jr., and Jeffrey H. Dyer, *Beyond Team Building. How to Build High Performing Teams and the Culture to Support Them*, 2019, Wiley, New York.

[9] Bruce W. Tuckman, "Developmental Sequence in Small Groups". *Psychological Bulletin*, 1965;63(6):384–399.

[10] Gerben S. Van Der Vegt., and J. Stuart Bunderson, "Learning and Performance in Multidisciplinary Teams: The Importance of Collective Team Identification". *The Academy of Management Journal*, 2005;48(3): 532–547.

[11] Meredith Belbin, *Management Teams: Why They Succeed or Fail?*, 2nd edn, 2004. Elsevier Butterworth-Heinemann.

[12] Barbara Oakley, "Coping with Hitchhikers and Couch Potatoes on Teams". *Journal of Student Centered Learning*, 2004;2(1):32–34.

Chapter 11
Interpersonal communication

Vítor Braga[1]

11.1 Introduction

Interpersonal communication is crucial for engineers and other professionals to collaborate effectively, manage projects, and advance their careers. This chapter examines the various aspects of interpersonal communication, offering techniques to improve personal and professional capacity to engage with other individuals.

Exchanging information is only one aspect of effective interpersonal communication [1]. In addition to comprehending the motivations and feelings that underline the exchange of information, interpersonal communication includes empathy, clear communication, active listening, and the capacity to adapt messages for various audiences [2]. These abilities are particularly important for professionals working on projects with a significant social impact and in interdisciplinary teams [3].

Effective communication with stakeholders, clients, and coworkers can make the difference between a project's success and failure. Engineers must be able to work with various teams and explain technical ideas to individuals unfamiliar with technical jargon. Therefore, interpersonal communication abilities are just as important as technical knowledge.

Interpersonal communication is important in engineering for a range of tasks, from attending team meetings and interacting with clients to settling disputes and establishing business partnerships. While the communication skills needed for each activity vary, they all attempt to make interactions constructive. These are some examples of different contexts where communication can occur.

- Teamwork: Coordinating activities, exchanging ideas, and achieving shared objectives depend on effective interpersonal communication. To finalize projects, engineers frequently collaborate in groups.
- Engineers must interact with clients to learn about their demands, clarify technical aspects, and provide progress reports on their projects—sympathetic and transparent communication facilitates building trust and satisfaction.

[1]School of Technology and Management School of Felgueiras, Instituto Politécnicao de Porto, Portugal

- Resolution of conflicts: Disagreements are unavoidable in a cooperative context. Resolving conflicts and preserving a cordial work environment require interpersonal communication skills such as active listening and negotiating.
- Networking: Developing a professional network is crucial to career advancement. Professionals with interpersonal communication skills can better interact with mentors, peers, and business executives.

In engineering, interpersonal communication can be difficult. Effective communication can be obstructed by several obstacles that engineers frequently encounter, such as:

- Technical jargon: The language and terms used in engineering are highly sophisticated. Technical jargon can be a barrier when speaking with non-technical audiences, even though it is helpful for accurate communication among peers.
- Complexity of information: Engineering projects frequently include intricate data and concepts that can be challenging to communicate concisely and clearly.
- Cultural differences: Engineers in multinational teams may encounter cultural differences that impact expectations and communication efficiency.
- Time restrictions: Because engineering projects are frequently time-sensitive, there is little opportunity for in-depth talks or prolonged explanations.

Therefore, engineers must acquire various interpersonal communication techniques and skills to overcome these obstacles. This chapter will offer suggestions and methods for improving communication in a variety of contexts, such as:

- Active listening: Methods for increasing listening to comprehension and ensuring that the team feels heard and understood.
- Empathy: Techniques for understanding and experiencing others' emotions that promote affinity and trust.
- Clear expression: Strategies for expressing ideas and thoughts concisely and clearly so that the message is understood.
- Conflict resolution: Methods for resolving disputes harmoniously while preserving good working ties.
- Cultural sensitivity: Techniques to interact with clients and coworkers from different cultural backgrounds.

Gaining practical interpersonal communication skills is an ongoing process that requires reflection and practice. Engineering students should actively seek extracurricular activities, internships, and coursework that help develop their communication skills. In addition to improving professional performance, good communication skills help students to grow and improve the engineering sector personally.

11.2 The basis of communication

11.2.1 Interpersonal communication types and definition

Interpersonal communication refers to sharing thoughts, feelings, ideas, and information. It is a complex process in which individuals communicate verbally and nonverbally to transfer meanings, emotions, and information [4]. Face-to-face communication is naturally dynamic and interactive, allowing participants to watch and read each other's body language and verbal signs. Thus, it depends on relationship construction and maintenance, direct communication of thoughts and feelings, and developing understanding and connection [5].

One of the primary forms of interpersonal communication is verbal communication, which involves exchanging written or spoken messages. This communication style is perceptible by directness and clarity, enabling people to express their thoughts and feelings clearly [6]. Verbal communication includes written correspondence such as emails or letters, and spoken activities such as speeches, presentations, and conversations. The consistency of the content, the clarity of the message, and the communicator's capacity for attentive listening and suitable response affect the effectiveness of verbal communication.

Another fundamental type of communication is nonverbal, encompassing various actions and signs that convey meaning without words. This includes posture, eye contact, facial expressions, gestures, and body language. Nonverbal signals often provide context and emotions to spoken words or can contradict them. For example, a smile can express warmth and openness, while crossed arms might suggest defensiveness or discomfort. Because nonverbal signals frequently reveal underlying attitudes and feelings that words alone may not convey, understanding and interpreting them are vital for successful interpersonal connections [7].

A key element of interpersonal communication is listening, more than simply hearing others say. To engage in active listening, one must give the speaker one's whole attention, attempt to comprehend what they are saying, respond thoughtfully, and retain the information or knowledge transmitted. Effective listening techniques show respect and empathy for the speaker and are essential for establishing connection and trust. People can create more meaningful and fruitful interactions by actively participating in the listening process.

Written communication differs from spoken communication because it uses written symbols to deliver messages despite overlaps. Examples of this type of communication include emails, texts, reports, and social media posts. Given its significant role in documenting and sharing information, written communication is essential in both personal and professional contexts. Its permanence and clarity promote thorough deliberation and editing, accurately conveying the intended message [8].

Visual communication utilizes visual aids such as charts, graphs, images, and videos to enhance information delivery [9]. By providing complex material with a visual context, this form of communication becomes easier to understand and more accessible. Visual aids reinforce and complement written and spoken communications, boosting their significance and engagement.

Therefore, successful interaction in various contexts requires awareness of the different forms of interpersonal communication and how to use them. Each type of communication—spoken words, nonverbal cues, written text, active listening, and visual aids—is essential for understanding and fostering connections between people.

11.2.2 The process of communication

Information, ideas, and feelings are exchanged through a complex and dynamic series of events known as the communication process. This process includes several elements and steps to ensure that the intended message is effectively conveyed and understood. Understanding the communication process is essential for improving interpersonal relationships and reducing misunderstandings.

The sender initiates the communication process as the original creator of the message. The goal is to share facts or ideas with another individual or group. In this initial step, the sender transforms ideas and emotions into a format that can be transmitted by converting the message [10]. This coding process may involve selecting appropriate words, tone, and nonverbal cues to convey the intended meaning.

After being encoded, a message is sent over a channel, which is the vehicle through which it passes from sender to recipient. Channels can be nonverbal, such as body language and facial expressions, or verbal, such as written or spoken words. Since different channels are appropriate for different messages and situations, channel choice can significantly impact communication effectiveness [11].

The recipient, the person or group to whom the communication is addressed, plays a role in the subsequent step. The receiver is responsible for decoding the communication—analyzing and deriving meaning from the information received [12]. The receiver's background, experiences, and perceptions influence the decoding process, which affects how the message is interpreted. Accurately interpreting the sender's intended meaning is essential for effective communication.

Feedback is essential to communication because it informs the sender whether the message was received and understood as intended. Feedback can be provided verbally or nonverbally, allowing the recipient to respond, clarify, or ask questions [13]. This interactive dialogue ensures that everyone is on the same page and that any misunderstandings can be quickly resolved.

Another important factor to consider in the communication process is noise. Any internal or external elements that obstruct the transmission or reception of a message are known as noise. Physical distractions, such as background noise or poor lighting, serve as examples of external noise, while psychological factors, like emotional states or preconceived notions, comprise internal noise. To ensure effective communication, minimizing noise is essential, as it can distort the message and result in misunderstandings [14].

Another crucial factor in the process is the environment in which communication occurs. Context encompasses social dynamics, cultural background, and physical surroundings that affect how messages are transmitted, received, and understood. By fully grasping the context, communicators can more effectively adapt their messages and identify potential barriers to successful communication [15].

The communication process involves a complex series of steps, including the sender encoding a message, sending it across a selected channel, the recipient decoding it, and the recipient responding. Context and noise are two other elements that may affect how effectively people communicate. By comprehending and controlling these elements, people can improve their interpersonal communication abilities and create meaningful and fruitful interactions.

11.2.3 Obstacles to successful interpersonal exchanges

Building and maintaining relationships requires effective interpersonal communication, but many obstacles can lead to miscommunication and conflict. These barriers impact communication differently and can be broadly categorized into cultural, psychological, and physical [3].

External factors that can complicate message transmission are known as physical barriers. These include environmental noises that hinder hearing and understanding such as loud or disruptive sounds. Physical distance can also be a barrier, especially when technology becomes necessary and in-person communication is not possible [16]. Uncomfortable seating arrangements, dim lighting, and other environmental aspects can also obstruct communication.

Internal psychological barriers also affect signal transmission, reception, and interpretation. These barriers include mental states, attitudes, and emotions that influence communication. For example, anger, worry, or stress can impair judgment and lead to misunderstandings of signals. Stereotypes and preconceived beliefs may also be obstacles, causing individuals to judge others based on limited information [17]. Furthermore, some individuals may struggle to express themselves honestly and freely due to a lack of confidence or fear of being judged.

Cultural barriers result from disparities in cultural backgrounds, values, and customs [18]. Because people may have different communication standards and expectations, these discrepancies may result in miscommunication and misinterpretations. A significant cultural barrier is language disparity, which can lead to misunderstandings and misperceptions [19]. Differences in accents, dialects, and colloquial expressions can also cause problems when people speak the same language. Body language and other nonverbal cues, like gestures, can differ significantly across cultures, which can also be the cause of misunderstandings [20]. This creates a perceptual barrier when individuals interpret the same situation differently based on their experiences, values, and beliefs. Each person may understand the message in a way that aligns with their viewpoint, and these differences in perception can lead to misunderstandings. For instance, depending on the recipient's perspective and past experiences, a comment meant as constructive criticism may be taken as a personal attack.

Jargon, technical terms, and complex vocabulary that some people may not comprehend serve as examples of language barriers that extend beyond variations in spoken language. Using overly technical language or industry-specific jargon can lead to confusion and misunderstanding, distancing those unfamiliar with these terms [21]. Simplifying language and avoiding jargon helps ensure that messages remain clear and accessible to everyone.

Emotional barriers refer to how emotions affect communication. Feelings such as excitement, anger, or frustration can impact how messages are sent and received [22]. For example, an angry individual might escalate conflicts and hinder clear communication by adopting a harsh tone or displaying hostile body language. However, by managing emotions and maintaining a calm demeanor, individuals can promote more productive interactions [23].

Attitudes and actions that hinder open and effective communication contribute to attitude barriers [4]. These barriers include insensitivity, defensiveness, and arrogance. It can be difficult for others to express themselves in a hostile communication environment created by someone who is condescending or dismissive. However, these obstacles can be overcome, and communication can be enhanced by treating people respectfully and positively.

Physical, psychological, cultural, perceptual, linguistic, emotional, and attitudinal factors can obstruct successful interpersonal communication [4,18]. Recognizing and eliminating these barriers are crucial for fostering greater understanding, cooperation, and communication among individuals.

11.3 Oral communication

11.3.1 Active listening

A key component of verbal communication is active listening, which involves giving the speaker full attention to understand what they are saying, providing a thoughtful response, and remembering what has been discussed [20]. Active listening requires a conscious effort to be present and engaged during the conversation, unlike passive listening, where the listener may hear the words but not fully understand or engage with them [21]. This skill is essential for effective interpersonal communication because it fosters understanding among individuals, enhances trust, and solidifies relationships.

The initial active listening stage requires dedicating focus to the speaker. This means setting aside electronics and other distractions to ensure the speaker has your undivided attention [22]. Individuals can demonstrate engagement by nodding, maintaining eye contact, and employing other nonverbal cues. By giving full attention, individuals foster a supportive environment that encourages honest and open dialogue [19].

A crucial element of active listening is grasping the message. This requires hearing the words and interpreting the underlying emotions and significance [16,24]. Listeners must observe the speaker's body language, tone, and facial expressions, as these nonverbal cues often add crucial context and insight into their intentions and feelings. One can better understand the overall message by reflecting on the speaker's words and mentally summarizing the key points.

An additional crucial aspect of attentive listening is providing a thoughtful response [25]. This involves offering comments demonstrating an understanding of the speaker's viewpoint and showing interest in the conversation. To confirm comprehension of the message, verbal affirmations such as "I see," "I understand," or "That makes sense" can be employed, along with asking clarifying questions. Summarizing or paraphrasing the speaker's remarks can also reflect attentiveness and deepen understanding of their message. For example, one might say, "So what you're saying is..." followed by a summary of the speaker's key points.

The final aspect of active listening is knowledge retention. This requires the ability to recall the main points of the discussion later. Taking physical or mental notes can help retain during lengthy or complex conversations [17]. Remembering the speaker's information can show commitment to the conversation and appreciation for their contributions. Over time, this can strengthen connection and build trust [16].

Patience and empathy are also essential elements of active listening. Even when disagreeing with the speaker, empathy allows them to understand their perspective and see things from their point of view [1]. This compassionate approach fosters a more supportive and encouraging communication environment. Patience is equally important, enabling the speaker to finish their thoughts without feeling rushed or interrupted. By allowing for pauses and silences, the speaker can gather their thoughts and continue the conversation at their own pace [4].

Active listening is an essential verbal communication ability that requires complete focus, comprehension, deliberate answers, and information memory. By practicing active listening, people can improve their interpersonal communication.

11.3.2 Unambiguous expression

Effective verbal communication requires precise language for the recipient to grasp the message quickly. It involves expressing ideas and concepts clearly, succinctly, and cohesively to minimize miscommunication and enhance overall communication. Gaining proficiency in clear communication is vital for both professional and personal interactions, as it fosters improved understanding and collaboration.

Clear ideas are essential for effective communication. It is important to arrange thoughts and identify the key concepts to be expressed before speaking. This process involves identifying the main idea and supporting information to ensure the message is clear [19]. By organizing thoughts, messages can be more effective and reduce the risk of misunderstandings.

Simplicity is another essential element of effective communication. Clear language enables a wider audience to understand the message. Avoiding jargon, technical terms, and complex vocabulary improves comprehension, particularly when addressing individuals who may not be familiar with the subject matter [26]. When technical terms are unavoidable, providing clear definitions and explanations can help bridge any gaps in understanding.

Another essential aspect of clear expression is conciseness. Succinct means using the fewest words to convey ideas without sacrificing completeness or clarity [10,27]. This eliminates unnecessary words, phrases, and repetition, making the main point more prominent. In professional settings, where time is often limited and effective communication is vital, concise communication is highly advantageous.

A coherent message is organized logically and flows smoothly from one idea to another. This involves structuring thoughts, using transitions to connect concepts, and ensuring each part supports the main objective. When a message is clear and easy for the recipient to follow, misunderstandings are less likely.

Tone and delivery also greatly influence precise expression. Speech rate, tone of voice, and focus on essential details can all affect how the message is understood [6]. Speaking moderately, using a confident and clear tone, and highlighting important topics can help convey the message. Furthermore, paying attention to nonverbal signs—such as gestures and facial expressions—can support and provide context for what is said.

Feedback is vital in effective communication, indicating whether the message has been understood [23]. Encouraging the recipient to ask questions and seek clarification helps identify any misunderstandings and clarify the message. By proactively seeking and being open to criticism, you can enhance the effectiveness of your communication and ensure a proper understanding of your message.

In summary, a clear presentation relies on effective tone and delivery, coherence, simplicity, conciseness, and clarity of thought. By mastering these aspects, individuals can enhance their verbal communication skills, ensuring their messages are clear and reducing the risk of misunderstandings. Accurate expression promotes effective interpersonal communication and builds strong, collaborative relationships.

11.4 Nonverbal communication

A key aspect of interpersonal relationships is nonverbal communication, which involves expressing oneself without words. This includes body language, posture, gestures, eye contact, facial expressions, and other physical actions [12,15,24]. Nonverbal cues often provide context and emotional nuance to verbal messages, reinforcing or contrasting them. Understanding and effectively utilizing nonverbal communication can greatly enhance the impact and clarity of interpersonal interactions.

Body language is a potent form of nonverbal communication that expresses ideas through physical actions. It includes elements like body orientation, posture, movements, and gestures. Body language can convey a wide array of emotions and attitudes, often signifying more than spoken words [24].

Hand gestures emphasize ideas, provide instructions, or convey specific meanings. For example, a wave can be used to greet or say goodbye, while a thumb-up indicates agreement or acceptance. How individuals use their hands and arms can also reflect their level of engagement and openness. In contrast, closed postures, like crossing one's arms, may suggest defensiveness or discomfort; meanwhile, open gestures, such as extending one's arms or displaying palms, often show receptiveness and honesty [15].

Another key element of body language is posture. Standing or sitting can show confidence, focus, and curiosity. Unlike slouching or leaning away, which can indicate insecurity or disinterest, an upright position with retracted shoulders and a held-high head often communicates assertiveness and self-assurance [12]. Additionally, mirroring someone else's posture can build rapport and a sense of connection while demonstrating understanding and empathy.

Body position and movements can further enhance communication. While turning away or creating physical distance may indicate disinterest or discomfort, leaning toward the speaker can convey interest and attentiveness [15,24]. The fluidity and pace of movements can also convey various emotions; for example, slow, deliberate movements might signify tranquility and control, while rapid, jerky movements may reflect agitation or anxiety.

11.4.1 Eye contact and face expressions

Facial expressions play a pivotal role in nonverbal communication, as the human face possesses remarkable expressiveness capable of conveying a wide array of emotions. Countless expressions can manifest on the human face, symbolizing a distinct emotion or reaction [16]. Notable examples of fundamental facial expressions encompass smiles, frowns, elevated eyebrows, and various movements of the eyes, lips, and forehead. For instance, a smile is a universal symbol of approachability, joy, and friendliness. It can create an atmosphere of positiveness and encourage open dialogue. Conversely, a frown or furrowed brow might indicate uncertainty, annoyance, or worry, suggesting something wrong. A person's emotional state and reactions during the conversation can also be inferred from subtle changes in their facial expressions.

Eye contact is another essential aspect of nonverbal communication [1]. It is crucial in guiding interactions, indicating interest and focus, and nurturing connections between people. Proper eye contact demonstrates engagement, self-assurance, and genuineness. It helps establish rapport and trust, thereby increasing the overall effectiveness of communication.

However, various cultures may hold differing perspectives regarding the appropriateness and nature of eye contact. Direct eye contact may be perceived as confrontational or impolite, depending on the cultural context. In certain cultures, it signifies respect and attentiveness. Understanding these cultural distinctions is essential to facilitating effective communication across diverse contexts.

Nonverbal signs, including body language, facial expressions, and eye contact, are essential for expressing emotions and messages. By being conscious of and skillfully utilizing these nonverbal clues, people can improve their interpersonal communication abilities, forge closer bonds with others, and promote greater understanding.

11.5 Written exchange of information

Written communication is a vital component of both personal and professional relationships, necessitating the transfer of messages through written symbols. This form of communication depends on clarity, accurate documentation of information, and the capacity to review and reference communications later [28]. Effective written communication requires clarity, a thorough understanding of the audience, and meticulous attention to detail.

11.5.1 Business-related emails

Professional emails are a commonly used method of written communication in the workplace. They are mainly used for sharing information, making requests, and maintaining professional relationships. Several essential components contribute to the effectiveness of a successful business email. The subject line should be clear and concise, providing a brief overview of the email's contents. This approach helps the recipient quickly understand the email's intent.

A courteous greeting and an introduction outlining the email's purpose should come first in the well-structured body of the message. The main information should be organized into concise, clear paragraphs, each focusing on a different topic. Complex information can be easily understandable using numbered lists or bullet points. Avoid overly informal language or jargon that the recipient might not grasp, and instead, opt for a polite and professional tone.

To ensure that the recipient understands their expectations, it is advisable to end the email with a polite closing and, if necessary, a call to action. Including your name, title, contact information, and a professional signature boosts the message's credibility and makes follow-up easier for the recipient [6].

11.5.2 Memoranda and reports

Reports and memoranda are formal written documents that share information, present updates, and offer organizational recommendations [4]. These documents require a clear structure and a logical flow of information to ensure that the message is understood and actionable [1].

Typically, the introduction outlines the report's goals and scope, followed by the main body containing the results, analysis, and supporting data. Headings and subheadings enhance content organization and help readers navigate the document. Additionally, charts and graphs serve as visual aids that improve data presentation and clarify complex concepts.

In contrast, memoranda [21] are concise documents designed primarily for internal communication. They often provide updates on ongoing initiatives or address specific issues. The core content of a memo should be organized into clear, succinct paragraphs that follow a well-defined purpose [13]. The tone should be direct and professional to ensure the message is both comprehensible and actionable.

11.6 Visual interaction

Visual communication involves using visual aids to improve comprehension and transmit information. This form of communication can complement both written and spoken messages by making complex information more accessible and engaging.

11.6.1 Visual aid utilization

Visual aids are effective communication tools that significantly enhance the impact and clarity of a message. They include various components such as charts, graphs, images, videos, and infographics, each serving a specific purpose in making information more accessible and engaging [4]. By transforming complex information into visually understandable formats, visual aids promote improved information retention and comprehension [19].

Graphs and charts are invaluable when presenting numerical data and illustrating patterns. They highlight essential details and simplify intricate information, helping the audience grasp the main message. Pie charts depict proportions, line graphs illustrate trends over time, and bar charts facilitate comparisons among different data sets. The choice of graph or chart depends on the type of data and the aim one wishes to convey. It is vital to ensure that these visuals remain uncluttered and use clear labels and legends to enhance understanding.

Videos and images enrich presentations by engaging the audience and providing visual illustrations. High-quality images can evoke emotions, clarify concepts, and increase the relatability of the information. Videos can captivate audiences with dynamic content, demonstrate procedures, and provide real-world examples. To maintain professionalism in the use of images and videos, it is crucial to ensure that this media is of high quality and relevant to the subject matter.

By integrating text, images, and data, infographics convey information in a visually attractive and easily understandable way. They are especially effective for emphasizing key points and summarizing information.

11.6.2 Creating powerful images

Imagination, clarity, and a deep understanding of the target audience are essential for creating visually striking content. The main goal is to produce images that improve the message's communication while also being aesthetically pleasing. This effort includes several fundamental design principles such as relevance, consistency, and simplicity [28].

Simplicity is vital in visual design. Overly complex visuals can confuse viewers and obscure the main point. Effective imagery emphasizes the central ideas and presents them clearly and directly [29]. This entails eliminating unnecessary elements, using straightforward lines and shapes, and maintaining a clutter-free image. Separating different components using white or negative space can enhance the clarity of a visual presentation [23].

Consistency in design significantly contributes to a cohesive and polished appearance. This entails maintaining uniformity in layout, font style, and color scheme throughout the image. The consistent use of color can guide the viewer's eye as it navigates the image [30] and helps differentiate between various types of information. Headings, subheadings, and body text should use easily readable fonts, while a logical structure must underpin the layout, ensuring components are evenly spaced and aligned.

Relevance is another crucial principle. The images must reinforce the main points and create a direct connection to the accompanying text [28]. This requires choosing pictures, charts, and other materials that are relevant to the topic and support the arguments presented. Decorative or unrelated elements can undermine the primary message and reduce the visual's effectiveness.

In addition to adhering to these principles, it is essential to consider the audience during the image creation process [6]. Understanding the audience's needs, preferences, and comprehension level can boost the graphics' effectiveness. A general audience may benefit from clearer visuals and illustrative language, while a specialized audience may value intricate graphs and detailed data.

Finally, two crucial stages in the design process are testing and feedback. Sharing the images with colleagues or a representative segment of the target audience can yield valuable insights into their effectiveness and possible areas for improvement. This iterative process guarantees that the final images are clear and engaging, effectively conveying the intended message.

Understanding the audience, ensuring relevance, maintaining simplicity, and upholding consistency are crucial aspects to consider when creating visually striking content. By adhering to these guidelines and embracing feedback, we can create images that enhance communication and make complex content more accessible and enjoyable.

11.7 Interaction in groups

11.7.1 Team meetings

Team meetings are a fundamental communication tool in the workplace. They promote decision-making, solve problems, and allow team members to share updates. Thoughtful preparation and facilitation are essential to ensure these meetings are both productive and inclusive [4].

It is essential that the organized agenda clearly distributes itself in advance to participants [1]. This helps maintain the meeting's focus. Additionally, allocating specific time slots for each item on the agenda is beneficial for effectively managing meeting time.

Encouraging open communication and active participation from all members is essential during the meeting [21]. Fostering a welcoming atmosphere where individuals feel empowered to share their thoughts and opinions will significantly enhance this process. The facilitator should guide the discussion, ensuring every voice is heard and the conversation remains focused.

After the meeting, summarizing the key concepts and outcomes ensures that all participants are aligned and aware of their next steps. Distributing meeting minutes that cover the topics discussed, decisions made, and action items identified is an essential next step. Accurate documentation also keeps accountability and offers a reference for future meetings.

11.7.2 Resolving conflicts

Conflict resolution is an essential aspect of team communication. Differing viewpoints, misunderstandings, and conflicting interests can cause conflicts [2]. To ensure effective conflict resolution, it is important to address these issues honestly and constructively while seeking solutions that meet the needs of all parties involved.

Conflict resolution requires active listening, empathy, and a willingness to compromise. Creating a respectful and safe environment where team members can voice their concerns and collaborate toward a resolution is also essential [13]. This process involves acknowledging the feelings and viewpoints of all individuals and understanding the underlying issues at play.

Additionally, establishing guidelines for civil discourse and encouraging problem-solving techniques are crucial aspects of conflict mediation. This method prioritizes addressing the issue instead of resorting to personal attacks while promoting collaboration to achieve a mutually beneficial solution [31]. In certain circumstances, the participation of an impartial third party, like a manager or a human resources representative, can help facilitate the resolution process.

Effective team communication involves transparent and open dialogue, active listening, and the ability to resolve conflicts amicably [1,4]. Fostering a collaborative and supportive environment enhances interpersonal relationships, creativity, and overall productivity.

11.8 Intercultural interaction

Interacting with people from diverse cultural origins is a component of cross-cultural communication. Effective communication in various contexts requires understanding and respecting cultural diversity. Cultural conventions, attitudes, and communication styles must be understood, as must the flexibility and adaptability to change course. This is analyzed in depth in Chapter 13, on intercultural understanding.

11.8.1 Recognizing cultural disparities

Cultural variations influence communication through linguistic elements, nonverbal cues, and various communication styles. A thorough understanding of cultural norms and expectations and awareness of these variations can help build deeper relationships and reduce misunderstandings [32].

One of the foremost challenges in cross-cultural communication is the limitations presented by language. To enhance the likelihood of mutual comprehension, it is advisable to use plain language, eschew jargon, and exercise patience. Additionally, it is advantageous to be aware of idioms and cultural references that may be unfamiliar to individuals from different backgrounds.

Nonverbal indicators, such as gestures and eye contact, also vary significantly across cultures and should be assessed within their cultural contexts [15]. For instance, in particular cultures, direct eye contact signifies respect and attentiveness; conversely, in others, it may be interpreted as aggressive or disrespectful. Acknowledging these cultural distinctions can help minimize misinterpretations and promote effective communication.

11.8.2 Working in global teams

Multinational teams require effective communication and collaboration across various time zones, languages, and cultural backgrounds. Building trust and understanding among team members is essential for achieving success [1].

Consistent communication, using technology to bridge geographical gaps, and acknowledging cultural differences can promote a cohesive and effective team [1]. This involves scheduling meetings at convenient times for all team members, utilizing video conferencing to improve face-to-face interactions, and being conscious of time zone differences.

Encouraging open communication and creating opportunities for team members to share their viewpoints can enhance collaboration and innovation. This method involves fostering an inclusive environment where all team members feel valued and respected, regardless of their cultural background. Providing cultural awareness training and resources can further support team members in understanding and appreciating the diverse perspectives within the team.

Cross-cultural communication requires recognizing and respecting cultural differences, adapting communication styles, and fostering an inclusive environment. Organizations can enhance collaboration, innovation, and overall success by cultivating trust and understanding within multinational teams [30].

11.9 Digital communication

Digital communication involves using technological devices and platforms to transmit information. In today's interconnected society, this type of communication is increasingly crucial, requiring careful attention to etiquette and online presence [23]. Effective digital communication necessitates adopting suitable tools, maintaining professionalism, and being aware of the impact of one's online interactions.

11.9.1 Email etiquette

Email etiquette is crucial for professional communication. This includes using a clear and concise subject line, appropriately addressing the recipient, and maintaining a formal tone throughout the correspondence. A well-crafted subject line offers a brief overview of the email's content, allowing the recipient to understand the purpose briefly [28]. It should be specific and relevant, steering clear of vague or overly generic terms.

The body of the email should begin with a polite greeting, addressing the recipient by their name and title, if applicable. This creates a respectful tone and recognizes the recipient's status. The introduction should clearly and briefly state the email's purpose, providing essential context for the following content. The main information should be organized into concise, understandable paragraphs, each focused on a specific topic. Complex information can be made clearer by using numbered lists or bullet points.

It is essential to maintain a professional tone while avoiding overly casual language or slang. The language should convey respect and courtesy, especially when addressing challenges or making requests [30]. Being mindful of the recipient's time and presenting necessary information in an organized manner are vital elements. This may include attaching relevant documents or providing links to additional resources as needed.

The email should conclude with a polite closing and, if necessary, a clear call to action to ensure that the recipient understands the expected actions. A professional signature that contains one's name, title, and contact information boosts credibility and facilitates follow-up communication from the recipient [6,23]. Furthermore, proofreading the email for errors and clarity before sending it is essential to maintaining professionalism and preventing misunderstandings.

11.9.2 Social media and online presence

Social media and online presence are vital in personal and professional communication in today's digital world. Social media sites such as LinkedIn, X (formerly Twitter), Facebook, Instagram, and others offer unique chances to connect with a worldwide audience, share knowledge, and establish a personal brand. A good online presence can increase your professional reputation, broaden your network, and open new prospects [23,28]. However, it also demands careful management to ensure that your online activities reflect positively on you and your firm.

11.9.2.1 Building a professional profile

A carefully crafted professional profile is the foundation of a successful online presence. This involves developing profiles on appropriate social networking platforms that highlight your skills, experiences, and accomplishments [6]. For example, LinkedIn is specifically designed for professional networking. A strong LinkedIn profile should feature a professional photograph, an engaging headline, a detailed overview of your work history, and a list of your skills and endorsements. Additionally, including references from colleagues and supervisors can boost the credibility of your profile.

Your biography should clearly convey your identity and professional pursuits on platforms like Twitter and Instagram. Using industry-specific keywords can help others find you and understand your professional focus. Consistency across different platforms is essential, as it aids in building a cohesive and recognizable personal brand.

11.9.2.2 Content sharing and engagement

Sharing relevant material is crucial for maintaining an active and engaging online presence. This includes posting articles, blog posts, industry news, and other content related to your field [30]. Expressing your thoughts and opinions on these topics can position you as a thought leader and encourage audience engagement. Finding a balance between sharing and curating content from various sources is essential, as it reflects your depth of knowledge and commitment to the broader industry.

Engagement is equally essential, involving linking, commenting on, and sharing posts within your network. Actively interacting with others' content nurtures relationships and broadens your reach. Participating in forums, joining relevant groups, and attending virtual events can strengthen your online presence and create networking and collaboration opportunities [23].

11.9.2.3 Professionalism and etiquette

Maintaining professionalism in your online interactions is crucial. This requires using respectful and considerate language and tone, even when discussing contentious or sensitive topics. Avoiding slang, overly informal language, and inappropriate content is essential for upholding a professional image [6]. Moreover, it is important to be aware of the potential impact of your posts, as they can be seen by a wide audience, including current and prospective employers, colleagues, and clients.

Responding swiftly and carefully to comments and messages suggests you are active and approachable. Acknowledging and praising people for their contributions and support is also crucial. Building a positive and polite online presence can enhance your reputation and develop stronger professional ties.

Managing an online presence also requires awareness of privacy and security threats. This includes adjusting privacy settings on social media accounts to control who can see posts and personal information. Regularly reviewing and updating these settings will help maintain privacy and ensure that online activities are visible only to the intended audience [23].

It is also vital to be cautious with the information you provide online. Avoid revealing essential or personal information that could be used maliciously. Using strong, unique passwords for your accounts and enabling two-factor authentication will help safeguard them from unauthorized access.

Personal branding is the practice of marketing yourself and your profession as a brand. It involves creating a unique value proposition, identifying your target audience, and consistently communicating your brand message across all your online activities. A strong personal brand can set you apart from others in your field and make you more appealing to potential employers, clients, and partners.

To establish a personal brand, start by defining your qualities, abilities, and areas of expertise. Consider what makes you unique and how you can provide value to your audience. Craft a clear, compelling brand statement that reflects your professional identity and ambitions. Utilize social media profiles, content sharing, and interactions to spread this message consistently. Social media offers valuable opportunities for collaboration and networking. You can build a strong professional network by engaging with professionals in your field, joining trade associations, and participating in online forums. Networking with influential individuals and opinion leaders can enhance your credibility and visibility.

Another significant advantage of social media is collaboration. Connecting with people on social media sites like LinkedIn and Twitter lets you find people who share your interests and objectives. This opens possibilities for cooperation on projects, studies, and other endeavors. Online conferences, webinars, and virtual events also offer more chances to connect with and work with professionals worldwide.

Analyzing social media activity and vigilantly observing one's online presence can yield critical insights regarding individual impact and efficacy. Most social networking platforms offer analytical tools for monitoring data, such as follower growth, engagement levels, and reach metrics. By methodically examining these indicators, one may discern the types of content that resonate with the audience and identify strategies to enhance online visibility.

Employing tools such as Google Alerts enables individuals to remain informed about discussions related to their names or enterprises on the internet. This proactive approach can be pivotal in maintaining a positive online reputation and enabling timely responses to criticism or misinformation.

The digital landscape is constantly evolving, with new features, platforms, and trends emerging daily. To remain relevant, one must stay abreast of these developments and adapt one's online presence accordingly. This adaptability includes following influential figures and thought leaders, remaining informed about industry developments, and embracing new strategies and technologies.

To stay ahead of the curve and capitalize on social media opportunities, one must be receptive to constructive criticism and continually refine one's online presence. This involves learning from past mistakes, adjusting one's strategies as needed, and being open to experimenting with various platforms, formats, and types of content.

Finding the right balance between personal and professional information is essential for creating an authentic and approachable online presence. While professionalism is important, sharing personal stories can humanize a brand and strengthen connections with the audience. This may include discussing personal achievements, interests, or insights into daily life. However, it is crucial to consider how such content might affect professional reputation, ensuring that each post stays relevant and aligns with the overall brand messaging.

In modern communication, a strong online presence and effective use of social media are vital. Creating professional profiles, sharing relevant content, interacting with others, and maintaining privacy and professionalism are key components in building a successful online identity. The primary advantages of using social media include networking, collaboration, and personal branding. These platforms offer opportunities to connect with a global audience and advance one's career. Ultimately, sustaining a significant and effective online presence necessitates staying updated with trends and changes, making necessary adjustments, and balancing personal and professional content. By engaging with social media mindfully, individuals can enhance their communication skills, cultivate a robust professional network, and attain their personal and professional objectives.

11.10 Case studies and practical illustrations

11.10.1 Effective interaction in technical projects

Successful engineering project case studies demonstrate that good communication is important to reaching project objectives. These illustrations offer helpful perspectives and takeaways. The following are some instances of effective communication in engineering projects:

11.10.1.1 Project A: development of smart cities

This project aims to develop smart city infrastructure in a medium-sized metropolitan area. The initiative integrates data analytics, Internet of Things (IoT) sensors, and sustainable technology to improve urban living conditions.

Techniques for communication:

- Frequent team meetings: Weekly meetings were held to plan future work, discuss progress, and resolve issues. Representatives from each department involved attended these sessions to ensure consensus. Facilitation techniques allowed all team members to participate. Online collaboration technologies, including project management platforms and video conferencing, made communication among team members in different locations easier.
- Clear documentation: All stakeholders received current, comprehensive project plans, schedules, and progress updates. This documentation facilitated the monitoring of developments and the early identification of potential issues. User and procedure manuals were also created to ensure that all team members could easily and quickly access essential information.
- Stakeholder engagement: Through public forums, publications, and social media, citizens, community leaders, and city officials receive regular updates. This interaction addressed various issues and helped build the community's trust. Feedback meetings were scheduled to gather residents' opinions and make any necessary adjustments to the project. Establishing trust and ensuring community support for the initiative required open communication and transparency.
- Results: Stakeholder satisfaction was high, and the project was completed on schedule and within budget. Effective communication was attributed to efficient collaboration and problem-solving throughout the project. The smart city infrastructure enhanced public safety, energy efficiency, and traffic control. Due to improved surveillance and emergency response, effective deployment increased public safety while also significantly reducing energy consumption and traffic congestion.

11.10.1.2 Project B: utilizing renewable energy

The primary goal of this project was to install solar panels and wind turbines to provide the community with sustainable energy, emphasizing the development of renewable energy solutions in a rural setting.

Communication challenges: The project team faced challenges in conveying the benefits and details of the renewable energy systems because of the local community's limited technical knowledge and some skepticism toward the new technology.

Strategies implemented:

- Community meetings: To inform the community about the project, its benefits, and its implementation, the project team organized regular community meetings. These sessions allowed residents to express their concerns and ask questions. Additionally, educational materials, including brochures and video presentations, were utilized to clarify complex technical concepts.

- Visual aids: To improve understanding of intricate technical subjects, the team employed visual aids such as models and diagrams. This approach made the content more accessible to non-technical audiences. Moreover, the renewable energy systems were highlighted through guided tours of installation sites and demonstration videos.
- Practical demonstrations: Practical demonstrations illustrated the operation and advantages of solar panels and wind turbines. This hands-on approach significantly increased community members' acceptance and trust. To cultivate a sense of responsibility and ownership, workshops were organized to instruct residents on the maintenance and operation of the systems.
- Outcomes: The project stimulated greater community involvement and support, leading to a sustainable and well-maintained energy system. Renewable energy solutions improved residents' quality of life by providing reliable and affordable electricity. Long-term sustainability was ensured through ongoing and effective system maintenance, supported by community acceptance and collaboration.

11.10.1.3 Project C: healthcare software development

Develop healthcare management software to improve hospital operations and manage patient data effectively. The software must comply with healthcare regulations and integrate seamlessly with the hospital's existing systems.

Techniques for communication:

- Agile processes: To promote continuous communication and feedback, the project team adopted agile methodologies, which featured daily stand-up meetings, sprint reviews, and retrospectives. This iterative approach enabled the swift identification and resolution of issues. Examples of agile project management tools used include Kanban boards and task-tracking software, which monitor progress and foster collaboration.
- User involvement: The development process incorporated the expertise of medical professionals, including physicians, nurses, and administrative staff. Their feedback was regularly sought to ensure that the software met their needs and remained user-friendly. Usability testing sessions were held to identify and address issues with the user interface, ensuring that the software was both practical and accessible.
- Training sessions: Training sessions were organized to help hospital personnel familiarize themselves with the new software. These workshops included hands-on practice and question-and-answer segments to address any concerns. User manuals and instructional videos were also made available to support ongoing training and troubleshooting.
- Results: A high-quality software solution developed through an iterative communication approach effectively met user needs. The program improved hospital efficiency reduced administrative burdens, and enhanced patient data management. Consequently, healthcare professionals could focus more on patient care, as the successful implementation significantly decreased data errors and the administrative workload.

11.10.1.4 Project D: improvement of infrastructure

An infrastructure upgrade initiative, which encompasses installing new traffic management systems and enhancing public transit, aims to modernize the city's transportation framework.

Techniques for communication:

- Public consultations: Public consultations were conducted to gather insights into the needs and preferences of residents and stakeholders. This process helped develop a transportation system addressing the community's concerns. During these discussions, surveys and focus groups were utilized to collect comprehensive feedback and identify areas needing improvement.
- Collaborative technologies: The project team utilized collaborative technologies such as project management software and shared online platforms to enhance communication and coordination among team members. Issues were resolved, and all participants were quickly informed through real-time communication tools, including video conferences and chat platforms.
- Frequent information: The public received consistent updates through community meetings, social media, and press releases. This transparency helped foster trust by keeping the community informed about the project's progress. Newsletters and progress updates were distributed regularly to ensure that all stakeholders were aware of the various phases and milestones of the project.
- Results: The project was completed successfully, significantly improving public transportation and traffic flow. Effective communication ensured that the project met the community's needs and aspirations. Modernization has resulted in greater satisfaction among public transit users, as commuting times and traffic congestion have notably diminished.

11.10.1.5 Project E: development of products

The creation of a new consumer electronics product entails several critical stages, including design, prototyping, and production.

Techniques for communication:

- Cross-functional teams: The established cross-functional teams included engineers, marketers, and manufacturers. Regular meetings were held to ensure collaboration and alignment across all functions. To encourage creativity and exchange of ideas, brainstorming sessions and group workshops were arranged.
- Consumer input: Focus groups, surveys, and beta testing were used to gather consumer feedback, which improved the product's features and design. Various data analysis techniques were used to analyze customer input and identify trends and preferences.
- Accurate specifications: Every team member received a detailed copy of the product requirements and specifications. This distribution fostered a mutual understanding of the product's objectives and specifications. The specifications were visualized and validated with prototypes and technical diagrams.

- Results: The product was developed and launched successfully, receiving positive customer reviews. Timely delivery and effective communication across all departments enabled the product to meet market demands. The successful launch resulted in strong sales and favorable reviews, highlighting the importance of clear communication in product development.

11.10.1.6 Project F: global cooperation

This proposal outlines the research, development, and testing phases of a collaborative project involving teams from multiple nations. The objective is to create innovative technology.

Techniques for communication:

- Team members participated in cultural sensitivity training to understand and appreciate cultural differences. The training sessions encompassed cultural conventions, communication styles, and conflict-resolution strategies, resulting in a more cohesive and collaborative team environment.
- Timetables for flexible communication: Flexible communication schedules were created to accommodate different time zones. Regular virtual meetings were conducted to discuss progress and tackle any challenges that arose. Despite the time zone differences, asynchronous communication methods, such as email and project management software, facilitated ongoing cooperation among team members.
- Clear documentation: Every team member had access to thorough documentation of project plans, progress reports, and technical specifications. This accessibility ensured clarity and consistency across the project. The documentation was regularly updated and available through a centralized online platform.
- Results: The project was completed successfully, resulting in the development of new technology that aligned with the project's objectives. Effective communication and cultural sensitivity helped overcome challenges and ensured smooth collaboration among multinational teams. This fruitful partnership produced groundbreaking technology that gained industry recognition and opened the door for future international initiatives.

11.10.2 Learning from communication errors

Ineffective communication can have serious consequences in both personal and professional settings. Its shortcomings may lead to misunderstandings, conflicts, decreased productivity, and strained relationships. However, these challenges also offer valuable learning opportunities. By examining and understanding the factors that cause communication breakdowns, individuals and organizations can develop strategies to improve their communication skills and avoid recurring issues.

One of the key lessons from communication failures is the necessity of precision and clarity in delivering messages. Ambiguous or unclear communication can result in misunderstandings and misinterpretations. For example, if a manager gives confusing instructions, team members may interpret expectations differently, potentially leading to inconsistent results. Therefore, it is vital to communicate clearly and precisely to prevent such situations. This involves articulating thoughts clearly, providing thorough explanations, and ensuring that the message is appropriate for the audience's understanding level.

Another critical lesson is the importance of active listening. Communication is a two-way process that involves both speaking and listening. When individuals do not engage in active listening, they often miss essential information, leading to misunderstandings. Active listening requires giving the speaker one's full attention, understanding the conveyed message, and responding appropriately. It involves avoiding interruptions, seeking clarification, and summarizing the information presented to ensure comprehension. Practicing active listening improves understanding of others and enhances the overall effectiveness of communication.

Nonverbal communication also plays a significant role in determining the success or failure of interactions. Cues such as body language, facial expressions, and eye contact can either reinforce or undermine spoken words. Misinterpretation of nonverbal signals can lead to communication breakdowns. For instance, a lack of eye contact, even when accompanied by a positive verbal message, may be perceived as dishonesty or disinterest. Awareness of and skill in nonverbal communication is essential for accurately conveying and understanding the intended message.

Cultural differences can complicate communication breakdowns, especially in multicultural and international settings. Misunderstandings may arise if various cultures' conventions, values, and communication styles are not properly understood and respected. For instance, while some cultures prefer a more indirect form of communication, others favor directness. Effective cross-cultural communication requires understanding these cultural differences and the ability to adapt accordingly. This involves keeping an open mind, appreciating diverse perspectives, and honoring cultural practices.

Emotional intelligence is crucial in preventing communication breakdowns. Emotions significantly impact how messages are conveyed and received. High levels of stress, anger, or frustration can cloud judgment and lead to ineffective communication. Cultivating emotional intelligence involves being aware of one's own emotions and those of others, managing these emotions effectively, and showing empathy to understand and respond to others' emotional states. Enhancing emotional intelligence enables individuals to communicate more clearly and build deeper connections with others.

Feedback plays a crucial role in effective communication and serves as a valuable tool for learning from mistakes. Both giving and receiving feedback promote constructive communication practices and help identify areas for improvement. Constructive feedback should be delivered in a kind and supportive manner, detailed, and focused on actions rather than personal traits. Individuals and organizations can continually enhance their communication strategies by fostering an environment that values open and honest feedback.

Technological barriers may also lead to communication breakdowns, particularly in remote or virtual environments. Challenges such as technical issues, inconsistent internet connectivity, and limited familiarity with communication technologies can impede effective communication. To address these challenges, it is crucial that all participants have access to reliable technology and receive training on using communication platforms. Establishing clear virtual communication protocols, including email etiquette and video conferencing guidelines, can further reduce technological barriers.

Lastly, biases and assumptions can distort our interpretation of communication, resulting in ineffective exchanges. For instance, assuming that others share the same knowledge or perspective may lead to unclear or insufficient communication. Effective communication demands awareness of and engagement with one's preconceived notions and biases. This involves actively seeking diverse perspectives, staying open to feedback, and asking questions for better understanding.

Communication breakdowns offer valuable lessons for improving communication practices among individuals and organizations. Important aspects include clarity and precision, active listening, effective nonverbal communication, cultural sensitivity, developing emotional intelligence, giving and receiving feedback, tackling technological challenges, and recognizing biases and assumptions. By learning from these experiences and applying strategies to address them, individuals can greatly enhance communication effectiveness and foster stronger, more productive relationships.

11.11 Conclusion

Effective communication is a multifaceted skill that includes written, spoken, nonverbal, visual, and digital forms. Throughout this chapter, we have examined the key elements and methods necessary for achieving proficiency in each mode of communication, highlighting its importance in both personal and professional contexts. As we are near the conclusion, it is crucial to emphasize the key lessons learned and offer practical recommendations for further development.

11.11.1 Principal concepts

Verbal communication refers to the exchange of messages conveyed through spoken or written language. Key components include clarity, attentive listening, and precise communication. Expressing ideas and actively listening are essential for successful verbal communication.

Nonverbal communication includes various forms such as body language, facial expressions, and eye contact. These nonverbal cues enhance understanding by providing additional information and emotional nuances that can either support or contradict spoken messages. Successful interactions can be more effective and transparent when nonverbal communication is recognized and used appropriately.

Written communication involves sharing information through written formats such as memoranda, reports, and emails. Key components of effective written communication include coherence, conciseness, and clarity. Understanding the audience and paying close attention to details is vital for optimal written communication.

Visual communication involves conveying information through visual aids such as charts, graphs, images, and videos. Effective visual communication requires the creation of engaging, relevant, and clear visuals that support the main messages.

Digital communication involves sharing information through electronic platforms and devices. Key elements include email etiquette, social media presence, and online professionalism. Maintaining a professional image and being mindful of privacy and security is essential for effective digital communication.

11.11.2 Recommendations for future development

Continuous learning: Communication skills can always be improved. Engage in continuous development by attending workshops, reading relevant literature, and seeking feedback from peers and mentors. Stay up to date with the latest trends and advancements in communication technologies.

Practice active listening: Aim to listen actively in every interaction. This means giving your full attention to the speaker, avoiding interruptions, and providing thoughtful responses. Active listening can strengthen relationships and improve mutual understanding.

Enhance nonverbal skills: Pay attention to your body language, facial expressions, and eye contact. Practice using nonverbal cues to strengthen your verbal messages and stay aware of how others might interpret your nonverbal signals.

Enhance writing skills: Strive to craft clear, concise, and coherent messages. Concentrate on organizing your thoughts logically and employing straightforward language. Carefully proofread your written communications to reduce errors and guarantee clarity.

Leverage visual aids: Visual aids can greatly enhance your presentations and written communications. Practice creating engaging and relevant visuals. Use tools like PowerPoint or Canva to develop professional-quality visuals.

Maintain a professional online presence: Be mindful of your online activities and their potential impact on you and your organization. Use social media strategically to build your professional network and share relevant content. Maintain professionalism in all online interactions.

Adapt to cultural differences: Develop an understanding of cultural differences in communication styles and norms. Adjust your communication strategies to show respect and accommodate these variations, especially in global and diverse environments.

Seek feedback: Regularly ask for feedback from peers, mentors, and supervisors about your communication skills. Use this feedback to pinpoint areas needing improvement and create a plan for enhancing your skills.

11.11.3 Roadmap for interpersonal communication

Figure 11.1 graphically summarizes the roadmap for interpersonal communication, explained in this chapter.

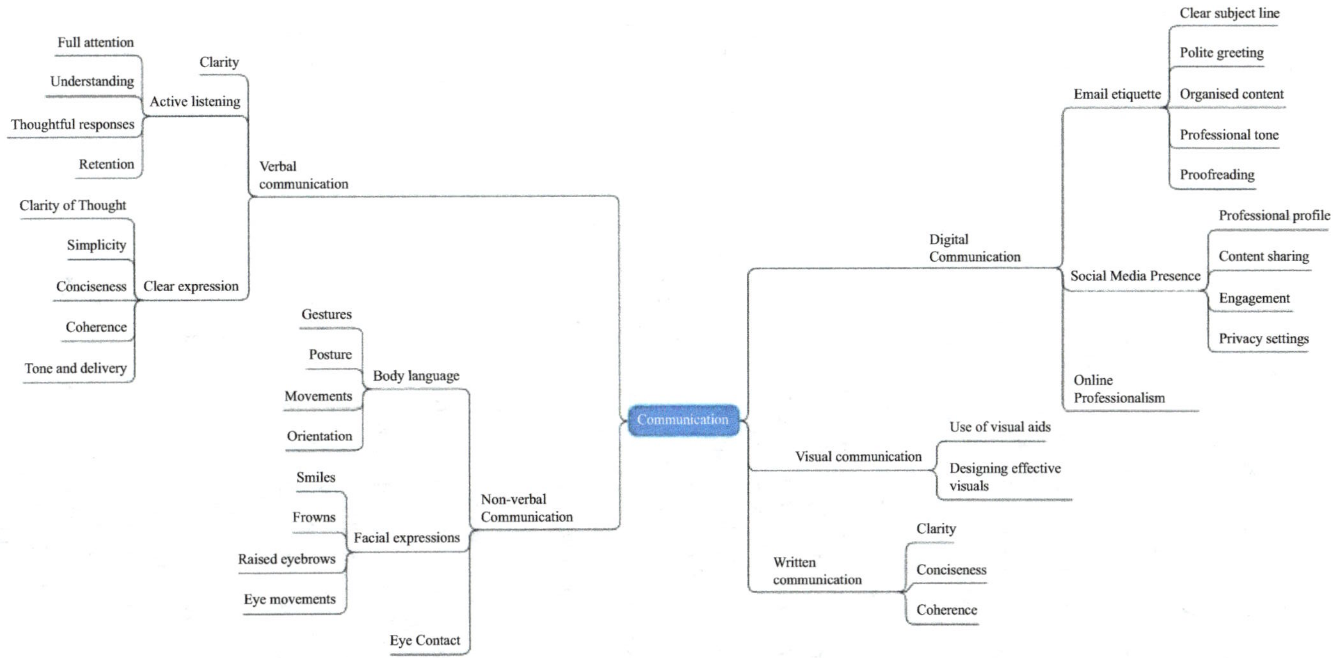

Figure 11.1 Roadmap for interpersonal communication

References

[1] R. B. Adler, and J. M. Elmhorst, *Communicating at Work: Principles and Practices for Business and the Professions*. New York, NY: McGraw-Hill Education, 2009. [Online]. Available: https://books.google.pt/books?id=ssxXAAAAYAAJ [Accessed 30 Oct 2024].

[2] M. E. Guffey, and D. Loewy, *Essentials of Business Communication*. Cengage Learning, 2015. [Online]. Available: https://books.google.pt/books?id=y9stBQAAQBAJ [Accessed 30 Oct 2024].

[3] J. D. Hoover, *Effective Small Group and Team Communication*. Harcourt College Publishers, 2002. [Online]. Available: https://books.google.pt/books?id=zaHYAAAAIAAJ [Accessed 04 Nov 2024].

[4] K. O. Locker, and S. K. Kaczmarek, *Business Communication: Building Critical Skills*. McGraw-Hill/Irwin, 2004. [Online]. Available: https://books.google.pt/books?id=NVXRngEACAAJ [Accessed 04 Nov 2024].

[5] R. Ludlow and F. Panton, *The Essence of Effective Communication.*in Essence of Management Series. Prentice Hall, 1992. [Online]. Available: https://books.google.pt/books?id=TBmPQgAACAAJ [Accessed 04 Nov 2024].

[6] C. Meinel and H. Sack, *Digital Communication: Communication, Multimedia, Security*. in X.media.publishing. Springer Berlin Heidelberg, 2014. [Online]. Available: https://books.google.pt/books?id=5O25BAAAQBAJ [Accessed 12 Nov 2024].

[7] N. N. Sharan, A. Toet, T. Mioch, O. Niamut, and J. B. F. van Erp, "The relative importance of social cues in immersive mediated communication," in T. Ahram and R. Taiar (eds.), *Human Interaction, Emerging Technologies and Future Systems*, Lecture Notes in Networks and Systems, vol. 319, Cham: Springer, pp. 491–498, 2022. doi: 10.1007/978-3-030-85540-6_62.

[8] C. Simons and B. Naylor-Stables, *Effective Communication for Managers: Getting Your Message Across*. Cassell, 1997. [Online]. Available: https://books.google.pt/books?id=3DJEAAAAYAAJ [Accessed 12 Nov 2024].

[9] E. R. Tufte, *The Visual Display of Quantitative Information*. Graphics Press, 2001. [Online]. Available: https://books.google.pt/books?id=qmjNngEACAAJ [Accessed 12 Nov 2024].

[10] N. Thompson, *Effective Communication: A Guide for the People Professions*. Bloomsbury Publishing, 2018. [Online]. Available: https://books.google.pt/books?id=WfJGEAAAQBAJ [Accessed 12 Nov 2024].

[11] N. Wang, A. Jajodia, A. Karpurapu, and C. Merchant, "Charisma and learning: Designing charismatic behaviors for virtual human tutors," in I. Roll, D. McNamara, S. Sosnovsky, R. Luckin, and V. Dimitrova (eds.), *Artificial Intelligence in Education - 22nd Int. Conf. AIED 2021*, Part II, Lecture Notes in Computer Science, vol. 12749, Cham: Springer, pp. 372–377, 2021. doi: 10.1007/978-3-030-78270-2_66.

[12] M. Wieland, L. Thevin, A. Schmidt, and T. Machulla, "Non-verbal communication and joint attention between people with and without visual

impairments: Deriving guidelines for inclusive conversations in virtual realities," in K. Miesenberger, G. Kouroupetroglou, K. Mavrou, *et al.* (eds), *Computers Helping People with Special Needs. ICCHP-AAATE 2022.* Lecture Notes in Computer Science, vol. 13341. Cham: Springer, pp. 295–304, 2022. doi: 10.1007/978-3-031-08648-9_34.

[13] D. J. Whalen, *I See What You Mean: Persuasive Business Communication*. SAGE Publications, 1996. [Online]. Available: https://books.google.com.bn/books?id=uzffaohN5lMC [Accessed 15 Nov 2024].

[14] P. Bayl-Smith, D. Sturman, and M. Wiggins, "Cue utilization, phishing feature and phishing email detection," in M. Bernhard, A. Bracciali, L. J. Camp, *et al.* (eds.), Financial Cryptography and Data Security: FC 2020 International Workshops, AsiaUSEC, CoDeFi, VOTING, and WTSC, Revised Selected Papers, Lecture Notes in Computer Science, vol. 12063, Cham: Springer, pp. 56–70, 2020. doi: 10.1007/978-3-030-54455-3_5.

[15] E. Peatfield, "Policing and non-verbal communication," in K. Corteen, R. Steele, N. Cross, and M. McManus (eds.), *Forensic Psychology, Crime and Policing*, Bristol: Policy Press, pp. 257–262, 2023. doi:10.51952/9781447359418.ch042.

[16] M. G. Frank and A. Solbu, "Nonverbal Communication: Evolution and Today," in R. J. Sternberg and A. Kostić (eds.), *Social Intelligence and Nonverbal Communication*, Cham: Springer International Publishing, pp. 119–162, 2020. doi:10.1007/978-3-030-34964-6_5.

[17] L. Ang, *Principles of Integrated Marketing Communications*. Cambridge University Press, 2021. doi:10.1017/9781108628662.

[18] Z. Kang, C. Indudhara, K. Mahorker, E. P. Bucy, and J. Joo, "Understanding political communication styles in televised debates via body movements," in A. Bartoli and A. Fusiello (eds.), *Computer Vision – ECCV 2020 Workshops*, Lecture Notes in Computer Science, vol. 12535, Cham: Springer, pp. 788–793, 2020. doi: 10.1007/978-3-030-66415-2_55.

[19] U. Hess, "Nonverbal communication," in H. S. Friedman (ed.), *Encyclopedia of Mental Health*, Elsevier, pp. 647–659, 2023. doi:10.1016/B978-0-323-91497-0.00133-8.

[20] D. O'Hair, H. Rubenstein, and R. A. Stewart, *A Pocket Guide to Public Speaking*. Bedford/St. Martins, 2019. [Online]. Available: https://books.google.pt/books?id=zU_cvAEACAAJ.

[21] G. Salim, *Business Communication*. Kaav Publications, Delhi, India, 2022. doi:10.52458/9789391842321.2022.tb.

[22] M. Sanaei, M. Machacek, S. B. Gilbert, C. Eubanks, P. Wu, and J. Oliver, "Behavioral coding for predicting perceptions of interactions in dyads," in A. Antona and C. Stephanidis (eds.), *Learning and Collaboration Technologies – 10th Int. Conf. LCT 2023* (HCII 2023 part I), Lecture Notes in Computer Science, vol. 14040, Cham: Springer, pp. 79–90, 2023. doi: 10.1007/978-3-031-34411-4_7.

[23] P. A. Argenti and C. M. Barnes, *Digital Strategies for Powerful Corporate Communications*. McGraw Hill LLC, 2009. [Online]. Available: https://books.google.pt/books?id=gbANmCLR9bMC [Accessed 15 Nov 2024].

[24] A. Gruber and R. Kaplan-Rakowski, "Verbal and nonverbal communication in high-immersion virtual reality for language learners," in *Intelligent CALL, Granular Systems and Learner Data: Short Papers from EUROCALL 2022*, Research-publishing.net, 2022, pp. 129–134. doi:10.14705/rpnet.2022.61.1447.

[25] K. Schlegel, "Inter- and intrapersonal downsides of accurately perceiving others' emotions," in P. A. Andersen (ed.), *Social Intelligence and Nonverbal Communication*, Cham: Springer International Publishing, pp. 359–395, 2020. doi:10.1007/978-3-030-34964-6_13.

[26] M. Markel and S. Selber, *Technical Communication*. Bedford/St. Martin's, 2017. [Online]. Available: https://books.google.pt/books?id=6Ds9DwAAQBAJ [Accessed 19 Nov 2024].

[27] J. Cornelissen, *Corporate Communication: A Guide to Theory and Practice*. SAGE Publications, 2020. [Online]. Available: https://books.google.pt/books?id=z0i3DwAAQBAJ

[28] L. A. Lievrouw and B. D. Loader, eds., *Routledge Handbook of Digital Media and Communication*. Routledge, 2020. doi:10.4324/9781315616551.

[29] L. Oliveira, A. Fleury, and M. T. Fleury, "Digital power: Value chain upgrading in an age of digitization," *International Business Review*, vol. 30, no. 6, 2021, pp. 3–4, doi:10.1016/j.ibusrev.2021.101850.

[30] E. Darics, *Writing Online: A Guide to Effective Digital Communication at Work*, London: Palgrave Macmillan, 2015. [Online]. Available: https://api.semanticscholar.org/CorpusID:64213373 [Accessed 27 Nov 2024].

[31] I. Montiel, J. Park, B. W. Husted, and A. Velez-Calle, "Tracing the Connections between International Business and Communicable Diseases," *Journal of International Business Studies*, vol. 53, no. 8, pp. 1785–1804, 2022, doi:10.1057/s41267-022-00512-y.

[32] S. Decock, J. De Wilde, S. Van Hoof, E. Van Praet, and B. De Clerck, "Professional Discourse in Multilingual Settings: Policies and Practices," *Multilingual Journal of Cross-Cultural and Interlanguage Communication*, vol. 37, no. 4, pp. 321–330, 2018, doi:10.1515/multi-2018-0023.

Chapter 12

Conflict management and negotiation

Pilar Pazos[1]

Engineering programmes and companies need to foster the development of non-technical and professional skills as part of preparing current and future engineers to succeed in the workplace. Accreditation agencies also expect academic programmes to provide students with opportunities to cultivate these skills. These nontechnical professional skills are considered critical to project success in technical settings. This chapter will explore conflict management and negotiation during the execution of technical projects and present approaches to promote the development of these skills.

We introduce the topics of conflict management and negotiation within project management environments, focusing on skill development. The first part of this chapter explores conflict at the team or project level by providing an operational definition and identifying different types of conflict that emerge in teams, as well as the impact that conflict can have on team outcomes. We will explore the most common management strategies that can be used to prevent conflict from emerging and to effectively address existing conflicts.

The second part discusses negotiation in project management settings, including approaches and best practices used in negotiation tasks.

12.1 Conflict in project-based settings

Conflict is a natural aspect of social interactions in organisations. It is a complex phenomenon occurring at multiple levels, including organisational, unit/departmental, team and individual levels [1]. At the organisational level, factors such as company policy changes, resource allocation or managerial decisions can trigger conflicts between employees or departments. Within teams, a lack of trust can create conflict in project teams. At the individual level, differences in work styles, work–life balance priorities or personality traits can also influence the emergence of conflict. Conflict emerging at different levels often requires different skills and approaches to manage them effectively. Organisational level factors are beyond the

[1]Department of Engineering Management and Systems Engineering, Old Dominion University, USA

scope of this chapter as they originate at the top levels of management, and their impact usually goes further the project team. This chapter will focus on conflict at the project team and individual levels.

Team conflict: *Conflict in a team setting is a disagreement between team members, typically arising from differences in goals, values, perspectives or work styles* [2]. *Conflict can take various forms; some types can be beneficial, whereas others can undermine team processes and performance.*

Conflict in project teams has powerful implications for team processes, outcomes and viability [3]. Different types of conflict have different impacts on team effectiveness. Some types of conflict can harm the team's performance, while others can benefit the team and their outcomes. Understanding the different types of conflict, including the drivers and consequences, is a crucial aspect of effective project management.

12.1.1 Types of conflict

Relationship conflict (RC) arises from personality differences or disagreements among group members about interpersonal issues, norms or values [4,5]. This type of conflict involves opposing relationships and interpersonal tension. When facing this type of conflict, members are likely to approach the situation as a win-lose, and they become less concerned about collective goals and more concerned about their personal gain or status.

Task conflict (TC) is a disagreement between people on ideas and opinions about the task such as disagreement regarding the content of a specific report or the approach to analysing data. Some research has shown that moderate levels of TC can be beneficial to team outcomes under certain conditions such as nonroutine [4] or innovative tasks [5]. TC focuses on the approach to executing the task, so it tends to be more tangible, more impersonal and less emotional charged than interpersonal conflict [4]. Some studies suggest that moderate levels of TC can make teams more effective by broadening their perspectives. Constructive criticism when addressing TC can help teams evaluate the quality of alternatives [6,7]. RC is linked to decreases in group performance and member satisfaction [5]. Simons and Peterson [8] noted that building trust is crucial in gaining the potential benefits of TC without suffering the downsides of RC in the context of top management teams.

Process conflict (PC) is a disagreement about the logistics and delegation aspects when managing a project team, including task assignments, sequencing of activities and delegation of activities. PC is considered a separate category from task and RC with differentiated effects on outcomes. Issues related to PC include planning and task delegation, while task-conflict issues relate to the execution and approach used to accomplish the task. For instance, in a task related to software testing and development, if developers disagree about the interpretation of testing data, they are experiencing TC. If they disagree about who is responsible for specific aspects of the coding or the description of user cases, that is an example of PC.

Team conflict and its types

Team conflict is a disagreement between team members, typically arising from differences in goals, values, perspectives or work styles.

Relationship conflict emerges from personality differences or disagreements among group members about values, norms or interpersonal interactions.

Task conflict is a disagreement among team members on ideas and opinions related to the task, such as disagreement regarding the content of a specific report or the approach to analysing data.

Process conflict involves disagreements about the logistics and delegation aspects when managing a project team, including task assignments, sequencing of activities and delegation of responsibilities.

12.1.2 Conflict profiles

The different types of conflict do not emerge independently in work teams. Instead, they present patterns or profiles, with specific combinations of conflict levels that replicate across teams and contexts [9,10]. These conflict profiles offer a more realistic approach to understanding the nature of team conflict than looking at each type of conflict separately. Team conflict profiles reflect a combination of different types and levels of conflict (relationship, task and process) that coexist and emerge as a result of team members' interaction.

Recent research has investigated the relationship between team conflict profiles and team outcomes. It has been found that some profiles lead to higher team functioning and performance levels, whereas others can lead to team dysfunction and dissolution [10,11]. Conflict profiles represent a novel and holistic approach to examining and managing conflict by helping us characterise different types of teams and design targeted approaches that collectively address the combined levels of conflict. These profiles demonstrate distinct team behaviours that can be addressed using conflict management strategies aligned with the issues at hand.

O'Neill's work on team conflict profiles [9,10] identified three conflict profile categories that showed differences in team outcomes: TC dominant, minor/midrange overall conflict and dysfunctional conflict. The *TC dominant profile* had almost negligible levels of interpersonal and PC, and moderate to high levels of TC. Teams with this profile focused on evaluating inputs and ideas from others and did not experience interpersonal tension between team members. The minor/midrange conflict profile had small to medium levels of interpersonal and PC and high levels of TC. The dysfunctional conflict profile had lower levels of TC than the other profiles, but high levels of RC and PC.

An evaluation of these teams' conflict profiles [10] suggests that teams can experience high TC and maintain low RC and PC. They provide evidence that the TC-dominant profile represents an ideal pattern of productive or constructive team conflict. This important discovery offers new and unique empirical evidence that TC can favour teams in certain conditions, beyond what prior research has found.

Cima and Pazos [11] successfully used team building and scaffolding activities for team training to manage the emergence of RC and PC. Their approach included a kick-off meeting with a team building and alignment activity, developing a team charter and a project plan. This study also identified three categories of team conflict profiles that differed in team effectiveness and outcomes. One profile had almost negligible levels of RC and PC with a small level of TC. The second had low levels of RC and PC and moderate TC. The third and last profile showed similar characteristics as the dysfunctional profile identified by O'Neil, with moderate to high levels of all the three conflict types. Teams with this high conflict profile reported more negative team experiences and outcomes than the other two profiles. In particular, they showed lower commitment to the task and lower satisfaction compared to the other two profiles [29].

The conflict profile examination in teams incorporates a holistic perspective that helps explore the relationship between team conflict and effectiveness. *The empirical evidence suggests that TC, occurring in the absence of both RC and PC, is the most effective, efficient and healthy profile* [9]. They showed that teams can thrive in a beneficial conflict profile by applying constructive controversy approaches.

12.1.3 Conflict management approaches

Most experts agree that conflict management in contemporary organisations should focus less on conflict resolution and more on conflict prevention [12]. Successful conflict management involves using effective strategies to minimise the emergence of dysfunctional conflict, deal with it when it arises and facilitate constructive conflict. Conflict management should go beyond the avoidance, reduction or termination of conflict. Best practices in conflict management involve using effective strategies to minimise the emergence of dysfunctional conflict and facilitate the emergence of constructive conflict to enhance learning and effectiveness in teams and organisations.

12.1.3.1 Preventing dysfunctional conflict

There is evidence of specific approaches and practices that can prevent the emergence of dysfunctional conflict [11] by focusing on the team formation stage of Tuckman's development model (see Chapter 8 on Teamwork). Some of the intra-team PC that can potentially emerge in teams can be prevented by focusing on preparation activities like goal alignment during the early stages of team formation, as well as through the use of collaborative project planning and team charters [11,13,14]. These preparation tasks help create a shared vision for the team, setting up the team for success by clarifying roles, establishing explicit team norms from the start and developing contingency plans to deal with potential team problems. When teams work on building a shared vision and collectively develop team norms and a project plan, they are less likely to face interpersonal conflict, and they can focus on tasks and issues. The resulting team environment is more conducive to cooperatively solving the differences in ideas. In these conditions, there is often an increase in the number and quality of ideas, where members are more likely to have an impartial exploration of differences in opinions. Increased openness to diverse ideas is fundamental in decision-making tasks or those requiring creativity and innovation such as new product development.

There is empirical evidence indicating the positive impact that team building, goal alignment and collaborative planning can have on preventing dysfunctional conflict from emerging [11,13,14]. Academic programmes and on-the-job training can support the development of conflict management and teamwork skills by incorporating some of these proven scaffolding activities into team projects.

Conflict management under ideal conditions

Successful conflict management approaches create an environment that is open to input from all members, with a focus on the benefit of the group and respect for an idea's originator. These conditions set the stage for developing conflict management approaches that are more cooperative and less competitive.

Some of the conditions that facilitate effective conflict are:

- A clear understanding of team members' preferences, constraints, strengths and weaknesses.
- A shared vision of the team and its purpose.
- Open communication based on mutual respect, focused on discussions around ideas and approaches to meet the team's goal.
- Clear roles and responsibilities.

There are different approaches to addressing existing conflict. The two most widely used are the competitive conflict management approach (e.g., individual team members treat their goals as mutually exclusive) and the cooperative conflict management approach (e.g., members have goals that align with those of their teammates) [15]. Teams that use a cooperative management approach tend to see their goals and motives as aligned with the goals and motives of their teammates; they see benefits from team success and participate in mutual support processes such as backing up others in the team when they face individual struggles [16,17].

Collaborative conflict management

There is strong empirical evidence that *collaborative conflict management* processes have a positive impact on affective outcomes. These types of outcomes are important for the long-term sustainability of teams.

When a team has developed a shared attitude of supportive cooperation, the damaging effect of RC can be controlled by the team. Thus, having teams participate in activities that promote the development of a shared vision and supportive attitude, as well as understanding each other's constrains, preferences, strengths and weaknesses, can be very effective in promoting a cooperative approach to conflict management. It is also key to have clearly defined roles and responsibilities so that accountability can be established early during the team formation.

There are essential tactics that a *team lead or project manager* can use to prevent interpersonal conflict or to address some of its root causes:

- Build a team environment that welcomes objections or opinions framed around ideas and approaches, objectively evaluating information from all parties and using active listening. In new teams, the expected behaviours that create this environment should be incorporated into the team's charter.
- Work with the team to develop collective team norms or agreements that prevent the use of language that is not task-related, or that can be perceived as a personal attack.

12.1.4 Common barriers to conflict resolution

Some common problems prevent teams from resolving escalating conflict, including cognitive bias, power dynamics and emotional barriers.

12.1.4.1 Cognitive biases

Biases can hinder successful negotiation by preventing the parties from objectively evaluating the evidence. Confirmation bias is a common type of cognitive bias that can have a negative impact on team conflict resolution. Individuals or teams that display this kind of bias favour information that confirms their beliefs while ignoring or downplaying opposing views of contradictory evidence. This type of bias prevents teams from reaching integrative, collaborative solutions.

12.1.4.2 Power dynamics

Power imbalances can also interfere with conflict resolution. If a team member has more authority or influence, others may feel reluctant to express their concerns, fearing retaliation or dismissal [4]. This imbalance stifles open communication and interferes with problem solving.

12.1.4.3 Emotional barriers

Emotional factors such as anger, fear or frustration can cloud judgement and interfere with rational decision-making. Team members caught in emotional states during conflict often focus on defending their perspective and winning the argument rather than finding a solution [18].

12.1.5 Skills and tools to support conflict resolution and negotiation

12.1.5.1 Emotional intelligence

Emotional intelligence (EQ) is the ability to understand and manage one's emotions as well as recognise and influence other's emotions [19]. In project teams, negotiations are frequent, whether over resources, timelines, responsibilities or differing perspectives. Emotional intelligence can help project managers and team members constructively navigate these interactions. Self-awareness and empathy are qualities of an individual's emotional quotient that can help support successful negotiation processes.

12.1.5.2 Communication skills

Communication is vital to effective negotiation and conflict resolution in project teams. Effective communication involves a transparent exchange of ideas, expectations and concerns among team members. Good communication skills involve the ability to articulate different parties' needs, understand others' perspectives and build consensus. Another critical aspect of successful communication in project teams is active listening.

Active listening involves giving full attention to the person communicating, minimising interruptions to those that are strictly necessary. This skill helps negotiators understand underlying concerns or motivations. Active listening fosters a collaborative environment in project teams where members feel valued and heard.

Another critical aspect of communication skills is using clear and precise language so that all parties involved in the negotiation fully understand the issues under discussion. Project teams often deal with complex tasks, and misunderstandings can lead to confusion, misalignment and conflict. A negotiator must have strong communication skills to present different viewpoints in a precise, clear and structured way that reduces ambiguity.

12.2 Negotiation in engineering project teams

Companies operate in uncertain and rapidly changing environments involving complex relationships. Some of these relationships are internal to the organisation (within a department, across departments, interpersonal), whereas others involve external stakeholders (suppliers, customers, government). Negotiation is an inherent element in forming those relationships, and it is necessary to manage them over time. Project-based organisations rely on teams at different levels to handle the negotiations with internal and external stakeholders.

Some examples of negotiations that may involve project teams include:

- Developing contracts to fulfil the requirements of a customer.
- Developing marketing plans for different products or services.
- Create product development contracts with external partners or subcontractors.
- Negotiating budgets with clients.

There are very few business transactions that do not require some element of negotiation between different parties. Possessing negotiation capabilities can help support many business interactions and improve the bottom line. Although negotiation skills are essential for engineers, they have historically received little attention in academic programmes [20,21].

12.2.1 Negotiation tasks and negotiation skills

We propose Cohen's definition of negotiation as a joint decision-making process of two or more parties collaborating to reach a mutually acceptable agreement on one or more issues [30].

There are some common characteristics of a negotiation task:

1. It involves two or more parties.
2. There is an information exchange between the parties.
3. The parties can be creative and cooperate in arriving at a joint decision.
4. The payoff depends either on the consequences of the joint decision or variables external to the negotiations.

In engineering settings, most transactions involve negotiations around quality, schedule and other project or product requirements. Cost and delivery time are the common elements in project negotiations, especially in project-based organisations such as construction companies or consulting firms. When project teams work on price negotiations, they must find a balance between the highest price the customer is willing to pay and the lowest price the offering company can accept. The range between these two prices is called zone of possible agreement (ZOPA).

Another approach used in negotiation strategies is best alternative to a negotiated agreement (BATNA) [22]. It refers to a party's most beneficial course of action if the negotiation does not reach an agreement. Knowing one's BATNA is critical because it provides a baseline to measure against any agreement or final decision. This approach reduces the chances that the negotiating party ends up with unfavourable terms out of desperation or lack of understanding. This approach requires comprehensive preparation, where negotiators generate and evaluate the available alternatives. The process must be informed by a thorough analysis of the conditions. BATNA can also be used as a walk-away argument. If the negotiated terms are worse than the BATNA, the rational choice is to abandon negotiations. This approach prevents negotiators from accepting unfavourable deals out of fear or pressure [23]. BATNA can be a valuable tool in negotiation. However, a rigid focus on one's BATNA might lead to missed opportunities for creative problem-solving or integrative solutions, where both parties could achieve better outcomes than their alternatives [24].

The importance of conflict prevention in negotiation: *Effective negotiation skills are not necessarily associated with mastering confrontational tactics. It is also not only about solving an unfolding crisis. Many crucial skills are linked to the prevention of the crisis in the first place. Providing students with the opportunity to participate in activities where they can tackle the problem at its root is a critical and necessary skill to prevent the initial conflict from arising.*

Approaches to developing negotiation skills in engineering students

The role-playing method is widely used to support the development of negotiation skills and is considered superior to the lecture method. This approach asks students to interact with others and assume roles in a hypothetical engineering negotiation. They receive some common information about the scenario and the issues to be resolved or optimised (e.g., the price of a material), and some confidential information unknown by others (e.g., their company negotiating position). Another successful approach is having authentic team projects embedded in the curricula using team activities that model best practices. These practices include establishing precise individual and collective goals, incorporating peer and expert evaluation of processes and outcomes and creating a transparent approach to planning and tracking team [11,13,14].

12.2.2 Negotiation strategies in project settings

There are different approaches to handling interpersonal team conflict based on two primary dimensions: concern for self and concern for others. Concern for self represents the degree to which individuals want to satisfy their own concerns or interests. Concern for others represents the degree to which people want to satisfy the concerns or interests of others. The combination of these two dimensions characterises the motivational orientations of an individual who faces conflict in a project team. This two-dimensional model is called the dual concern model, and it is associated with five identifiable conflict-handling styles: integrating, obliging, compromising, dominating and avoiding [12,25]. Figure 12.1 illustrates the five approaches to conflict.

1. **Integrating** is an approach to conflict reflecting a high concern for both self and others. This approach involves openness to new ideas, critical evaluation of ideas and information, the examination of alternatives and the achieving of solution acceptance. Integrating styles have been linked to a high commitment to the solution and a more comprehensive and fair evaluation of alternatives among the parties involved. It is considered suitable for dealing with complex problems requiring a group of experts to reach a solution, as it supports the

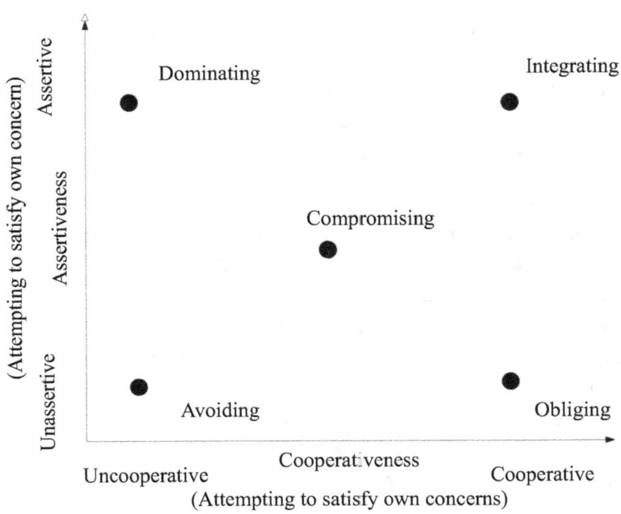

Figure 12.1 Conflict management styles [12]

effective use of all project team members' skills, information and resources. Integrating has also been successful in addressing social conflict within teams.

2. **Obliging** is a style reflecting low concern for self and a high concern for others by attempting to play down differences in order to satisfy the other party's concern. This approach is less appropriate when the individual has evidence to support their concern or when the issue is essential to them. This approach can result in groupthink if systemic pressure exists to agree with a decision without the proper supporting evidence. However, this approach could be appropriate when the individual does not have complete information or is in a position of weakness. It may also be used when a party is willing to give up something with the hope of getting some benefit from the other party when needed.
3. **Dominating** is a style with high concern for self and low concern for others. This approach pushes the other party to agree with one's position. A supervisor may use this style if the issues involve routine matters or require a speedy decision. A supervisor may have to use it to deal with subordinates who are very assertive or do not have the expertise to make technical decisions. This style is also effective in dealing with the implementation of unpopular courses of action. In these scenarios, some pushback can be expected in the project team if the courses of action are not justified or understood.
4. **Avoiding** is a style characterised by low concern for self and others. This approach is associated with withdrawal or detachment from the issue or the task. Although it is considered one of the less effective approaches, it can be used to deal with trivial or minor issues, or when a cooling-off period is needed before complex problems can be effectively addressed. However, this style is not recommended when issues are critical to one of the negotiating parties involved or when a quick resolution is required [12].
5. **Compromising** is a style that sits at the centre of all four remaining styles. It requires all parties to make concessions in order to reach a mutually acceptable decision. This style can be helpful when the goals of the different members are mutually exclusive, when both parties have similar negotiating power (e.g., workers and management) and have reached an impasse in their negotiation process or when other styles are ineffective.

A style is considered appropriate if its use leads to the effective formulation of a solution to the problem. In this respect, the predominant view is that the integrating style is generally appropriate for managing conflict in most situations [26]. In general, integrating and, to some extent, compromising styles are the most appropriate for dealing with strategic issues. The remaining styles can be useful for handling tactical or day-to-day problems, and the approach should consider the context and stakeholders [12]. Some recent studies suggested that the situation's context and characteristics are essential considerations when selecting a conflict management approach. Evidence suggests that effective negotiators adopt this contingency approach to conflict management. The approach considers the context and stakeholders when addressing conflict in a team situation and selects the most appropriate style based on those conditions.

12.2.3 Groupthink and the absence of group conflict

Groupthink is a psychological phenomenon that can occur when the desire for consensus, harmony or conformity overrides the group's ability to critically evaluate alternative ideas, options or courses of action as part of decision-making [27]. *Groupthink* leads members to suppress dissenting opinions and overlook potential risks, resulting in flawed decision-making. In environments characterised by groupthink, little to no conflict occurs. Historical failures such as the Challenger disaster and the Bay of Pigs invasion have been cited as examples of groupthink in action [28]. Groupthink creates conditions where the priority becomes maintaining group unity rather than making well-informed, rational decisions.

Proposed group activity: case study analysis (group work or individual)

Instructions:

1. **Form teams**: Students will be divided into groups of 4–5 members.
2. **Review case study**: Each group will be given a case study based on a real-world team conflict scenario.
3. **Identify conflict types**:
   - As a team, discuss the case and identify and describe the different types of conflict involved (TC, PC, RC).
   - Analyse the causes of the conflicts and how they impact the team's dynamics and project success.
4. **Propose solutions**:
   - Discuss and propose specific strategies for resolving each type of conflict.
   - Include negotiation techniques (e.g., compromise, collaboration) and communication strategies (e.g., active listening, reframing) to develop practical solutions.

Deliverables:
A brief (2–3 pages) written report summarising your case study analysis, the identified conflicts and your proposed solutions.

References

[1] Korsgaard MA, Soyoung Jeong S, Mahony DM, and Pitariu AH. A multilevel view of intragroup conflict. *Journal of Management*. 2008;34(6):1222–52.

[2] De Dreu CK, and Weingart LR. Task versus relationship conflict, team performance, and team member satisfaction: A meta-analysis. *Journal of Applied Psychology*. 2003;88(4):741.

[3] DeChurch LA, Mesmer-Magnus JR, and Doty D. Moving beyond relationship and task conflict: Toward a process-state perspective. *Journal of Applied Psychology*. 2013;98(4):559.

[4] Jehn KA. A multimethod examination of the benefits and detriments of intragroup conflict. *Administrative Science Quarterly*. 1995;40(2):256–82.

[5] De Dreu CK. When too little or too much hurts: Evidence for a curvilinear relationship between task conflict and innovation in teams. *Journal of Management*. 2006;32(1):83–107.

[6] Amason AC. Distinguishing the effects of functional and dysfunctional conflict on strategic decision making: Resolving a paradox for top management teams. *Academy of Management Journal*. 1996;39(1):123–48.

[7] Schweiger DM, Sandberg WR, and Rechner PL. Experiential effects of dialectical inquiry, devil's advocacy and consensus approaches to strategic decision making. *Academy of Management Journal*. 1989;32(4):745–72.

[8] Simons TL, and Peterson RS. Task conflict and relationship conflict in top management teams: The pivotal role of intragroup trust. *Journal of Applied Psychology*. 2000;85(1):102.

[9] O'Neill TA, and McLarnon MJ. Optimizing team conflict dynamics for high performance teamwork. *Human Resource Management Review*. 2018;28(4):378–94.

[10] O'Neill TA, McLarnon MJW, Hoffart GC, Woodley HJ, and Allen NJ. The structure and function of team conflict state profiles. *Journal of Management*. 2015;44(2):811–36.

[11] Cima F, and Pazos, P. Conflict profiles and team outcomes in cross-disciplinary teams: An integrated latent profile analysis and natural language processing approach. *Proceedings of the American Society for Engineering Management Conference*, 2024.

[12] Rahim MA. Toward a theory of managing organizational conflict. *International Journal of Conflict Management*. 2002;13(3):206–35.

[13] Magpili NC, and Pazos P. Self-managing team performance: A systematic review of multilevel input factors. *Small Group Research*. 2018;49(1):3–33.

[14] Pazos P. Conflict management and effectiveness in virtual teams. *Team Performance Management: An International Journal*. 2012;18(7/8):401–17.

[15] Maltarich MA, Kukenberger M, Reilly G, and Mathieu J. Conflict in teams: Modeling early and late conflict states and the interactive effects of conflict processes. *Group & Organization Management*. 2018;43(1):6–37.

[16] Zhou Z, and Pazos P. How teams perform under emergent and dynamic situations: The roles of mental models and backup behaviors. *Team Performance Management: An International Journal*. 2021;27(1/2):114–29.

[17] Tjosvold D. Cooperative and competitive goal approach to conflict: Accomplishments and challenges. *Applied Psychology*. 1998;47(3):285–313.

[18] Ayoko OB, Callan VJ, and Härtel CE. Workplace conflict, bullying, and counterproductive behaviors. *The International Journal of Organizational Analysis*. 2003;11(4):283–301.

[19] Der Foo M, Anger Elfenbein H, Hoon Tan H, and Chuan Aik V. Emotional intelligence and negotiation: The tension between creating and claiming value. *International Journal of Conflict Management*. 2004;15(4):411–29.

[20] Hindriks K, Jonker CM, and Tykhonov D, editors. The benefits of opponent models in negotiation. *2009 IEEE/WIC/ACM International Joint Conference on Web Intelligence and Intelligent Agent Technology;* 2009 Sept 15–18; Milan, Italy/New York: IEEE; 2009. pp. 439–44.

[21] Smith ML. Planning your negotiation. *Journal of Management in Engineering*. 1992;8(3):254–60.

[22] Fisher R, Ury WL, and Patton B. *Getting to yes: Negotiating agreement without giving in*. New York: Penguin; 2011.

[23] Lewicki RJ, Barry B, and Saunders DM. *Essentials of negotiation*. McGraw-Hill Education; 2011.

[24] Neale MA, and Bazerman MH. Negotiator cognition and rationality: A behavioral decision theory perspective. *Organizational Behavior and Human Decision Processes*. 1992;51(2):157–75.

[25] Rahim MA, Garrett JE, and Buntzman GF. Ethics of managing interpersonal conflict in organizations. *Journal of Business Ethics*. 1992;11:423–32.

[26] Mouton JS. *The managerial grid: Key orientations for achieving production through people*. Houston, TX: Gulf Publishing Company; 1964.

[27] Janis IL. Groupthink. *IEEE Engineering Management Review*. 2008;36 (1):36.

[28] Esser JK, and Lindoerfer JS. Groupthink and the space shuttle Challenger accident: Toward a quantitative case analysis. *Journal of Behavioral Decision Making*. 1989;2(3):167–77.

[29] Tekleab AG, Quigley NR, and Tesluk PE. A longitudinal study of team conflict, conflict management, cohesion, and team effectiveness. *Group & Organization Management*. 2009;34(2):170–205.

[30] Cohen SP. *Negotiating skills for managers*. New York: McGraw-Hill; 2002.

Chapter 13
Intercultural understanding

Morgane Lamoureux[1]

This chapter focuses on the perception of intercultural understanding as a challenge in international communication. The key question explored is "Why is intercultural communication important?" Under this main idea, various aspects will be developed, including barriers to intercultural communication, cultural shock, cultural identity, strategies for working effectively at an international level, and the significance of this essential soft skill in today's globalized context.

13.1 Intercultural understanding: communicating from our own culture

13.1.1 What is intercultural communication?

In intercultural communication, the central focus is culture, a concept with numerous definitions. Generally, culture is defined as the distinctive traits – spiritual, material, and emotional – that characterize a society or social group. It also encompasses elements such as literature, music, and art. However, in the context of intercultural communication, scholars like Geert Hofstede [1] and Erin Meyer [2] emphasize shared values, and behaviors as key components of culture.

One of the most widely used models of understanding culture is Edward T. Hall's Iceberg Model [3]. This model shows that we typically only perceive about 20% of another person's culture, which would be the visible portion. This visible part provides information about aspects of verbal and nonverbal communication such as opinions and behaviors. It represents the conscious elements of culture which are often mistakenly interpreted as encompassing the entirety of an individual's or group's cultural identity. Figure 13.1 depicts this model.

However, the remaining 80% of culture lies beneath the surface, comprising deeper and less visible aspects. These include values, beliefs, thought patterns, and underlying concepts. This "hidden" part of culture takes time and effort to uncover and understand.

The iceberg can be compared to the infamous iceberg that sank the Titanic: if approached without preparation, awareness, or understanding, it can lead to

[1]Department of Psychological Sciences and Education, Haute École de la Province de Liège, Belgium

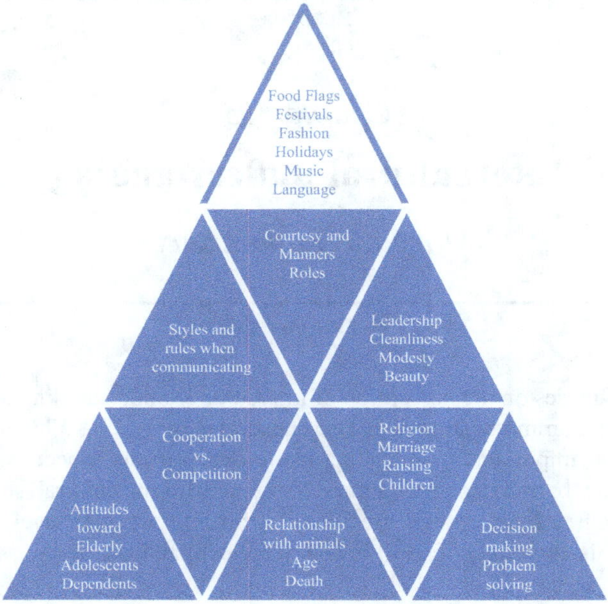

Figure 13.1 The cultural iceberg (in white, the surface culture, over the sea level; in dark, the deep culture, under the sea) [3]

miscommunication and conflict. Proper preparation and a willingness to explore the deeper layers of culture are essential to avoid "crashing" into cultural misunderstandings.

13.1.2 Why is intercultural communication important?

Understanding one another is a fundamental priority, especially in communication. To communicate effectively, it is essential to comprehend the communication process and be aware of our own communication style, as well as the impact it may have on relationships with others.

Even within the same culture, communication can be challenging; between individuals from different cultures, it becomes significantly more complex. Intercultural communication involves exchanging messages between people of distinct cultural backgrounds. In today's interconnected world, this type of communication is more frequent than ever, facilitated by various factors such as the rise of the internet, the proliferation of low-cost airlines, the growing job opportunities abroad, the reception of refugees, and the movement of workforces across borders. These developments underscore the importance of understanding and navigating intercultural communication effectively.

One of the main challenges in intercultural communication is navigating differences in language, but the obstacles extend beyond that. Cultural barriers, such as differing ideological constructs, also play a significant role. Unfortunately, cultural differences are not limited to surface-level customs, such as politeness or initial contact behaviors what might be referred to as "first impression customs".

These differences influence much deeper aspects of interaction, including communicate styles, perception of agreements, the role of family, and broader group dynamics in business context. When these deeper cultural perceptions are overlooked, misunderstandings can persist or even intensify over time. This is especially true during negotiations, where differing cultural interpretations and values may lead to increasing friction, even as the parties become more familiar with one another on a personal level. Recognizing and addressing these cultural nuances is essential for fostering mutual understanding and effective collaboration in intercultural settings.

Having the ability to communicate effectively across cultural boundaries is essential for the success of any intercultural or multinational endeavor. It not only ensures smoother collaboration but also fosters mutual understanding and respect among individuals from diverse cultural backgrounds.

Moreover, this ability significantly enhances relationships by facilitating open and constructive conversations, breaking down barriers, and building trust. By bridging cultural gaps, it helps create more inclusive environments where people can work together harmoniously, leveraging their differences as strengths. Some tips could be:

- To know each other: understanding and appreciating different cultural perspectives foster mutual respect and awareness.
- To increase insurance: developing confidence in navigating cultural differences enhances communication and collaboration.
- To break down the barriers: removing cultural misunderstandings and prejudices helps establish stronger connections.
- To build trust: establishing reliable and respectful communication builds confidence among individuals and groups.
- To open new horizons: engaging in intercultural communication broadens perspectives and encourages innovative ideas.
- To develop interpersonal skills: enhancing communication skills improves empathy, adaptability, and emotional intelligence.
- To find common ground: identifying shared values and goals strengthens collaboration and unity.

To successfully enter a new market, companies and their representatives must be ready to adapt to the local habits, customs, and mindset of the population. This requires not only an understanding of cultural norms but also the ability to interpret and navigate them effectively.

For exporters, having *intercultural decoding tools* is crucial to ensure smooth communication and establish trust with partners and customers in the target market. It is equally important to recognize that culture exists in various forms and influences interactions on multiple levels. These include:

- Corporate culture: the unique values, practices, and behaviors within organizations.

- Professional culture: the shared norms and expectations within specific professions or industries.
- Religious culture: beliefs and practices rooted in religious traditions.
- Regional culture: the customs and values specific to geographic areas.

Understanding and respecting these cultural layers can significantly enhance a company's ability to succeed in international markets.

13.1.3 Barriers to intercultural communication

There are several barriers to effective intercultural communication. These can arise from language difficulties, misinterpretations of the verbal or nonverbal message, or stress. Barriers stemming from negative attitude include stereotypes, prejudice, xenophobia, ethnocentrism, discrimination, and racism. Let us analyze some of them.

13.1.3.1 Stereotypes

We all use simple (but convenient) descriptions of national cultures and behaviors: stereotypes. These are generalizations, clichés, that allow us to summarize complex traits in a way that is both useful and dangerous at the same time. For example, Belgians are said when they speak that "they eat French Fries". This expression refers to a humorous stereotype that reflects how the French perceive the Belgian accent or way of speaking. It is often said that Belgians "speak as if they have French fries in their mouth". This saying stems from the friendly rivalry and jokes between the French and Belgians, particularly regarding the pronunciation and the Belgian accent when speaking French, which sometimes sounds different due to regional variations. Moreover, Belgium is famous for its French fries (*frites*), which reinforces the metaphor.

The stereotype is a simplified understanding of another culture, as we often synthesize information too quickly and oversimplify what we receive from it.

13.1.3.2 Discrimination

Unia is a Belgian independent public institution that fights against discrimination and defends equality. It defends the equal and inclusive participation of all in all sectors of society and ensures respect for human rights. Unia [4] lists 19 discrimination criteria:

1–5. The five so-called "racial" criteria: race, skin color, nationality, ancestry (e.g., Jewish), and national or ethnic origin;
6. Disability;
7. Philosophical or religious convictions;
8. Sexual orientation;
9. Age;
10. Wealth (in other words, financial resources);
11. Civil status;
12. Political beliefs;
13. Trade union membership;

14. Health status;
15. Physical or genetic characteristics;
16. Birth;
17. Social origin;
18. Sex;
19. Language.

Discrimination constitutes an action. When the attitude of rejecting or denying another culture turns into an action, this is called "discrimination", which refers to treating an individual unfairly based on factors such as nationality, gender, age, or profession. The most terrible form of discrimination is that which occurs between one group and another.

13.1.3.3 Xenophobia

Xenophobia can lead to discrimination, exclusion, and even violence against people of different nationalities, ethnic origins, or cultures. It hinders social cohesion, intercultural dialogue, and the promotion of diversity.

13.1.4 Cultural shock

Cultural shock "refers to feelings of uncertainty, confusion, or anxiety that people experience when they are transplanted into a society different from their own. Culture shock occurs when people vacation, do business, attend school, or move to another city or country. The syndrome arises from an individual's unfamiliarity with local customs, language, and acceptable behavior" [5].

Culture shock refers to the reactions people have when living in a new culture. Some common symptoms include strain, a sense of loss, feeling rejected, confusion, anxiety, helplessness, and obsession with hygiene. There may also be physical symptoms like headaches, sleeplessness, overeating, or excessive consumption of alcohol.

The first anthropologist who has analyzed these reactions was the Canadian Kalervo Oberg (Figure 13.2). He has identified different stages of culture shock [6].

The first stage is the honeymoon period. Everything looks exciting and feels fascinating. There is a sense of euphoria about new things and discovering foreigner people and customs.

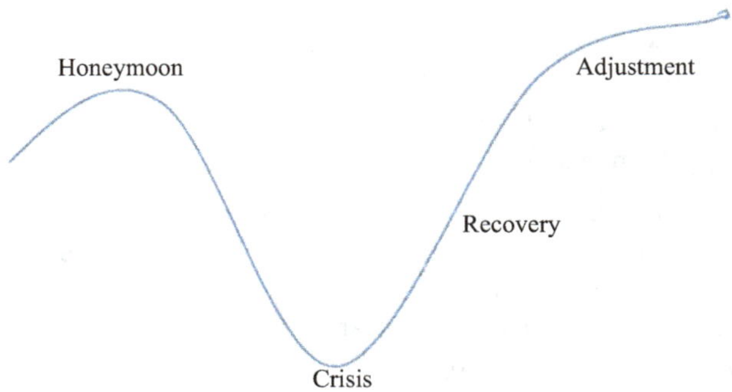

Figure 13.2 The stages of culture shock

Then, in the second stage, the reaction changes. It is called "crisis": habits are coming and enters in a period of disappointment. Negative aspects of the new culture appear. Misunderstanding with local people occurs, and everything becomes difficult to endure. It is the crisis stage.

In the third stage, recovery stage, individuals accept that there is a gap between expectations and reality. They begin to find a balance, increasing their sense of control and feeling less lost in this new culture.

The last stage, adjustment stage, allows individuals to achieve their goals. They learn to appreciate both the positive and negative aspects of the new culture, integrate into the new environment, and begin to see its positive aspects.

Some authors have analyzed reverse culture shock [7] which occurs when individuals return home after spending time abroad. They have determined that people must reacclimatize to their own culture upon returning. They may encounter difficulties in resuming their routines and feel sadness about leaving behind new friends. While abroad, individuals often experience a sense of freedom, and they may feel a sense of loss upon leaving it behind. They also often feel misunderstood by people who have not had the same experiences. Common problems associated with reverse culture shock include social withdrawal, depression, anxiety, stress, and anger.

13.1.5 Cultural identity

Cultural identity is the sign of recognition for a human community, which can be social, political, corporate, national, religious, or otherwise. It includes shared values, ideas, languages, traditions, beliefs, and common experiences. Identity provides a secure refuge from the social disorders and insecurities of life, but it can be adaptable to other communities. Traveling and interacting with people from different cultures allow us to engage with new perspectives, lifestyles, traditions, and languages. These interactions can influence our own identity allowing us to adopt new practices, modify existing ones, revive forgotten traditions, and become more open-minded.

13.2 Culture map by Erin Meyer

Erin Meyer is an American author and professor in intercultural management. She has developed an eight-scale model that represents the extremes of cultures to which managers must pay attention [8].

The theory suggests that when working with multicultural teams, one must analyze these eight dimensions:

- Communicating: low context versus high context
- Evaluating: direct negative feedback versus indirect negative feedback
- Persuading: principle first versus applications first
- Leading: egalitarian versus hierarchical
- Deciding: consensual versus top-down

- Trusting: task-based versus relationship-based
- Disagreeing: confrontational versus avoid confrontation
- Scheduling: linear time versus flexible time

Analyzing these axes helps to improve effectiveness in managing others.

13.2.1 Communicating: low context versus high context

In certain cultures, communication is based on low-context, explicit messages. There is a classification of countries between low-context communication and high-context communication. In low-context communication, the characteristics are that communication is direct, explicit, and relies heavily on verbal messages. Messages are clear and unambiguous, with little reliance on nonverbal cues or shared context. Countries like the US, Australia, Germany, and the UK are placed in this category.

In high-context communication, the characteristics are that communication is indirect, nuanced, and relies heavily on nonverbal cues, shared context, and relationships. Messages are often implied and require interpretation based on the social and cultural context. Countries like Japan, China, Korea, and Saudi Arabia are placed in this category.

For example, a direct and explicit communication style, typical of low-context cultures, may be perceived as rude or insensitive in high-context cultures where indirect communication is more common. Conversely, a more indirect and nuanced communication style, common in high-context cultures, may be seen as evasive or unclear in low-context cultures.

Among the cultures with explicit communication and low context are North American cultures (the US and Canada), along with Germanic cultures (Germany, Switzerland, and Austria), and cultures of Scandinavian countries, Australia, and New Zealand.

At the opposite end of the scale, we find Japan, where context plays a significant role. For example, rules of politeness govern speech, with the manner of speaking shifting across more than 20 subtly different registers depending on the age, gender, and social position of the conversation partner, as well as the relative social positions of the speaker and listener. The word “no” is almost absent from the Japanese vocabulary: a “yes” in certain circumstances can, in fact, mean “no”.

13.2.2 Evaluating: direct negative feedback versus indirect negative feedback

This suggests how different cultures approach evaluation or feedback, specifically negative feedback. There is a two-end axe:

- Direct negative feedback: This end of the spectrum represents cultures where negative feedback is given bluntly, honestly, and without softening or positive wrapping.
- Indirect negative feedback: This end signifies cultures where negative feedback is delivered more subtly, diplomatically, and often cushioned with positive messages.

The countries are positioned along this axis based on their perceived preference for direct or indirect negative feedback. Direct negative feedback: countries like Russia, Israel, and Germany are closer to this end, implying a more direct approach to giving negative feedback. Indirect negative feedback: countries like China, Japan, and Indonesia are closer to this end, suggesting a preference for indirect feedback.

Some cultures, typically low context and explicit, may deliver criticism indirectly and in a coded manner. Conversely, high-context and implicit cultures can express criticism in an explicit and direct way. On the right side of the spectrum, we find Asian countries, including India and Thailand, which are among the most critical.

Concept of context and (in)direct feedback can be crossed. Cultural communication styles can significantly impact how negative feedback is delivered and received across different countries. In low-context, direct cultures like the US, Canada, and Australia, individuals tend to be straightforward and assertive when providing feedback, often using explicit verbal messages. Conversely, low-context, indirect cultures such as Germany, Denmark, and the Netherlands appreciate directness but prefer a subtler approach, incorporating nonverbal cues and implied meanings. In high-context, indirect cultures like Japan, Thailand, and China, the emphasis is on maintaining harmony; thus, feedback is often communicated in a subtle manner to avoid confrontation. Meanwhile, high-context, direct cultures, though less common, include countries such as Brazil and Israel, where context and shared understanding are crucial, and feedback may still be direct. It is important to note that these categorizations are generalizations, and individual communication styles can vary widely within each country. Additionally, as cultural norms evolve, understanding these differences can enhance cross-cultural communication and help prevent misunderstandings, particularly when delivering or receiving negative feedback.

13.2.3 Persuading: principle first versus applications first

Under this dimension, countries are located on an axis with two extreme points:

- Applications first: Individuals from cultures in this category tend to begin with facts, statement, or opinion. They prefer to start with a practical, concrete approach, often using executive summaries or bullet points. Theoretical or philosophical discussions are generally avoided in business settings. Countries listed in this category include Italy, Russia, Germany, Argentina, Sweden, Netherlands, Australia, France, Spain, Brazil, Mexico, Denmark, UK, Canada, and the US.
- Principles first: Individuals from these cultures tend to prioritize the development of a theory or complex concept before presenting a fact, statement, or opinion. They prefer to build a theoretical argument before reaching a conclusion. Conceptual principles underlying each situation are highly valued.

It could have some implications like misunderstandings: differences in communication styles can easily lead to misunderstandings between individuals from these two categories. Someone from an applications-first culture might perceive someone from a principles-first culture as being overly theoretical or impractical. Conversely, someone from a principles-first culture might find someone from an applications-first culture to be too focused on details and not considering the broader context.

Adapting communication by understanding these cultural differences is crucial for effective communication. When interacting with someone from a different cultural background, it is helpful to adapt your communication style to align with their preferences. For example, if you are from a principles-first culture, you might want to start a conversation by explaining the underlying theory or concept before sharing specific examples. Similarly, if you are from an applications-first culture, you might want to begin with a concrete example or fact before delving into the theoretical implications.

The left part of the axis at persuading dimension represents cultures that are more principles first compared to those on the right. However, it is important to consider the concept of cultural relativity.

According to this theory, Asian cultures fall outside the persuading scale and do not appear on it. Their worldview is so distinct from that of Europe that they operate within an entirely different frame of reference.

Asians tend to have more holistic thought patterns, while Westerners adopt more specific approaches. For example, Chinese individuals often think from macro to micro, whereas Westerners typically think from micro to macro.

13.2.4 Leading: egalitarian versus hierarchical

This dimension provides leadership styles across different cultures. It categorizes countries into two main groups based on their preferred leadership approach: egalitarian and hierarchical. This concept refers to the extent to which a society accepts power inequalities.

In egalitarian cultures, the ideal distance between a boss and a subordinate is low and the best boss is a facilitator who works alongside their team members. The organizational structure is flat and flexible, with communication often bypassing hierarchical lines. It can be quoted countries like Denmark, Canada, and Finland.

In hierarchical cultures, the ideal distance between a boss and a subordinate is high and the best boss is a strong, directive leader who takes charge. The organizational structure is multi-layered and rigid, with communication following strict hierarchical lines. It can be quoted countries like France, Poland, and Japan.

13.2.5 Deciding: consensual versus top-down

Deciding dimension presents the concept of decision-making styles, which can be broadly categorized into two main types: consensual decision-making and top-down decision-making.

In consensual decision-making countries, like Japan, Sweden, Netherlands, Germany, or the UK, the decisions are made collectively within a group, and everyone involved needs to reach a unanimous agreement. This style emphasizes collaboration, inclusivity, and building consensus among all stakeholders.

In top-down decision-making countries, like the US, France, Brazil, Italy, Russia, India, China, or Nigeria, the decisions are made by individuals, typically those in positions of authority or leadership. This style is often associated with hierarchical structures, where decisions flow from the top down.

The positioning of countries on the scale reflects cultural norms and preferences regarding decision-making. It is important to note that countries do not strictly adhere to one style or the other. Many organizations and situations may involve a mix of both approaches. The choice of decision-making style can be influenced by various factors such as the urgency of the decision, the complexity of the issue, and the level of expertise required.

There are two different ways to make decisions that impact in project chronology. In a consensual culture, decision-making takes time because all stakeholders are consulted. However, once the decision is made, execution is very fast. In a top-down culture, the decision is delegated to one person, usually the boss. But all decisions are subject to revision.

To avoid cultural shock when making decisions, the best solution is to organize dialogue and find a solution that works for everyone.

13.2.6 Trusting: task-based versus relationship-based

Countries could be categorized along a spectrum ranging from "task-based" to "relationship-based" trust, taking into account how trust is built in different cultures.

In task-based trust countries, like US, Denmark, Germany, UK, Poland, Netherlands, or Finland, the trust is built through business-related activities and performance. Work relationships are formed and dissolved based on practicality and the situation. Reliability and consistent good work are key factors in building trust.

In relationship-based trust countries, like France, Italy, Mexico, or Brazil, the trust is built through shared experiences, social interactions, and personal connections. Work relationships develop slowly over time and are often based on deep-level understanding and shared personal moments. Trust is often reinforced by knowing others who trust the individual.

Task-based cultures prioritize performance and reliability, while relationship-based cultures prioritize social interactions and personal connections.

13.2.7 Disagreeing: confrontational versus avoid confrontation

This seventh dimension compares how different cultures approach disagreement and debate. It proposes two-end axes:

- Confrontational: as in countries like Israel, Germany, and the US. This indicates that people from these cultures tend to be more direct and open about expressing disagreement, even if it might lead to some tension. They believe that debate and disagreement can be healthy and even beneficial for the group or organization. Open confrontation is considered appropriate and not harmful to relationships.
- Avoids confrontation: as in countries like Sweden, India, and Indonesia. People from these cultures generally prefer to avoid direct confrontation and disagreement. They believe that harmony and maintaining positive relationships are more important than expressing differing opinions openly. Open confrontation is considered inappropriate and could damage relationships or group harmony.

To understand how your own culture navigates disagreement, consider this question: "If someone in my culture strongly disagrees with my idea, does that suggest they are disapproving of me personally or just the idea?" In confrontational culture, it is natural to challenge the idea, not the person who expresses it. In avoiding cultures, the two are often perceived as interlinked, with disagreement sometimes interpreted as personal disapproval.

13.2.8 Scheduling: linear time versus flexible time

Countries can be classified into cultural dimension related to scheduling, specifically focusing on linear-time and flexible-time orientations. Let us examine what each means:

- In linear-time orientation cultures, tasks are approached sequentially, one at a time, and in a step-by-step manner. Interruptions are minimized, as there is a strong focus on completing one task before starting another. Meeting deadlines and adhering to a strict schedule are of utmost importance. Punctuality and meticulous planning are highly valued.
- In contrast, flexible-time orientation cultures adopt a more adaptable approach to tasks, with less emphasis on strict sequencing. Interruptions are more readily accepted, and it is common to switch between tasks as needed. This orientation places a greater value on adaptability and spontaneity over rigid scheduling.

For example, countries like Germany, Switzerland, Denmark, and the Netherlands typically align with linear-time orientation, when punctuality, deadlines, and structured planning are cultural priorities. However, countries such as Italy, Brazil, and India are often associated with flexible-time orientation, emphasizing adaptability, flexibility, and spontaneity, while placing less importance on adhering to fixed schedules.

The way time is managed is deeply rooted in cultural norms. While some cultures view delays as unacceptable, others find them more tolerable. Time management in a cultural context is often analogous to how queuing is organized in different countries, revealing distinct approaches and values.

13.3 Keys to work at an international level

When working at an international level, it is important to be open-minded about others and their culture. It is essential not to assume that our way of thinking is the best. It is also crucial to be aware of our own culture in order to better understand others: "How are we reacting?" [9].

Listen to others and try to communicate as clearly as possible. Do not hesitate to ask if you were understood clearly. The ability to embrace change, adjust to new situations, and work in different time zones and work environments is essential for success in international roles.

Do not be afraid to make mistakes, everyone makes them. Do not hesitate to ask for help, advice, or explanations.

It is important to learn about the culture before meeting people and to respect local practices.

How would you go about giving a business presentation to an international audience? It is recommended to have a cultural checklist:

- What kind of language should you use? Formal or informal?
- What kind of structure should you use? Explicit or implicit?
- What kind of content should you use? Detailed or general?
- What kind of delivery is expected? Text reading or improvised?
- Do you need to respect timing? Fixed or flexible?
- What kind of dress code should you use? Formal or informal?
- What kind of behavior is expected? Serious or relaxed?

So, it is important to take the time you need to adapt and try not to make value judgments about the culture of others. Doing so could prevent you from appreciating all the richness of their culture.

To avoid misunderstandings, try to establish friendly relationships with the local population and ask them to enlighten you about some of their practices that may seem shocking or incomprehensible to you.

The conclusion of this chapter could be the traveling quote by Henry Miller, an American novelist: “One’s destination is never a place, but a new way of seeing things”.

References

[1] Geert Hofstede web page, https://geerthofstede.com/ [Accessed 17 Dec 2024].

[2] Erin Meyer –web page, https://erinmeyer.com/ [Accessed 17 Dec 2024].

[3] Edward T. Hall, “The Cultural Iceberg”, https://bccie.bc.ca/wp-content/uploads/2020/09/cultural-iceberg.pdf [Accessed 26 Feb 2024].

[4] Unia, “List of 19 Discrimination Criteria”, https://www.unia.be/en [Accessed 26 Feb 2024].

[5] Troy Segal, “Culture Shock Meaning, Stages, and How to Overcome”, https://www.investopedia.com/terms/c/culture-shock.asp, 2024 [Accessed 17 Dec 2024].

[6] Kalervo Oberg, “Culture Shock”, https://citeseerx.ist.psu.edu/document?repid=rep1&type=pdf&doi=c787fddd7e1557f100b9703a41f6f5ae2dac0412 [Accessed 26 Feb 2024].

[7] Kevin F. Gaw, “Reverse culture shock in students returning from overseas”, *International Journal of Intercultural Relations*, 2000;24(1):83–104. https://doi.org/10.1016/S0147-1767(99)00024-3.

[8] Erin Meyer, *The Culture Map: Decoding How People Think, Lead, and Get Things Done Across Cultures*, Public Affairs, 2016.

[9] Robert Gibson, *Intercultural Business Communication*, Oxford University Press, 2002.

Part IV

Cognitive skills

Chapter 14

Analytical thinking

Karolina Szturo[1]

Analytical thinking in scientific literature is defined as the ability to break down complex problems into smaller, more comprehensible components to facilitate detailed analysis, understanding, and resolution [1]. It is characterized by a systematic approach to information analysis, which includes data collection, pattern identification, and the formulation of conclusions based on logical examination of evidence. Analytical thinking is a cognitive process that enables the evaluation of situations through the application of logical methods such as deduction, induction, and other problem-solving techniques.

The literature emphasizes that analytical thinking requires intellectual discipline, precise planning, and objectivity, distinguishing it from critical thinking, which focuses more on evaluating and questioning information. In educational and professional contexts, analytical thinking is essential for data-driven decision-making as well as in scientific and engineering processes, where precise analysis of technical and systemic problems is required. Analytical thinking—devoid of emotions and based on data—facilitates well-reasoned decision-making, particularly in contexts that demand meticulous calculations and a structured approach to details.

Analytical thinking serves as a foundation for other cognitive processes, including critical thinking, aesthetic thinking, decision-making, creative thinking, problem-solving, scientific reasoning, and coordination-based thinking. It is a complex competency that develops over time through life experiences and conscious decision-making. The more deliberately we approach problem-solving, the more factors—both advantageous and detrimental—can be identified and considered when seeking the optimal decision. The key to analytical thinking lies in avoiding stereotypical reasoning and automatic responses.

The process of analytical thinking can be divided into several stages, beginning with defining the objective of the analysis and identifying key issues, followed by determining possible explanations for the problem, gathering information from various sources and analyzing it, and ultimately selecting the most effective solution, as summarized in Table 14.1.

[1]Department of Safety Engineering, Faculty of Technical Sciences, University of Warmia and Mazury in Olsztyn, Poland

Table 14.1 Stages of analytical thinking

Stage	Description	Methods
Problem identification	Defining the essence of the problem, its scope, and context.	Root cause analysis, comparative studies/benchmarks
Formulating hypotheses	Developing preliminary assumptions and potential solutions based on available information.	5W1H Technique (Who, What, When, Where, Why, How), brainstorming, Ishikawa diagram (cause-and-effect analysis)
Data collection	Gathering information, facts, and evidence to confirm or refute the proposed hypotheses.	Research, statistical analysis, surveys, observations, case studies, document analysis
Conducting analysis	Detailed examination of collected data to identify patterns, relationships, and anomalies.	SWOT analysis, trend analysis, comparative analysis, regression analysis, data mining algorithms, mathematical models
Testing hypotheses	Verifying assumptions by examining whether the data supports the hypotheses.	Experiments, modeling, statistical tests, scenario analysis, computer simulations, correlation analysis
Drawing conclusions	Final interpretation of analytical results and their application in decision-making.	Deductive and inductive reasoning, reporting findings, data visualization, developing recommendations

Employers seek employees with analytical skills to examine key company challenges or successes—this analysis serves as the basis for taking concrete actions to eliminate factors contributing to problems or to continue those that drive success. Analytical thinking is crucial for gaining a deeper understanding of and improving a company's operations.

Individuals with an analytical mindset prioritize facts and logic over emotions, favor rationality and structured approaches, and make well-considered decisions that they systematically oversee. They are independent and non-aggressive, carefully weighing all alternatives while remaining steadfast in their pursuit of goals. Their distinguishing traits include emotional detachment, a business-oriented approach, and perseverance. Additionally, they demonstrate discipline and allow others to take the initiative in group settings.

Conversely, individuals with a low level of analytical competence tend to make impulsive decisions based on emotions, intuition, and a limited scope of information. As a result, they may overlook critical factors, fail to consider key data, and ultimately arrive at decisions that are unsatisfactory for all stakeholders. Furthermore, the inability to identify alternative options and assess their advantages and disadvantages can lead to significantly unfavorable or even detrimental outcomes.

14.1 Characteristics of analytical thinking

Analytical thinking is a method of processing information that relies on evidence, logical structure, and a systematic approach to problem-solving. It is a conscious process that requires precise organization of thought, enabling a thorough understanding of situations and the ability to make optimal decisions. Its key characteristics make it indispensable for solving complex problems. It also plays a vital role in supporting rational, fact-based decision-making.

One of the most important traits of analytical thinking is its reliance on evidence rather than emotions or intuition. The analytical process involves questioning assumptions and continuously seeking answers to the question "Why?" This approach fosters a deeper understanding of a given situation, leading to more informed decision-making while minimizing biases and cognitive errors. Basing analysis on facts also requires a critical evaluation of information sources, which further enhances the reliability of conclusions.

Another essential characteristic of analytical thinking is methodical precision and attention to detail. The analytical process is not chaotic; it demands a systematic approach and a logically structured thought process. It involves a meticulous examination of every aspect of a problem, allowing for better comprehension and identification of the most significant factors influencing the situation. Such an approach helps eliminate errors resulting from hasty conclusions and enables the creation of a comprehensive picture of the issue under analysis.

Sequential reasoning and decisiveness are also key aspects of analytical thinking. Information analysis follows a structured sequence, preventing disorder and ensuring a systematic progression toward a solution. This method ensures clarity in the thought process and reduces the tendency to become distracted by irrelevant details or alternative scenarios that might introduce unnecessary complications. As a result, analytical thinking becomes more efficient and goal-oriented.

An inherent feature of this thinking style is breaking problems down into their constituent parts. Complex issues are rarely solvable through intuition alone, making it essential to deconstruct them into smaller, more comprehensible components. This approach facilitates the identification of relationships between various factors and enables a detailed examination of their influence on the overall situation. Such structural analysis not only helps pinpoint the root causes of a problem but also allows for the discovery of effective solutions.

Thus, analytical thinking is a systematic, evidence-based process characterized by a methodical approach, logical sequencing, and the ability to decompose problems into smaller elements. These attributes make it highly effective in analyzing complex situations, making rational decisions, and avoiding errors resulting from emotional influences or hasty conclusions. It is an indispensable tool in any field that requires logical and precise reasoning.

14.2 Pitfalls of analytical thinking

Contrary to popular belief, analytical thinking is not an innate ability; its effective application requires the right approach and awareness of potential errors. One of the key challenges is the confirmation bias—the tendency to interpret information in a way that uncritically confirms pre-existing assumptions. To counteract this bias, it is advisable to apply the principle of falsification, which involves actively seeking evidence that could refute a hypothesis, as well as counterfactual analysis, which considers what would happen if the accepted assumptions were incorrect. A useful technique in this regard is the Devil's Advocate approach, where one deliberately argues against their own assumptions, as well as the inclusion of alternative hypotheses, enabling a more comprehensive and objective assessment of the situation.

Another cognitive flaw in analytical thinking is the first-impression syndrome, which refers to the difficulty of modifying initial beliefs when confronted with new data. An effective countermeasure is iterative analysis, which involves repeatedly reviewing data and updating conclusions accordingly. Another useful strategy is the "blank slate" technique, where analysis is periodically restarted from scratch without previous assumptions, preventing first impressions from unduly influencing subsequent reasoning. Additionally, critical evaluation of sources and team-based discussions—where diverse viewpoints are confronted and subjected to objective scrutiny—can help mitigate this issue.

A common mistake in analytical thinking is premature closure, where conclusions are drawn too quickly without gathering sufficient information. To avoid this, it is beneficial to adopt a gradual inference approach, where hypotheses are formulated in stages rather than based on initial intuitions. Another effective strategy is multi-criteria analysis, which evaluates the problem from multiple perspectives, thereby reducing the risk of one-sided conclusions. Using standard analytical procedures, such as decision checklists or Ishikawa diagrams, enhances the precision of analysis. Additionally, the "5 Why's" technique, which involves asking "why?" five times to identify the root cause of a problem, is a valuable tool for deeper investigation.

The limited capacity of working memory poses another significant constraint in analytical thinking, making it difficult to process multiple alternative explanations simultaneously. To address this limitation, data visualization tools—such as diagrams, mind maps, and tables—help organize information and identify patterns. Another effective approach is to divide analysis into stages, enabling a step-by-step evaluation of hypotheses rather than attempting to process all information at once. Analytical tools such as Excel, Power BI, R, and Python can further enhance the efficiency of data analysis. Additionally, the scenario method, which considers multiple possible future outcomes instead of a single reasoning pathway, provides a more comprehensive approach to problem-solving.

Analytical thinking can be more reliable and resistant to cognitive biases when conscious error-avoidance strategies, a systematic approach that considers alternative hypotheses, and tools supporting analysis are applied. Team collaboration and debates also play a crucial role, as confronting different perspectives helps eliminate individual cognitive errors. Through these rigorous procedures, analytical thinking becomes more efficient, reliable, and objective.

One of the key aspects of analytical thinking is hypothesis formulation, which involves developing provisional explanations for a given situation and subsequently verifying them through an analysis of available data. There are several ways to formulate hypotheses—one may rely on situational logic, base the analysis on theoretical assumptions, refer to similar past cases, or delve into a detailed data analysis. After formulating various hypotheses, selecting the most justified one is necessary. Different approaches can be adopted in this regard. A commonly used method is selecting the first hypothesis that seems sufficiently convincing and then adjusting the further analysis process to fit it. While this strategy allows for quickly reaching conclusions, it may lead to ignoring information that could challenge the adopted assumption, as well as overlooking alternative explanations. Another approach is creating a hypothesis that combines several similar concepts, allowing for a broader context. Sometimes, hypothesis selection is guided by sociotechnical considerations, consisting of finding a solution that gains acceptance from all participants in the analysis. Another possible strategy is basing the hypothesis on experience and analogy to previous cases, assuming that solutions that have worked in the past may also be effective in a new situation.

Different ways of handling information are also essential in analytical thinking. Three primary approaches can be distinguished: the data-driven approach, the theory-based approach, and the mosaic theory.

The *data-driven approach* involves applying a specific explanatory model, which is then used to analyze the available information. Its advantage lies in its clarity and predictability since the necessary conditions for decision-making are clearly defined. However, problems arise when there is too little available data, which limits the effectiveness of this method and restricts the areas in which it can be applied.

In the *theory-based approach*, when sufficient information is lacking, the analyst uses a selected theoretical concept to interpret the problem. This solution provides significant flexibility and allows for the analysis of even highly unusual situations. However, it carries the risk of subjectivity, as different analysts, relying on different premises, may arrive at completely different conclusions.

The *mosaic theory* assumes that analytical thinking resembles the process of reconstructing an image based on its fragments. In principle, individual pieces of information contribute to a larger whole that the analyst should be able to recreate. Although this concept appears attractive, in practice, cognitive psychology shows that different people, even when given the same data, may arrive at different conclusions.

To make analytical thinking as effective as possible, it is necessary to apply tools and procedures that reduce the risk of errors and help maintain the objectivity of analysis. A key element of this process is the continuous verification of assumptions. Many analytical errors do not result from inaccuracies in the data itself but from its misinterpretation. Therefore, at every stage of the analytical process, it is crucial to reflect on both the applied methods and the conclusions drawn from them, which helps avoid one-sided thinking and remain open to alternative solutions.

A common cognitive error is the mirror syndrome—the tendency to attribute one's way of thinking and evaluating reality to others. As a result, an analyst may fail to recognize significant differences in approaches to certain issues, leading to incorrect predictions and conclusions. To avoid this, it is advisable to consider alternative explanations for a given situation and take different perspectives into account.

One of the greatest challenges in analysis is recognizing the moment when a complete change of approach is necessary. New information can entirely undermine previous conclusions, which is why flexibility and a willingness to reconsider the problem are extremely important. Equally crucial is creative thinking, which helps uncover unconventional solutions and detect hidden relationships that might be overlooked in standard analysis.

14.3 Theories of the analytical thinking process

Formal methods supporting analytical thinking play a crucial role in information processing, knowledge modeling, and decision-making. One of the fundamental approaches is ontological theory, which utilizes logical structures to organize and model knowledge. This allows for hierarchical and interconnected information structuring, facilitating the analysis of relationships between different elements and enabling more effective reasoning. This method is commonly used in artificial intelligence, databases, and knowledge management systems.

Another significant method is contextual logic, which incorporates context into analytical processes. Traditional data analysis approaches often overlook the environment in which information is interpreted. Contextual logic enables a more precise understanding of problems by considering the variable conditions under which relationships occur. This is particularly useful in natural language processing, expert systems, and business analytics, where context plays a key role in decision-making.

The third important approach supporting analytical thinking is statistical semantics, which focuses on the numerical analysis of message content. It employs statistical methods to examine content, identify patterns, and assess the significance of conveyed information. It is widely used in natural language processing, big data analysis, and machine learning, where large volumes of data require a quantitative approach to interpretation.

These methods serve as invaluable tools supporting analytical thinking across various fields of science, technology, and business. Their application enables more precise, logical, and effective information analysis, leading to better decision-making and a deeper understanding of complex problems.

Formal methods applied in analytical thinking are primarily based on the principles of formal logic, mathematics, and, to a lesser extent, linguistics. Computer science has also significantly influenced their development, playing a crucial role in shaping these approaches. The application of computational technologies is one of the most important aspects of formal theories, as it allows for the rapid and precise processing of vast datasets—a task that would be extremely difficult to accomplish even with large teams of analysts. Some methods, such as the ontological approach, emerged directly from advancements in programming and artificial intelligence research.

The development of a formal approach to analytical thinking arose from the need to efficiently integrate data from various sources and techniques. Another key objective was to create an analytical tool that provides a clear insight into the analysis process and ensures a consistent understanding of results across different individuals.

14.3.1 Ontological theory

The ontological theory does not merely aim to describe the practice of analytical thinking but strives to establish a new, more precise analytical approach. Its purpose is to ensure a uniform understanding of analytical results for both the author and recipients, eliminating potential interpretational ambiguities.

In this context, the term ontology differs from its classical philosophical meaning. The primary function of ontology is to model a specific domain of knowledge or the structure of discourse language. Its key components typically include classes, which define categories of objects; attributes, which describe their characteristics; and relationships, which specify the connections between elements within a given class. Each of these components has a precise definition that determines its meaning and establishes constraints to ensure logical consistency in its application [2].

Ontology provides an agreed-upon set of concepts that can be used to model a given domain, encompassing object types, their properties, and interdependencies. In this sense, ontology serves as a tool—a structured language—supporting the development of information systems, databases, and software. Due to its abstract nature, ontology can be applied in various contexts without direct association with specific implementations, enabling flexible knowledge modeling and structuring.

The development of ontologies plays a crucial role in enhancing analytical thinking, as they support the systematic organization of knowledge, problem analysis, and informed decision-making. By leveraging ontologies, it becomes possible to navigate the stages of the analytical process more precisely and efficiently, significantly improving the quality of results obtained.

One of the most important aspects of supporting analytical thinking is problem identification, meaning a precise diagnosis of the issue. An analytical approach requires the ability to grasp the core of the problem and its context, allowing a focus on actual challenges rather than superficial symptoms. Ontologies facilitate logical and structured organization of information, making it easier to identify gaps in knowledge and areas requiring further analysis. In practice, this leads to greater accuracy in problem definition, resulting in more effective solutions.

Another essential element of analytical thinking is identifying the causes of problems through cause-and-effect analysis. Ontologies aid in tracking relationships and influences between various factors, enabling an accurate determination of root causes. The application of causal analysis methods, such as Ishikawa diagrams or root cause analysis, helps in identifying the mechanisms behind problems and formulating effective countermeasures.

An inseparable component of analytical thinking is also the presentation of alternative solutions, which allows for exploring different possible courses of action. By systematically structuring knowledge, ontologies support the design of alternative scenarios, the evaluation of their consequences, and the selection of the most optimal solution. In decision-making environments, where anticipating the effects of various actions is essential, such an approach reduces the risk of erroneous assumptions and enhances decision-making efficiency.

Another crucial factor supporting analytical thinking is consensus building and teamwork. Effective analysis often requires collaboration among individuals with diverse expertise and specializations. Ontologies provide standardized terminology and a unified knowledge model, minimizing the risk of misinterpretation and improving communication within analytical teams. As a result, team members can exchange information more efficiently, work with shared datasets, and reach more accurate conclusions.

One of the most significant advantages of applying ontologies in analytical thinking is knowledge sharing and reuse. Analytical thinking relies on past experiences, learning from previous analyses, and applying proven methodologies. Ontologies enable systematic knowledge storage and organization, ensuring that it can be accessed and applied in new contexts. This prevents analysts from repeating the same mistakes, accelerates pattern recognition, and allows for the implementation of best practices in future projects.

The aforementioned benefits of ontology stem primarily from replacing natural language with a structured artificial language and unifying terminology. In practice, this enhances information management and improves data analysis processes.

Ontologies serve as a powerful tool supporting analytical thinking by facilitating problem diagnosis, cause analysis, exploration of alternative solutions, team collaboration, and knowledge management. Their implementation significantly enhances the precision of analyses, improves decision quality, and enables efficient utilization of informational resources.

In today's dynamic world, where the volume of available data is rapidly increasing, the ability to systematically organize and utilize knowledge is becoming crucial for effective analysis and decision-making.

14.3.2 Contextual logic

Analytical thinking is a process that involves systematically analyzing problems, gathering data, and drawing logical conclusions. However, no data or facts exist in isolation—their interpretation is closely linked to the context in which they appear. This is why contextual logic plays a crucial role in effective analytical thinking, enabling the consideration of various conditions and the dynamic nature of situations in the decision-making process [3].

Contextual logic assumes that the meaning of information and the validity of conclusions depend on the context in which they are analyzed. Unlike classical formal logic, which operates based on fixed, immutable rules, contextual logic allows for the variability of conditions and their influence on the thinking process.

In analytical thinking, this means that conclusions should be drawn not only based on hard data but also by taking into account variable factors such as social, economic, cultural, or technological circumstances.

A practical example of the application of contextual logic in analytical thinking is market trend analysis. Sales data for a given product may indicate a decline in interest, but without considering the context—such as changes in consumer preferences, economic conditions, or competitor actions—it would be impossible to draw accurate conclusions. Only by analyzing the broader background can one understand whether the decline in sales results from a temporary crisis, a flawed marketing strategy, or the emergence of innovative solutions in the market.

Contextual logic is also useful in solving organizational problems. Suppose there is a drop in productivity within a company's department. An analytical approach allows for examining data related to employee performance, but only by considering the context—such as changes in company policies, team conflicts, or excessive workload—can the true cause of the issue be understood. Without factoring in these additional elements, decisions may be superficial and ineffective.

Contextual logic is crucial in data interpretation and presentation. Analytical thinking often involves statistics and mathematical models, but numbers alone do not provide a complete picture of a situation. The same indicator may have different meanings under different circumstances. For example, if the unemployment rate increases by 2%, it might be interpreted as a negative economic signal. However, in the context of increasing automation and labor market restructuring, this could be a natural consequence of long-term technological shifts.

In practice, contextual logic also helps avoid cognitive biases. In data analysis, considering different perspectives of stakeholders is key to avoiding misinterpretation of their intentions and decisions.

Today's world is characterized by high dynamics of change and a large number of variables affecting decision-making processes. Contextual logic allows for a more flexible approach to analysis and the adaptation of strategies to evolving conditions. In business, politics, or organizational decision-making, this means continuously updating assumptions and adjusting strategies to new circumstances.

For example, in crisis management, it is crucial not only to analyze available data but also to dynamically respond to new information, forecast possible scenarios, and adjust action plans to changing conditions. In this context, contextual logic becomes an essential tool for effective risk management and adaptation to unpredictable situations.

Thus, contextual logic is an inseparable element of effective analytical thinking. It enables the consideration of changing conditions, better interpretation of data, and avoidance of incorrect assumptions resulting from a lack of broader perspective. Thanks to contextual logic, it is possible to formulate more accurate conclusions, make more effective decisions, and avoid pitfalls stemming from superficial analysis. In a rapidly changing world, the ability to incorporate context into problem analysis and decision-making becomes an indispensable tool for anyone seeking to function effectively in a complex informational environment.

14.3.3 Statistical semantics

Analytical thinking is based on the systematic analysis of data and the ability to draw accurate conclusions from available information. In an era of massive data volumes and rapidly evolving technologies, language analysis and meaning interpretation have become key aspects of this process. Statistical semantics is a method that supports analytical thinking through a quantitative approach to content analysis, enabling the discovery of hidden patterns and relationships in language.

The foundation of statistical semantics lies in identifying patterns in the occurrence of specific concepts and analyzing their meaning within the overall content [4]. Once this analysis is conducted, it is verified by another analyst or an entire analytical team using the same methodological framework. This process enhances objectivity and reduces potential errors arising from individual interpretations.

To fully understand the essence of statistical semantics, it is important to consider two key aspects of its application:

1. Statistical content analysis is used to draw conclusions about non-verbal aspects of communication. The text itself is not the primary focus of study but serves as an intermediary carrier of information, allowing for an understanding of social, psychological, or cultural contexts.
2. The frequency of words alone is insufficient for accurate conclusions. It is crucial to combine at least two streams of information: verbal and non-verbal. This means that the analyzed text must be examined not only in relation to the words it contains but also in the situational context, communicative intent, tone, and purpose of the message. Only such a multifaceted analysis allows for a comprehensive understanding of content and its meaning.

In the context of analytical thinking, statistical semantics plays a crucial role in objectifying text data interpretation. Traditional language analysis methods can be subjective and prone to cognitive biases, whereas a statistical approach enables precise measurement and structured comparison of content.

Statistical semantics is a powerful tool that supports analytical thinking by enabling objective, quantitative research on communication content and the discovery of hidden patterns in textual data. By analyzing the frequency of specific concepts and their relationships, it becomes possible to gain deeper insights into social, political, and psychological contexts of statements.

Modern analyses require consideration of both verbal and non-verbal aspects, which is why the effective application of statistical semantics demands a multidimensional approach and the integration of various data sources. As a result, this method facilitates more accurate reasoning, reduces cognitive biases, and enhances data-driven decision-making.

14.4 Developing analytical thinking skills

Analytical thinking is a skill that can be systematically developed through conscious actions and daily practice. It requires not only knowledge of analytical tools but also openness to new information, a critical approach to data, and a willingness to explore alternative solutions. There are many ways to enhance this competency, and applying them regularly leads to more effective problem-solving and better decision-making.

One of the most effective ways to develop analytical thinking is engaging in intellectual challenges regularly. Exercises such as puzzles, Sudoku, logic games, or chess stimulate the mind by requiring analysis, pattern recognition, and anticipating the consequences of decisions. Additionally, engaging in case study analysis helps examine real-world problems and find optimal solutions.

Problem-solving can be practiced both individually and in groups. Teamwork allows for diverse perspectives, enriching the analytical process and fostering skills in argumentation and communication. Discussions and exchanging viewpoints often lead to the discovery of new aspects of a problem that might be overlooked in an individual analysis.

Analyzing real-life situations, whether from daily life or professional settings, helps translate theory into practical skills. It is crucial to draw lessons from each situation, whether a success or failure. After making a decision, it is useful to reflect: *What worked well*? *What could have been done differently*? Reviewing personal experiences allows for refining thinking processes and recognizing patterns that can be helpful in the future.

Developing analytical thinking also requires regular reflection on one's decision-making process. Allocating time for reviewing conclusions, analyzing mistakes, and seeking improvement opportunities are essential.

One effective method is keeping a journal to record thoughts, analyses, and insights from past decisions. This helps track personal progress and identify thinking patterns that may need adjustment. Regular self-evaluation enhances cognitive processes and prevents the repetition of past mistakes.

A good practice is also asking reflective questions, such as:

- What information did I consider in my analysis?
- Were my assumptions valid?
- Did I take alternative perspectives into account?
- What other possible solutions existed?

Knowledge is the foundation of effective analytical thinking. Reading books, scientific articles, and market analyses strengthens data interpretation and analytical skills. Particularly valuable are materials on logic, research methods, critical thinking, and data analysis, as they provide essential tools for daily analytical work.

Additionally, participating in courses and training helps acquire practical skills and learn new analytical techniques. Online courses and in-person workshops often provide hands-on experience working with real-world data and case studies.

Exploring different perspectives and approaches to problems is also essential. Reading materials that present various viewpoints helps avoid one-dimensional thinking traps and broadens analytical horizons.

Improving analytical thinking requires deliberate engagement and practice. Training the mind through solving logic puzzles, analyzing case studies, learning from experiences, and reflecting on decisions gradually strengthen this skill.

Likewise, it is important to continually seek out information, participate in training, and explore different perspectives. These activities not only enhance analytical skills but also contribute to more effective and informed decision-making in both professional and everyday life. Analytical thinking is a lifelong competency that can and should be nurtured continuously to improve problem-solving abilities and make better, data-driven decisions.

14.5 Conclusions

Analytical thinking is a fundamental cognitive process that enables the systematic analysis of information, the decomposition of complex problems, and the formulation of logical conclusions. This chapter explores its importance in scientific, professional, and decision-making contexts and distinguishes it from other cognitive processes.

A key feature of analytical thinking is its structured nature, with an emphasis on logical reasoning, objective evaluation of data, and deductive and inductive analysis. Essential characteristics such as reliance on empirical evidence, methodological precision, sequential reasoning, and problem decomposition enhance its role in decision-making and problem-solving.

The limitations of analytical thinking, including cognitive biases and premature conclusions, are outlined in this chapter, along with strategies for mitigating them, such as iterative analysis, hypothesis falsification, and the use of analytical tools, including data visualization and statistical modeling.

In addition, formal approaches—ontological theory, contextual logic, and statistical semantics—support structured knowledge organization, efficient relationship modeling, and improved decision accuracy.

Developing analytical skills requires ongoing intellectual engagement, case study analysis, and critical reflection. Effective enhancement involves applying cognitive tools, critically evaluating information and considering multiple perspectives.

In an increasingly complex world, analytical thinking remains essential for making rational decisions, optimizing cognitive processes and solving problems in scientific, technological, and organizational domains.

References

[1] The Peak Performance Center "Analytical Thinking", Available at: http://thepeakperformancecenter.com/educational-learning/thinking/types-of-thinking-2/analytical-thinking/. [Accessed 05 Feb 2025].

[2] Symons, J. Ontology and methodology in analytic philosophy. In: Poli, R., and Seibt, J. (eds.), *Theory and Applications of Ontology: Philosophical*

Perspectives. Springer, Dordrecht, pp. 155–174, 2010. https://doi.org/10.1007/978-90-481-8845-1_16

[3] Dau, F., and Klinger, J. From formal concept analysis to contextual logic. In: Ganter, B., Stumme, G., and Wille, R. (eds.), *Formal Concept Analysis.* Lecture Notes in Computer Science, vol. 3626. Springer, Berlin, pp. 120–133, 2005. https://doi.org/10.1007/11528784_4

[4] Sikström, S., and Garcia, D. *Statistical Semantics: Methods and Applications*. Springer International Publishing, Cham, 2020. https://doi.org/10.1007/978-3-030-37250-7

Chapter 15
Creative thinking

Manuel J. Fernández Iglesias[1]

In an ever-evolving world, the ability to think outside the box and find creative solutions is increasingly valued, both at work and at home. Whether solving job-related problems, seeking new perspectives in our personal relationships, or exploring opportunities for personal or professional growth, creative thinking has become an essential skill.

In this chapter, we explore the fundamentals of creative thinking and propose practical tools and effective strategies to stimulate creativity. We will begin by laying the groundwork for creative thinking, exploring concepts such as the fluidity of ideas, mental flexibility, and the ability to see things from different perspectives. As we progress, we will discuss specific techniques to expand our ability to generate original ideas, overcome obstacles, and challenge pre-established biases and assumptions. Our goal is to pave the way toward a creative mindset that will allows us to improve our ability to solve problems, promote collaboration, lead with creativity, and maximize our creative potential.

15.1 Myths and facts of creative thinking

Exploring preconceived ideas and misconceptions about creativity is critical to overcoming biases and fostering a broader and more accurate understanding of this ability. Many people believe that creativity is an innate gift or is limited to certain areas, which in turn limits their creative potential. By challenging myths such as these, self-imposed barriers can be broken down, enabling individuals to develop and apply their creativity across different aspects of their lives and professions. Understanding that creativity is not exclusive to certain people or disciplines fosters confidence in our own creative ability. This encourages a willingness to approach problems from multiple perspectives, empowering innovative solutions. Demystifying the belief that creativity is chaotic or unpredictable helps us recognize and use structured methodologies that foster innovation, some of which will be explored in this manual. Understanding that creativity is a developable skill, rather

[1]atlanTTic Research Center, Departamento de Enxeñaría Telemática, Universidade de Vigo, Spain

than an innate talent, facilitates the application of specific and practical techniques to enhance creative thinking.

15.1.1 Creativity is innate and cannot be learned

Contrary to the belief that creativity is innate and cannot be learned, creativity is an ability that can be developed and improved. Creativity involves various skills, such as divergent thinking, problem reformulation and ideation, which are techniques that can be learned and trained.

In fact, history offers numerous examples demonstrating that creativity can be cultivated. Thomas Edison, the inventor of the electric light bulb and other devices like the phonograph, worked tirelessly to develop and refine his inventions throughout his life. Leonardo da Vinci, recognized as one of the most creative geniuses in history, was an artist, inventor, scientist, and visionary. The evolution of his creative abilities is reflected in the evolution of his artistic production, in his studies of anatomy, or in his machine designs. Maria Sklodowska-Curie, the first woman to receive a Nobel Prize in two different scientific fields (Physics and Chemistry) for her pioneering research on radioactivity, developed creativity through her innovative thinking and applications. Her creativity was reflected not only in her fundamental discoveries but also in her ability to see beyond the state of the art and to explore the possibilities of radioactivity in various scientific and medical domains.

15.1.2 Creativity is just for artists and design professionals

Contrary to the myth that creativity is exclusive to artists and designers, reality shows us that it is applicable in virtually any field and is fundamental for innovation across discipline. In the case of engineering, we have examples such as 3D printing technology, which revolutionized product design and manufacturing. The ability to create three-dimensional objects by layering material has enabled significant advancements in customized manufacturing, medicine (like printing prostheses), and the aerospace industry. Engineering innovation has also driven progress in renewable energy technologies, including solar, wind, and tidal power. Creative design improvements in more efficient solar panels and more powerful wind turbines have expanded the capacity and viability of these energy sources. Similarly, creativity in the field of biomedical engineering has led to the development of advanced prosthetics, sophisticated medical diagnostic equipment like MRI scanners, and precision medical therapies that were unimaginable just a few decades ago.

In the case of economics and finance, creativity in applying technology has transformed how business transactions are conducted. The creation of e-commerce platforms, such as Amazon, eBay, and Alibaba, revolutionized the way in which we buy and sell goods and services, generating new business models and global markets. Creativity in the financial sector has also led to innovations like decentralized storage technologies based on blockchain and cryptocurrencies such as Bitcoin and Ethereum. These technologies are now being applied in many other areas and are transforming financial systems by enabling secure, decentralized transactions, introducing new forms of investment and financing.

15.1.3 Creativity is unpredictable and cannot be systematized

Creativity can be systematized and structured. In fact, there is evidence of several techniques and methodologies that foster creativity and innovation, some of which will be discussed in this chapter.

Structured processes and methodologies are key to fostering creativity in engineering, and many examples illustrate this. For instance, the Panama Canal expansion project aimed to accommodate larger ships and increase efficiency. Engineering professionals adopted a systematic approach to innovation, integrating advanced technologies such as modern lock systems, hydraulic gates, and computerized control systems. This approach enabled them to develop a constructive solution that expanded the canal without disrupting its normal operation. High-speed rail systems, such as the Japanese Shinkansen, the French TGV, and the Spanish AVE, are another example of systematic engineering creativity. Engineers combined aerodynamic designs, advanced propulsion systems, and track technologies to develop high-speed trains. Using innovative solutions, such as new electric traction motors and optimized aerodynamic designs, increased the speed and efficiency of transportation. Similarly, the design and construction of the Burj Khalifa building in Dubai required meticulous planning and systematic construction approaches. Engineers used innovative materials, advanced structural systems, and wind-resistant designs to overcome challenges such as horizontal loads at extreme heights. These challenges, exacerbated by sandstorms in desert areas, demanded a combination of creative problem-solving and systematic planning to achieve a structure unique in its field.

15.2 Toward creative thinking: convergent and divergent thinking

The creative process is a dance between two complementary ways of thinking, both essential for innovation and idea generation: convergent thinking and divergent thinking.

Convergent thinking focuses on finding a single, optimal solution to a problem. It requires us to analyze diverse ideas in depth, and challenges us to select, refine, and integrate concepts. This process transforms a diversity of ideas into a tangible solution, moving from the abstract to the practical using logic, experience, critical evaluation, and informed decision-making.

Convergent thinking is applied, for example, in situations where there is a right or wrong answer such as exams or scientific problem-solving. It is an important skill for success in studies, work, and in everyday life. Convergent thinking helps us solve problems, make decisions, and learn from mistakes by allowing us to weigh the pros and cons of the different options to eventually choose the best one.

This skill can be trained and improved with practice such as by solving puzzles, playing logic games, or engaging in activities that require problem-solving.

Divergent thinking, however, invites us to explore multiple paths, expand limits, and seek new possibilities. It encourages us to challenge existing rules, observe from unusual angles, and embrace uncertainty as a creative ally. Diverge thinking focuses on generating multiple solutions to a problem, relying on creativity, imagination, and the ability to think outside the box.

It is often used in situations where there is no right or wrong answer such as brainstorming or problem-solving. This type of thinking requires the ability to improvise, generate ideas quickly and fluidly, see the world in different ways, and think creatively to generate new solutions.

Like convergent thinking, divergent thinking is also useful in studies, professional activity, and in everyday life, as it can help us to be more creative, innovative, and effectively in problem-solving. For example, we use divergent thinking to paint a picture, write a story, design a new product, or innovate in various ways. Like convergent thinking, divergent thinking can be improved with practice by engaging in creativity-related activities such as free writing, seeking analogies or metaphors, role-playing or dramatization, or solving puzzles or riddles.

Between these two types of thinking emerges critical thinking. Like convergent thinking, critical thinking involves a deep and systematic analysis to solve problems or evaluate situations. Both convergent and critical thinking require evaluating information, data, or arguments to reach a conclusion, and both are based on logical reasoning and rationality to reach a solution or develop a plan. While both are essential in decision-making and problem-solving, they differ in focus and primary objective.

Convergent thinking focuses on finding a single optimal solution or conclusion, while critical thinking is more directed toward in-depth analysis and the objective evaluation of arguments, information, or situations. In other words, convergent thinking tends to be more structured and focused on finding specific answers, while critical thinking is more open to exploring diverse perspectives and considering multiple solutions. Critical thinking centers on evaluating and analyzing existing information, while convergent thinking seeks to bring together information and data to arrive at a defined solution.

Creative thinking is a metacognitive process that generates new or useful associations to develop solutions, models, patterns, or products previously unconsidered. From this perspective, it is common to find in the literature that creative thinking is presented as synonymous with divergent thinking [1,2]. In any case, creative thinking should be supported by convergent thinking to increase the ability to refine creative ideas and refine solutions through motivated decision-making and problem-solving, as appropriate [3].

In short, we can say that these two thinking approaches provide the foundations for the creative process. Divergent thinking nurtures imagination, expands horizons, and unleashes originality, while convergent thinking shapes, structures, and brings ideas to life.

15.3 The four stages of creative thinking

We can define creative thinking as a way of thinking that involves the generation of original, novel, and valuable ideas, characterized by the ability to find innovative solutions to problems, develop new perspectives, connect seemingly unrelated concepts, and produce exceptional results. Creative thinking thus involves thinking unconventionally, challenging established assumptions, and exploring new approaches based on mental flexibility, fluency of ideas, and openness to new possibilities. Moreover, creative thinking is not limited to a specific area but can be applied both personally and professionally, in many different fields such as the arts, science, technology, or business.

Creative thinking is original in the sense that it allows us to generate novel and unique ideas that have not been previously explored. It also allows us to adapt and change our perspective, exploring different approaches and possible solutions using skills such as the ability to generate a large number of ideas in a short period of time, without worrying about their initial quality or feasibility; the ability to connect seemingly unrelated concepts or ideas to create new associations; the ability to generate multiple solutions or answers to a problem instead of settling for a single solution; and the ability to evaluate and analyze the ideas generated, selecting the most promising ones and refining them.

In 1926, British psychologist and sociologist Graham Wallas postulated that creative thinking could be organized into four stages. In his book, Wallas named the stages of creative thinking as preparation, incubation, illumination, and verification [4] (cf. Figure 15.1), thus establishing a conceptual structure for understanding how creativity develops and evolves in people. Since then, these stages have been widely analyzed and studied.

The preparation stage lays the groundwork for creative thinking and involves information gathering, research, and immersion in the topic or problem at hand. In this phase, an open mind is required, along with curiosity and a willingness to explore different perspectives and sources of inspiration. Preparation also involves identifying existing challenges or roadblocks and establishing an environment conducive to fostering creativity.

During the incubation stage, the creative thinker consciously steps back from the problem or challenge being addressed. They allow the information and ideas gathered in the preparation stage to combine and connect. Incubation involves a period of reflection, rest, and detachment from the problem, with the goal of allowing divergent thinking and unexpected associations to emerge naturally. It is common for creative solutions to appear suddenly, when least expected, during this stage.

Figure 15.1 The four stages of creative thinking (pictures by Austin Distel, Jeffrey Erhunse, Tool Inc. and Warren in Unsplash)

Illumination is the stage where the famous "eureka" moment or sudden creative idea occurs, as the subconscious connections and associations generated during the incubation stage emerge into consciousness clearly and suddenly. They may present themselves as a flash of inspiration, a seemingly obvious solution, or a new perspective that was previously overlooked. Enlightenment may result from combining seemingly disparate ideas or reevaluating previous assumptions.

The verification stage consists of developing and evaluating the creative idea that emerged in the illumination stage by critically analyzing it and assessing its feasibility. Tests, additional research, or prototypes may be conducted to verify the validity and applicability of the creative solution. Verification also involves overcoming possible obstacles, adjusting and improving the initial idea, and finally putting it into practice in the relevant context.

Note that these stages do not always occur in a linear and sequential manner, as creative thinking is a fluid and flexible process where the stages may overlap or repeat at different times. However, understanding these stages can help structure and enhance the overall creative process.

15.3.1 Preparation

The preparation phase in creative thinking can be seen as the initial stage of the creative process, where the foundations are laid to generate innovative ideas and solutions. During this phase, information is gathered, research is conducted, the problem or challenge to be addressed is reflected upon, and an environment conducive to fostering creativity is created. The preparation phase is key because it provides the knowledge and resources needed to tackle the creative challenge in an informed and effective manner. During this phase, we gain a thorough understanding of the problem, identify objectives, and define limits and constraints.

Research is the fundamental element at this stage. To do research, we may use the usual resources, such as books, articles, studies, interviews, and online sources, to obtain information on the issue or problem addressed. Gathering different perspectives, background information, and knowledge will provide us with a solid basis for generating creative ideas. In this sense, we may also explore fields that are not directly related to our problem or challenge. Combining ideas from different fields can lead to new connections and creative approaches.

We may also use analogies and metaphors to explore the problem from different angles, looking for similarities among seemingly unrelated situations or concepts. Analogies and metaphors can open new avenues of thought and trigger creative ideas. However, exposing ourselves to visual and sensory stimuli may also help foster creativity. We can do this by enjoying works of art, listening to music, interacting with nature, or visiting inspiring places. Like analogies and metaphors, these stimuli can trigger new ideas and creative approaches.

In addition to the above techniques for stimulating idea generation and exploring different perspectives, a variety of tools can be used during the preparation stage in creative thinking. For example, empathic interviewing [5] is a type of interaction in which the interviewer actively engages in understanding and emotionally connecting with the interviewee's experience and perspective. It is about putting oneself in the other person's shoes, showing empathy and generating an atmosphere of trust and openness so that our interlocutor feels heard and understood. During an empathic interview, the interviewer seeks to understand not only the words and content of what the other person is expressing but also the underlying emotions, needs, and experiences. Attention is paid to nonverbal cues, open-ended questions are asked to invite the person to share more details, and judgment and interruptions are avoided.

Initially proposed by Tony Buzan [6], mind or concept maps are diagrams widely used in both education and business to represent ideas or concepts related to each other through a keyword or central idea. They are built around a key idea represented by a word, phrase, or short text. This idea is placed in the center of a sheet of paper or a board. Taking this central idea as a reference, we add other related ideas around it, clockwise, and connected them to the central idea with lines (cf. Figure 15.2). The purpose of a mind map is to stimulate creativity and facilitate the generation and development of ideas. They are used to capture and organize concepts, keywords, images, and other visual elements in a non-linear way. The structure of the mind map is very flexible and allows for free and fluid exploration of information.

A mood board is a visual tool that allows the association of images, colors, textures, typographies, and other visual elements related to a specific project or concept. They are created by gathering images, clippings, color swatches, text, and any other inspirational visual elements related to the creative project. These elements can be obtained from magazines and other printed material, websites, videos, social media, and many other sources of visual inspiration (cf. Figure 15.3). The

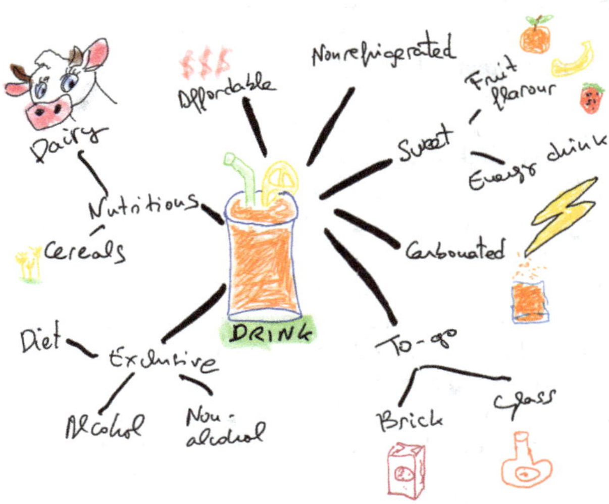

Figure 15.2 Mind map for a new drink

Figure 15.3 A mood board for T-shirt design (photo: June Lui at Unsplash)

collected elements are reviewed, and those that best represent the mood, style, aesthetics, and message to be conveyed are selected and organized into categories or groups according to their subject matter and characteristics. Like mind maps, their use was popularized by psychologist and writer Tony Buzan [7]. Although he is not exclusively credited with the creation of this technique, Buzan was a leading figure in the dissemination and promotion of inspiration boards as an effective tool in the creative process.

Brainstorming is an idea generation technique that encourages creativity and free, unrestricted thinking. It is conducted in a collaborative environment, where participants propose ideas spontaneously and without judgment, with the goal of generating as many ideas as possible, even those that may initially seem unconventional or impractical. It is commonly attributed to Alex Osborn, an American publicist and writer, and co-founder of the advertising agency Batten, Barton, Durstine & Osborn (BBDO). Osborn formally introduced the brainstorming technique as a structured approach to generating ideas creatively and collaboratively [8].

Affinity diagrams, also known as KJ diagrams or KJ-method diagrams, help visualize and group related concepts or ideas, allowing for the identification of patterns, connections, and common themes. They were proposed by Jiro Kawakita, a Japanese engineer and consultant, in the 1960s. Kawakita developed this technique as part of the KJ method, used for the organization and grouping of ideas in a collaborative context [9]. To construct an affinity diagram, we start with the result of a brainstorming session and write down each idea on a separate card or sticky note. Next, we review all the ideas generated looking for similarities or connections between them, grouping those that have affinity or that are related in terms of theme, characteristics, or common concepts. Finally, cards or sticky notes with related ideas are grouped together (cf. Figure 15.4).

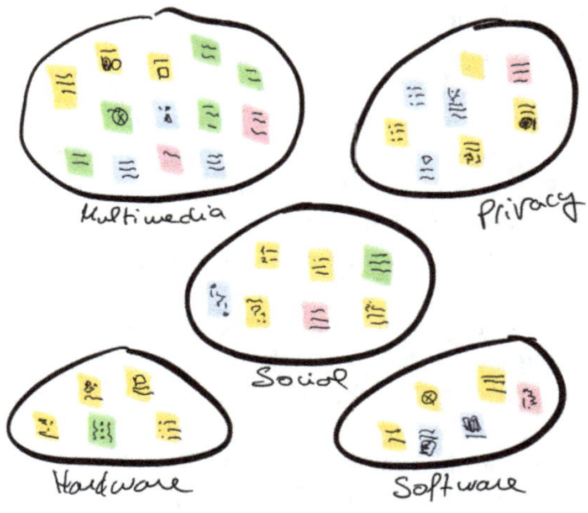

Figure 15.4 *Example of an affinity map for a new information and booking system at a mall*

15.3.2 Incubation

Once the preparation phase has been completed, the creative thinker enters in a process of incubation, which takes place mainly subconsciously. During this phase, we try to divert our attention away from the problem or challenge that we are addressing, allowing the subconscious mind to process the information gathered during the previous stage. In other words, during incubation, the creative thinker takes a mental break and actively detaches from the problem.

The reason for this conscious detachment is that it allows us to spend time in other activities, temporarily removing us from the creative challenge. The goal is to give the subconscious mind the space to work independently, making unexpected connections between different ideas and concepts. During this incubation period, the brain continues processing information in the background, even though we are not consciously focused on it. Often, associations, connections, or creative ideas occur spontaneously and suddenly, without conscious effort.

The incubation phase can be especially helpful when we face mental blocks, prejudices, or other difficulties to finding creative solutions. By detaching ourselves from the problem, we allow the mind to relax and open to new perspectives and approaches. Incubation also encourages the release of restrictive thought patterns, stimulating creativity and allowing divergent thinking to take center stage.

The incubation phase has no fixed duration, as it can vary widely depending on the individual and the project. It can last from a few minutes to several days, depending on the complexity and nature of the creative challenge, as well as the ability to abstract from the problem and allow the subconscious mind to work. As we return to the project in the next stage of enlightenment, new ideas, perspectives, or unexpected solutions may emerge, leading to breakthroughs in creative thinking.

Given the nature of the incubation phase, no specific tools are used as in the preparation stage, since the main objective is to relax and create space for ideas to arise spontaneously. However, there are some strategies that can facilitate the incubation process. For example, adequate rest and the ability to relax are essential during this stage, as activities that reduce stress and promote relaxing such as practicing meditation and other relaxation techniques [10]. This helps to relax the conscious mind and separate us from convergent thinking, allowing the subconscious mind work more effectively. Additionally, spending time without distractions, doing nothing in particular, can also be beneficial. This is often effective during activities such as taking a shower, driving, or simply sitting quietly (cf. Figure 15.5).

Moving to a different physical environment may also stimulate the mind and facilitate the incubation process. Taking walks outside the workplace, visiting new places, or simply changing location exposes us to different places, which can trigger new associations and facilitate the generation of creative ideas. However, engaging in creative activities unrelated to our project, such as drawing, painting, playing a musical instrument, or making crafts, can help free the mind, allowing ideas to emerge spontaneously.

Although the incubation phase is based on the subconscious process, it is useful to have available a notebook or device to record ideas that may arise spontaneously. This allows us to capture flashes of inspiration during incubation, without judgment or evaluation at that time, so they can be revisited in the next stage.

In summary, the incubation stage is a time of rest and relaxation where the objective is not to actively search for answers or solutions. The key is to enjoy this

Figure 15.5 During incubation, devote your time to activities that allow you to disconnect from your project (photo: Etienne Girardet in Unsplash)

time, and trust that creative ideas will emerge naturally, and be ready to engage in conscious thinking when transitioning to the next phase.

15.3.3 Illumination

This stage, often referred to as the "moment of revelations" or "eureka moment", occurs when a sudden revelation or insight emerges during the creative process. During the illumination phase, a sudden leap of understanding happens, offering a new perspective or solution to the problem or challenge. Unlike other stages, illumination cannot be forced or scheduled; it marks the end of the incubation stage, signaling that the creative process has reached a breakthrough.

In this phase, the subconscious connections and associations that were being built during the previous stages of preparation and incubation suddenly come to the surface, becoming conscious. At this moment, the result is often a revelation, a profound insight, an innovative solution, or original approach to addressing the creative challenge. The "eureka" moment is typically experienced as a sense of clarity, excitement, or surprise, and it may occur at any time, even when you are not actively focusing on the problem.

This stage can lead to highly creative and effective ideas and solutions as result of subconscious processing of information captured during the preparation phase and the combination of seemingly unrelated concepts during the incubation phase. In doing so, mental blocks, biases, and conventional assumptions are overcome, allowing for the emergence of fresh and surprising ideas and innovation.

Once the illumination moment occurs, it is crucial to capture these ideas immediately, as they often fade quickly from memory. The next step in the creative process is the verification stage, where the feasibility and effectiveness of the illuminating ideas will be evaluated and developed for practical application.

There are no specific tools that can be systematically applied during the illumination phase, as this moment is characterized by the sudden emergence of creative ideas and the unexpected connection between concepts. However, certain techniques can help encourage and enhance these moments of revelation. For example, as in the incubation phase, it is important to have a means to capture creative ideas as they arise. Keeping the same notebook, note-taking application, voice recorder, or any other medium used during incubation phase, we can ensure that they are not lost or forgotten before the next phase of development.

We can also use techniques such as brainstorming, perspective shifting, or creative evaluation techniques. Although brainstorming is usually associated more with idea generation in the preparation stage, it can also be useful during the illumination phase. We can brainstorm to develop inspirational ideas and write down all ideas that come up to mind, no matter how far-fetched they may seem. Similarly, adopting different perspectives or points of view can help to unblock thinking and generate creative ideas. For example, we can put ourselves in the shoes of a child, an extraterrestrial being, an expert from a different field, or a person from the future. By doing so, we allow ourselves to observe the problem from previously unexplored points of view. Creative visualization involves imagining scenarios, situations, or images related to the problem at hand, visualizing different possible solutions, exploring mental scenes, or using images or visual metaphors to stimulate generation of ideas during the illumination phase.

Note that the illumination phase is an internal and subjective process, and not all tools will work the same for everyone, as each person has their own of thinking and creative response style. Therefore, it is essential to experiment with different approaches and find the tools that work best for each individual, understanding that there are no specific tools for this process.

15.3.4 Verification

In this stage, also known as evaluation or validation, the ideas generated during the previous stages are analyzed and evaluated to determine their feasibility, effectiveness, and relevance to the problem or challenge addressed. The main objective of the verification phase is to ensure that the selected creative ideas are feasible and have the potential to solve the problem or achieving the established objectives.

During the verification phase, the generated ideas are evaluated to determine whether they are directly related to the problem or challenge, and whether they effectively address the issue or meet the established objectives. Additionally, their potential impact is assessed in relation to the expected results, analyzing whether the proposed solutions can generate significant improvements, positive changes, or competitive advantages. This process ensures that the proposed solutions are relevant and address the essential aspects of the problem.

We also assess whether the ideas can be practically implemented, considering factors such as the availability of sufficient resources, the maturity of required technologies, and any logistical constraints that could hinder their realization. Along these, the potential risks associated with their implementation are considered by identifying obstacles such as operational risks, financial limitations, or any other challenges that could affect the successful execution of the creative ideas.

During the verification phase, feedback is gathered from other relevant individuals, such as colleagues, experts or potential users, to obtain alternative perspectives, generate additional ideas, or identify areas for improvement in the creative proposals.

Based on the evaluation conducted during this phase, the most promising ideas are selected to further implementation, keeping in mind that some ideas may require adjustments or modifications before being fully accepted. Conversely, some ideas may not pass the verification stage and should be discarded.

Several tools can support the verification process, with some of the most relevant being strengths, weaknesses, opportunities, and threats (SWOT) analysis [11], rapid prototyping techniques [12], the think-aloud panels [13], and the comment capture box (cf. Figure 15.6).

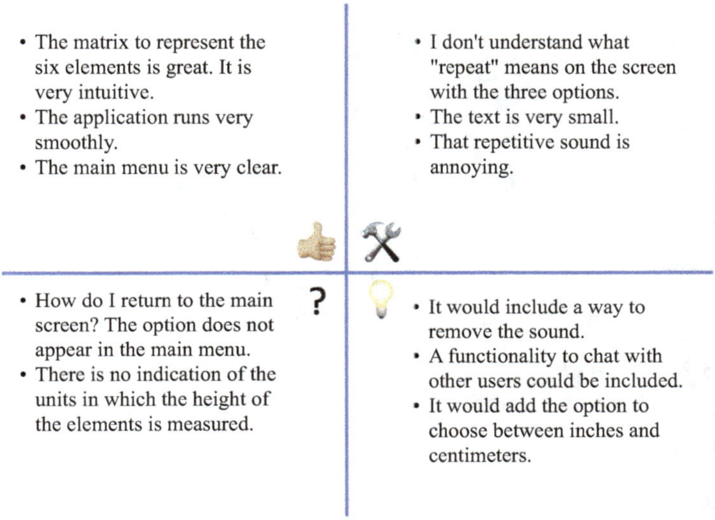

Figure 15.6 Comments capture grid to analyze a prototype of a mobile app

15.4 SCAMPER or posing the right questions to think creatively

Substitute, Combine, Adapt, Modify, Purpose, Eliminate and Reorganize (SCAMPER) [14] is a creative technique that prompts different questions to stimulate the generation of innovative ideas by modifying, adapting, or combining existing elements. Each letter in SCAMPER represents a key word in English which suggests a specific question to guide the creative process: Substitute, Combine, Adapt, Modify, Purpose or Purify, Eliminate, and Reverse or Rearrange (cf. Figure 15.7).

The **Substitute** phase involves exploring the possibility of replacing, changing, or substituting an element, material, process, or component with another that can perform a similar function or improve the result. Some relevant questions for this phase include:

- Can we replace an element or part of the process with another?
- What would happen if we changed a material, part, or process?
- Can we substitute one technique or method with a more modern or advanced alternative?
- What if we replace one component with another that works differently but achieves the same result?

An example of this strategy could involve replacing packaging with biodegradable materials such as those made from corn starch.

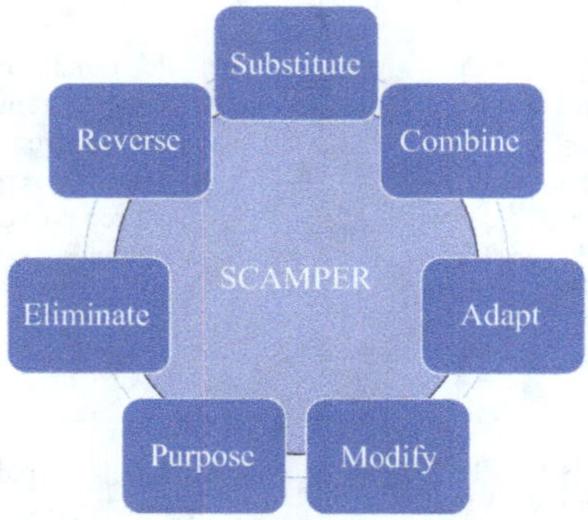

Figure 15.7 SCAMPER is the acronym of Substitute, Combine, Adapt, Modify, Purpose or Purify, Eliminate, and Reverse or Readapt

The **Combine** stage focuses on merging existing elements, ideas, or concepts to create something new or innovative. Some questions that might arise in this stage include:

- What happens if we combine two or more elements?
- Can two different ideas or elements be merged to create something new?
- Can we mix methodologies or approaches to address a problem from a different perspective?
- Can different concepts be combined to create a more complete solution?

For example, consider the idea of combining a smartwatch with Bluetooth headphones to create an all-in-one device that offers both health tracking and music playback, eliminating the need for additional devices.

The **Adapt** stage involves exploring ways to adjust, change, or modify an existing item to meet new needs or contexts. Some questions to consider during this phase include:

- How can something be adjusted or modified to work differently?
- Can its form, function, or focus be changed?
- Can this idea or product be adapted to serve a different demographic?
- Can this item be reconfigured to meet new requirements or standards?
- What changes are necessary to make it viable in a specific market or environment, different from the original?
- What adjustments could improve the effectiveness of this process in a particular situation?

For example, consider how a language learning application could be adapted to make it more accessible for visually impaired users.

The **Modify** phase focuses on examining how an existing element, process, or idea can be changed, adjusted, or altered to create something new or improved. Some relevant questions to consider during this stage include:

- What if we scale something up or down?
- Can we change the shape, size, color, or structure to improve its functionality or appeal?
- Can technological advancements or recent innovations be applied to modify the product or process?

An example associated with this phase could be modifying the design of a bicycle by integrating solar panels into the frame to charge electronic devices while pedaling.

The **Purpose** stage involves exploring how an existing item, product, or idea could be used in a different way or for an alternative task. Some key questions to consider during this stage include:

- Can this be used for a purpose different from the original?
- Are there other ways to use this product, tool, or idea?
- Are there other contexts or industries where this product or idea could be applied?
- What happens if we use this resource in a completely different way than intended?

For example, virtual reality systems originally designed for video games could be assessed for therapeutic applications such as helping patients with anxiety overcome their fears.

The **Eliminate** stage focuses on identifying elements, processes or parts that can be removed or simplified to improve, innovate, or solve a problem. Some useful questions for this stage include:

- What happens if we eliminate something?
- What happens if we simplify or remove unnecessary elements?
- What happens if we expand or add more detail?
- What happens if we reduce the size, complexity, or quantity of something?

For example, in a production process, evaluating which elements can be removed could improve efficiency and reduce waste of resources.

The **Reverse** stage involves considering the reversal of a process or a specific aspect of a product, idea or situation, starting at the end and working back to the origin. Key questions to consider during this stage include:

- What happens if the order of the steps is reversed?
- Could this be done in reverse order?
- What happens if we go in the opposite direction?

This strategy can also include rearranging, which involves changing the arrangement, sequence, or structure of something to achieve a different result. Questions to consider in this context include:

- Can parts be rearranged to improve efficiency or function?
- What happens if we redesign or adjust the layout of components?

A concrete example of this could be to assess what happens if we reverse the order of the steps in a production process to improve efficiency and reduce the total time required.

In short, **SCAMPER** is a flexible and versatile tool that can be applied to different contexts, from problem-solving to generating ideas for new products or services. It stimulates creativity by challenging traditionally perspectives and uses, encouraging the exploration of new possibilities. In fact, its versatility lies in its ability to enhance creative thinking, by providing a structured framework for generating innovative ideas and solutions. By applying the strategies that make up SCAMPER, every aspect of a concept can be systematically and provocatively explored. This methodology fosters innovation by stimulating the imagination, connecting seemingly unrelated concepts, and encouraging the reconfiguration, alteration, and transformation of existing elements. SCAMPER challenges conventional thinking, making it an essential tool for creativity and idea generation.

15.5 The six creative thinking hats

This technique, created by Edward de Bono [15], allows approaching a problem or situation from different perspectives, metaphorically represented by six hats of different colors. Each hat symbolizes a specific approach (cf. Figure 15.8):

- The **White Hat (facts)** represents objectivity and data. When wearing this hat, the focus is on available information, facts, and objective data.

Figure 15.8 The six thinking hats. Each hat represents a different perspective to face a challenge or problem.

- The **Red Hat (feelings)** represents emotions, intuitions, and feelings. This hat allows the expression of emotional reactions without the need for justification.
- The **Black Hat (caution)** represents critical thinking and caution. It is used to identify risks, weaknesses, or potential problems.
- The **Yellow Hat (benefits)** represents positive, optimistic, and constructive thinking. This hat is used to highlight benefits, strengths, and opportunities.
- The **Green Hat (creativity)** represents creativity and the generation of new ideas. With this hat, alternative solutions, possibilities, and innovative proposals are explored.
- The **Blue Hat (process)** represents control of the thinking process. The wearer of this hat moderates and supervises the use of the other hats, managing the flow of thought to ensure that the rules are respected, and progress is made effectively.

By wearing these hats, we can explore a problem from multiple angles, encouraging creativity, reflection, objective analysis, and the consideration of emotions, which in turn enables more holistic and balanced decision-making. This technique is particularly effective for teams seeking to approach a problem from different perspectives and reach more comprehensive and robust solutions. For example, imagine a team considering launching a new product to the market. To assess this possibility, they use the six thinking hats. The moderator or team leader wears the Blue Hat to ensure that time is properly allocated to each hat, guides the flow of discussion, and helps synthesize the conclusions reached from each perspective. To start, the team wears the White Hat of neutral thinking and reviews objective data such as market analysis, production costs, sales projections, and feedback from potential customers regarding the product to be launched.

Next, the team members wear the Red Hat of emotions. Each team member expresses their feelings, intuitions, and emotions about the project. Some may be delighted with the idea of seeing the new product for sale, while others may have concerns or doubts about its feasibility.

After putting on the Black Hat of caution, the team critically examines the potential risks of the product launch such as market competition, possible product failures, or financial challenges.

Having expressed their critiques of the project, team members now wear the Yellow Hat of benefits, to focus on the positive aspects of the product such as its potential impact on the company's customer base, the benefits it could bring to shareholders, and its expected profitability.

Finally, with the Green Hat on creativity, the team generates creative ideas for improving the product, exploring new features or marketing strategies that could make it more attractive or differentiated in the marketplace.

By using these thinking hats sequentially, the team can approach decision-making from multiple angles, considering objective data, emotions, pros, cons, and different options. This allows for a more complete and balanced view of the product launch, helping them make a more informed and well-rounded decision.

Like SCAMPER, the thinking hats offer a structured and systematic approach to creative thinking, where each hat represents a different mental stance: White for objective data, Red for emotions, Black for criticism, Yellow for optimism, Green for creativity, and Blue for process control. This allows the team to explore a problem or idea from different perspectives. This methodology facilitates the generation of broader and more varied ideas by encouraging the consideration of emotional, analytical, creative, and optimistic aspects, and it provides a framework for approaching a problem from multiple angles and triggering innovative solutions.

15.6 Creative thinking and design thinking

The design thinking methodology [16] proposes a structure similar to Wallas' creative thinking model described in this chapter, but with a more focused approach to problem-solving and innovation. Design thinking is a methodology for solving complex problems that brings an innovative and human-centered approach, fostering creativity throughout the process of finding solutions. One of its distinctive features is its holistic approach, which integrates empathy with user needs, multidisciplinary collaboration, and iterative experimentation. This methodology is based on deeply understanding users, challenging conventional biases and seeking solutions through co-creation and rapid prototyping. Flexibility in the problem-solving process and openness to redefining the initial challenge are other key features, facilitating adaptation as the process learns and evolves during the creative journey. Design thinking focuses not only on the functionality of the product or service but also on the overall experience of the real people who will use the new product or service, seeking solutions that are not only effective but also meaningful and emotionally impactful.

Design thinking is applicable across a wide range of fields, especially in the design of innovative products and services. In business, it is used to address strategic challenges and improve customer experience, fostering customer loyalty and satisfaction. In product design, it is applied to create people-centered solutions, incorporating user needs and desires from the earliest stages of development. In the technology sector, design thinking is essential for creating intuitive interfaces and engaging user experiences. Furthermore, in education, it helps develop more effective teaching methods, adapted to students' needs and learning styles. Likewise, in healthcare, it plays a key role in improving the accessibility and efficiency of medical services by focusing on the patient experience. In summary, design thinking has proven to be a versatile methodology, transcending productive sectors and being successfully applied in problem-solving and the generation of innovative solutions.

Design thinking is a process, a creative journey, typically organized into five stages (cf. Figure 15.9). These stages are designed to empathize with users, define the problem to be solved, generate solutions, prototype these solutions, and test them with the users and other stakeholders involved in the project. While the stages

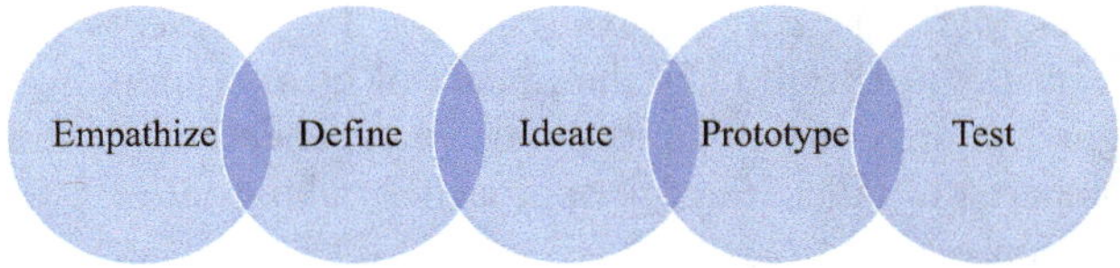

Figure 15.9 The design thinking stages

of creative thinking focus on the individual mental process and the internal flow of ideas, design thinking emphasizes user-centered problem-solving and iteration through collaboration and prototyping. Design thinking prioritizes empathy and a deep understanding of user needs from the beginning, continuously integrating feedback from user personas through prototyping and testing, while creative thinking is more focused on idea generation by the individual.

In terms of their organization in stages, despite the differences in their number, both models share some common elements. First, both models emphasize the importance of preparation and research to thoroughly understand the challenge or problem. Second, both models recognize that creative thinking often involves moments of inspiration or epiphany. Third, both models include an evaluation or verification phase to ensure that proposed solutions are feasible and effective.

The main differences between the two models lie in the incubation and illumination phases of creative thinking and the prototyping and testing stages of design thinking. As discussed, the Wallas model includes a specific incubation stage, where the individual stops consciously thinking about the problem. Design thinking does not include a specific incubation stage but assumes that creative thinking often involves moments of inspiration or epiphany that occur subconsciously. The creative thinking model treats illumination as a unique and unrepeatable stage, whereas the design thinking methodology views illumination as a continuous process that can occur at any point throughout the creative journey.

In contrast, design thinking model dedicates two specific stages to the construction and testing of prototypes. The Wallas model, however, does not include a specific prototyping stage but recognize the importance of validating ideas through feedback from users.

In summary, both creative thinking and design thinking are applicable to a wide range of creative problems, including design, engineering, social issues, or business problems. Both models are valuable for understanding the problem-solving process through creativity and innovation, and both emphasize the importance of preparation, research, inspiration, and evaluation.

15.7 To learn more

Below, we enumerate a small selection of works related to creative thinking that may offer new perspectives on this chapter, as well as extend the concepts discussed.

In *Creative and Critical Thinking Skills in Problem-based Learning Environments*, by Bengi Birgili [17], the impact of creative thinking on problem-based learning is explored, including its philosophy, general characteristics, and its uniqueness compared to other learning approaches, taking into account its advantages and limitations.

In *On the Path Toward the Science of Creative Thinking*, by Giovanni Corazza and Sergio Agnoli [18], the fundamental elements underpinning the science of creative thinking are enunciated. The authors review the theoretical models of the creative thinking process, the minimum number of stages, the concept of optimal originality, and the most important existing models, outlining their characteristics and limitations.

Creative Thinking and Problem Solving, by Michele Fisher [19], is an inspirational presentation aimed at ARUP laboratory staff to introduce them to different creative problem-solving skills and their application in the laboratory environment. The presentation includes many interesting quotes and examples of creativity throughout history.

Creative Thinking: A Field Guide to Build Your Strategic Core, by Angus Fletcher [20], serves as a guide for U.S. Army officers and advanced enlisted personnel, providing strategies and tools to become more creative. Creativity is presented as a tool that emphasizes adopting different perspectives, situational openness, and cooperative teamwork, making it an excellent resource for fostering individual freedom and diversity in a controlled and disciplined way for the public good.

Creative Thinking and Learning Project, by Anni McTavish [21], describes a project conducted at Sheringham Nursery School in the London borough of Newham, designed to inspire creativity within the educational community. The project is focused on enhancing children's learning opportunities in expressive arts and design, as well as encouraging creative thinking and learning as core features of effective education.

The articles by Michael D. Mumford and collaborators on the processes of creative thinking [22,23] propose and discuss eight fundamental processes related to creative thinking: (1) problem definition, (2) information gathering, (3) concept selection, (4) conceptual combination, (5) idea generation, (6) idea evaluation, (7) implementation planning, and (8) adaptive monitoring. These processes can be mapped to the initial and final stages of Wallas' methodology, introduced in this chapter.

Mastering the Power of Creative Thinking, by Anshul Sonak [24], a regular contributor to government agencies and educational institutions worldwide, discusses his experiences in promoting creative thinking and fostering creative thinking skills in young people. Increasingly, youth are using digital technologies, including generative artificial intelligence, to develop creative solutions to solve everyday problems facing their communities. The author argues that anyone can learn to think creatively.

References

[1] Hargrove, R. A. (2013). Assessing the long-term impact of a metacognitive approach to creative skill development. *International Journal of Technology and Design Education* 23, 489–517. DOI:10.1007/s10798-011-9200-6.

[2] Baker, M., Ruddy, R., and Pomeroy, C. (2001). Relationships between critical and creative thinking. *Journal of Southern Agricultural Education Research* 51(1), 173–188.

[3] American Management Association (2019). *AMA Critical Skills Survey: Workers Need Higher Level Skills to Succeed in the 21st Century*. Technical report. American Management Association.

[4] Wallas, G. (2014). *The Art of Thought*. London: Solis Press.

[5] Nelsestuen, K., and Smith, J. (2020). Empathy interviews. *The Learning Professional*, 41(5), 59–62.

[6] Buzan, T., and Buzan, B. (2009). *The Mind Map Book: How to Use Radiant Thinking to Maximize Your Brain's Untapped Potential*. Plume.

[7] Buzan, T. (2010). *Use Your Head. How to Unleash the Power of Your Mind*. London: BBC Active.

[8] Osborn, A. F. (1953). *Applied Imagination: Principles and Procedures of Creative Problem-Solving*. Charles Scribner's Sons.

[9] Scupin, R. (1997). The KJ method: a technique for analyzing data derived from Japanese ethnology. *Human Organization* 56(2), 233–237. DOI:10.17730/humo.56.2.x335923511444655.

[10] Davis, M., Eshelman, E. R., and McKay, M. (2020). *The Relaxation and Stress Reduction Workbook*. New Harbinger Publications.

[11] Puyt, R. W., Lie, F. B., and Wilderom, C. P. M. (2023). The origins of SWOT analysis, *Long Range Planning*, 56(3), 102304. DOI:10.1016/j.lrp.2023.102304.

[12] Gómez, G., and Lopez-Leon, R. (2019). Impossible design: fostering creativity by quick and dirty prototyping. In: Börekçi, N., Koçyıldırım, D., Korkut, F. and Jones, D. (eds.), *Insider Knowledge, DRS Learn X Design Conference 2019, 9–12 July*, Ankara, Turkey. DOI:10.21606/learnxdesign.2019.14026.

[13] Ramey, J., Boren, T., Cuddihy, E., *et al.* (2006). Does think aloud work? how do we know? In: *CHI '06 Extended Abstracts on Human Factors in Computing Systems (CHI EA '06)*. Association for Computing Machinery, New York, NY, USA, 45–48. DOI:10.1145/1125451.1125464.

[14] Eberle, B. (1996). *SCAMPER Let Your Imagination Run Wild*. Prufrock Press, Inc.

[15] De Bono, E. (2017). *Six Thinking Hats: The Multi-Million Bestselling Guide to Running Better Meetings and Making Faster Decisions*. Penguin UK.

[16] Cuiñas, I., and Fernández-Iglesias, M. J. (2023). *Design Thinking for Engineering. A Practical Guide*. London (United Kingdom): The Institution of Engineering and Technology. DOI:10.1049/PBME024E.

[17] Birgili, B. (2015). Creative and critical thinking skills in problem-based learning environments. *Journal of Gifted Education and Creativity* 2(2), 71–80. DOI: 10.18200/JGEDC.2015214253.

[18] Corazza, G. E., and Agnoli, S. (2016). "On the path towards the science of creative thinking". In: *Multidisciplinary Contributions to the Science of Creative Thinking*, G. E. Corazza and S. Agnoli (eds.). Singapore: Springer Singapore, pp. 3–19. DOI:10.1007/978-981-287-618-8_1.

[19] Fisher, M. (2014). *Creative Thinking and Problem Solving*. ARUP Laboratories.

[20] Fletcher, A. (2021). *Creative Thinking: A Field Guide to Build Your Strategic Core*. US Army – Command & General Staff College.

[21] McTavish, A. (2017). *Creative Thinking and Learning Project*. Sheringham Nursery School and Childern's Centre.

[22] Mumford, M. D., Martin, R., and Elliott, S. N. (2019). Creative thinking processes: managing innovative efforts". In: Aldag, R. (ed.), *Oxford Research Encyclopedia of Business and Management*. Oxford University Press, pp. 346–363. DOI:10.1093/acrefore/9780190224851.013.172.

[23] Mumford, M. D., and McIntosh, T. (2017). Creative thinking processes: the past and the future. *Journal of Creative Behavior* 51(4), 317–322. DOI: 10.1002/jocb.197.

[24] Sonak, A. (2022). "Mastering the power of creative thinking". In: Johri, L. M., Corich, K., and Haskins, G. (eds.), *Mastering the Power of You*. Routledge, p. 179.

Chapter 16
Critical thinking

Daniela Firoiu[1] *and George H. Ionescu*[2]

Critical thinking (CT) refers to the mental process of analyzing and evaluating information, reflecting on the meaning of available information, examining the evidence and reasoning provided, and forming judgments based on the facts. CT triggers complex cognitive processes, starting with information gathering and ending with decision-making.

This way of thinking does not develop automatically from birth, but requires consistent practice to master. To develop it, we need to overcome certain innate human characteristics such as egocentrism and sociocentrism. Through this chapter, we aim to provide an overview of the whole process of CT while offering students and others practical insights into the abilities underpinning CT. Additionally, it presents tools and tips to improve their CT skills.

16.1 Critical thinking – definition and importance

CT is nowadays regarded as an essential skill in education, professional environments, and everyday life. It plays an important role in developing the ability to analyze and evaluate information in a rational and objective way. This skill involves not only the acquisition of knowledge and the ability to reflect on one's own ways of thinking but also the identification of reasoning errors and the ability to make informed decisions based on sound arguments and relevant evidence.

Definitions of CT offer varied perspectives and reflect its importance in educational, professional, and personal contexts. These perspectives often depend on the experiences and specialization of the experts. For instance, the philosophical view of John Dewey (1910) defines CT as a continuous process of reflection and analysis, while Richard Paul and Linda Elder (2014) describe CT as the process of evaluating information necessary for making rational decisions.

Regardless of the perspective we adopt, CT reveals its multifaceted nature that allows a deeper understanding of complex subjects. It fosters the development of

[1]Department of Commerce, Economic Integration and Business Administration, Romanian-American University, Romania
[2]Department of Finance, Credit and Accounting, Romanian-American University, Romania

autonomous and rational thought, equipping individuals to address the ongoing challenges of the economic and social environment by generating innovative solutions.

Moreover, to address the challenges of the 21st century, it is essential to adopt learning methods that promote CT. Such methods not only improve academic performance but also contribute to the development of active and autonomous thinking, which is necessary in contemporary society [1,2].

CT stimulates curiosity and analytical thinking, as well as creates an interactive educational environment where students are actively involved in the learning process. It contributes to the development of higher cognitive skills such as problem-solving and argumentation, transforming learning into a collaborative process that fosters the personal and professional growth necessary to meet the challenges of an ever-changing world [3,4].

In order to cultivate a critical attitude and CT, the following fundamental concepts can be addressed across various educational and professional contexts.

16.1.1 What is critical thinking?

Paul and Elder [5] emphasize that CT is a process of analyzing and evaluating information and ideas in order to reach rational and objective conclusions. This definition emphasizes the importance of the process of evaluating arguments and assessing premises, which are essential for developing clear and logical thinking. In this context, CT is viewed not only as the ability to understand information but also as the competence to evaluate it and draw coherent and informed conclusions.

John Dewey defines CT as an active and continuous process of reflection on experience, involving questions and the search for rational solutions. This approach highlights the role of reflective thinking and learning from previous experiences. Dewey underscores the importance of CT in education, especially in context where learning is an active process of exploring and adjusting one's perspective on the world. Dewey's definition promotes a way of thinking that does not stop at the first solution but always seeks deeper understanding and continual improvement in reasoning [6].

Paul and Elder promote the idea of "rational and coherent decisions" and the careful evaluation of evidence. Their definition enhances our understanding of CT, emphasizing that CT is a process in which not only arguments are analyzed but also relevant information and evidence. This concept is essential in education and professions where decisions are based on facts and evidence [7].

Facione analyzes CT from an educational perspective, and emphasizes the importance of clear, logical, and rational thinking, as well as the ability to identify reasoning errors. It is important to emphasize that, in education, CT is not only about acquiring information but also about evaluating it. In this context, students are encouraged not only to absorb external information but also to form their own judgments and opinions. The ability to identify assumptions and errors in reasoning is vital for the development of autonomous thinking [8].

Halpern approaches CT from a psychological perspective, viewing it as a process of self-reflection on one's own thoughts and beliefs in order to adjust and improve them. This definition emphasizes self-evaluation and the adjustment of personal reasoning. It is a key concept for the development of self-CT which is essential in personal decision-making and managing one's own beliefs and biases. Halpern emphasizes the idea that CT not only helps to improve decisions but also allows continuous self-improvement [9].

Ennis defines CT as "reflective and rational thinking" which focuses on analyzing and evaluating information in order to reach reasonable conclusions. This definition highlights the process of critically assessing ideas and arguments, which is crucial in a world in which we are often exposed to vast amounts of information. The definition is particularly useful in contexts where distinguishing between valid and invalid arguments is essential [10].

From analyzing the above definitions, we highlight the fact that CT is a complex process that involves not only understanding information but also evaluating, analyzing, and making informed decisions based on sound logic and clear evidence. Although there are a differences between the approaches, each definition reflects the need for deep reflection and active engagement in evaluating ideas and arguments. Thus, CT is not just a cognitive skill but also an attitude that involves being open to new ideas, constantly evaluating one's own thinking, and adapting one's reasoning in light of new information.

Analyzing the concept of CT involves different types of thinking, including rational, objective, and self-reflective thinking, all of which are fundamental and absolutely essential for processing information, formulating opinions, and making decisions.

CT is also a process of formulating hypotheses about possible answers to questions, with its rationality of this thinking is based on the fact that the hypothesis is not random but justified in some sense. While thinking can be rational even when we do not have a high degree of confidence in the truth of an idea, it must be consistent with logical reasoning (e.g., thinking about "p" must involve thinking about "q," if "p" implies "q"). So CT does not depend directly on our beliefs or our confidence in them; in many situations, what we think is an active choice, not a pure reaction to evidence or beliefs [11].

This is because *rational thinking* refers to the use of logic and coherent reasoning to arrive at valid conclusions without being influenced by emotions, feelings, or prejudices. Thus, rational thinking becomes essential in CT, as it allows the identification and elimination of erroneous or illogical reasoning [10].

Objective thinking refers to approaching a situation from a balanced and unbiased perspective and is based on clear evidence and verifiable facts, thus eliminating different subjective viewpoints [5].

Self-reflective thinking refers to the ability of individuals to reflect on their own thoughts and beliefs [9].

As observed, these three traits generate a solid foundation for effective CT that is absolutely essential for solving complex problems and developing accurate, and informed judgment.

It is also important to note that a clear distinction must be made between CT and other forms of thinking (e.g., creative, intuitive thinking), as CT relies exclusively on reasoning, unbiased approaches, and objective analysis based on concrete examples.

16.1.2 The importance of critical thinking

CT is essential in today's world, which is undergoing a process of globalization, technologization, and digitalization. Moreover, we live in a world where we are exposed daily to vast amounts of information and also a diversity of viewpoints.

Therefore, in a society strongly influenced by technology, media, and social networks, the ability to analyze, evaluate, and interpret information in a rational and objective way becomes essential for facing daily and professional challenges.

Consequently, we can state that CT provides individuals with the ability to distinguish between credible and dubious sources of information, identify valid arguments, and reject manipulation or misinformation.

Another significant point is that CT plays an important role in the educational process. By acquiring a CT attitude, individuals may be able to think independently, ask informed questions, and objectively analyze different points of view.

Therefore, these skills not only improve academic performance but also prepare individuals to make informed decisions in their careers and personal lives.

Equally important to emphasize is the fact that CT contributes to the development of a democratic society, as it enables individuals to think critically, participate actively and responsibly in the democratic process, understand the implications of policy decisions, and contribute to the solution of complex social problems.

In conclusion, CT is not just an intellectual skill but an active attitude that influences all aspects of our lives. Developing this skill is essential to navigating today's complex world, making informed decisions, and contributing to a more equitable and rational environment (Figure 16.1).

Figure 16.1 The importance of critical thinking

Moreover, the lack of CT skills is often linked to the technological and social changes of the modern age. Many believe that modern technology has contributed to this decline by encouraging immediate gratification and superficial engagement at the expense of deep reflection and analysis. Additionally, current social norms are seen as diminishing the importance of CT, emphasizing conformity, speed, and pragmatism over the development of rigorous and independent thought.

Other factors, such as deficiencies in the educational system, suggest systemic inadequacies in the teaching of CT. These may reflect an overemphasis on machine learning or standardized testing at the expense of stimulating creativity and independent thinking. Low parental involvement during childhood is also mentioned, indicating the possibility that children may not receive sufficient guidance or encouragement to develop CT at home. Historical factors are less frequently cited, suggesting that respondents believe that the problem is predominantly caused by contemporary challenges rather than long-standing patterns. Overall, the figure underlines the perception that the structures and behaviors of modern society play a significant role in diminishing CT skills.

16.2 The elements of critical thinking

In a world where society is constantly assaulted with messages and information from various sources, it is important to develop the ability to analyze and correctly judge what is being conveyed through these messages. Therefore, elements of CT, such as clarity, analysis, evaluation of arguments, and logical reasoning, help to pick out valid arguments and errors in reasoning, base beliefs on solid evidence, and draw well-supported conclusions.

Not to be neglected to mention in this context is also the fact that we are currently identifying a number of artificial intelligence (AI)-based applications that have a positive impact on the development of CTs, contributing to increased self-confidence, openness, and maturity. These applications not only help to improve skills but also aid in developing essential 21st-century competencies [12].

It is thus evident that, regardless of the context, CT not only supports individuals in making rational decisions but also encourages them to be more open, reflective, and attentive to their own thoughts and beliefs.

Among the specific elements of CT, we mainly identify logical and deductive reasoning, which play an essential role in its development, as they help to formulate clear, coherent, and well-grounded arguments.

These skills are fundamental to decision-making and active participation in reasoned discussion. Equally important to emphasize are common errors in reasoning, which can be identified and avoided to maintain clear and rational thinking.

Therefore, we emphasize the following as essential to consider when defining the specific elements of CT:

1. **Logical and deductive reasoning** are essential in CT, providing a framework for reaching correct conclusions based on clear and correct premises. In deductive reasoning, the conclusion must inevitably follow from the premises

if they are true. For example, the syllogism "All humans are mortal. Socrates is human. Therefore, Socrates is mortal" is correct deductive reasoning because the conclusion follows directly from the premises [13]. However, deductive reasoning can be erroneous if the premises are false or misinterpreted. For example, "All cats are pets. Felix is a pet. Therefore, Felix is a cat" is faulty deductive reasoning because the premises do not sufficiently support the conclusion [14].

In this context, we identify the Logic-Muse model as an example of a method for applying logical and deductive reasoning. This is an intelligent tutoring system that helps to improve deductive reasoning skills through an adaptive approach. The system utilizes a Bayesian network (BN) to assess and predict knowledge levels in real time. This allows accurate diagnosis of skills and adjusts lessons according to progress, providing personalized feedback. The BN analyzes the relationships between different variables, which makes it possible to predict students' responses to exercises and adapt learning dynamically. Experimental studies have shown that Logic-Muse model can predict with about 85% accuracy, how individuals will respond to various deductive reasoning exercises. It helps tutors adjust instructions and support learners more effectively based on continuous assessment of their abilities [15].

2. **Argument analysis** involves assessing the structure, reasoning, and validity of the claims made to support a position or conclusion. For example, in an argument in favor of online education, the premises might be that technology allows access to education for students in remote areas and online courses are more flexible. The conclusion in this case is that online education should become a more widespread practice [16].

   As demonstrated, analyzing arguments is not limited to evaluating the logical coherence of reasoning, but also includes examining the evidence, identifying unsupported assumptions, and testing the alternative. Therefore, a well-constructed argument will have sound premises, clear reasoning, and relevant and credible evidence to support them [17].

   Another example in this context is the educational experiment with pre-service teachers, which aimed to improve mathematical argumentation skills by using Toulmin's framework (1958/2003) to solve cryptarithmetic problems. Participants moved from making unjustified assertions to constructing complete deductive arguments with clear requirements, proofs, and explicit reasoning, demonstrating a significant development in their reasoning skills. The results showed that structured activities, including case analysis and collective argumentation, can support the development of CT and the ability to make logical inferences. The study also recommends a sequence of activities that can help mathematics educators strengthen the argumentation skills of future teachers, thus preparing them to teach students to solve complex mathematical problems through logical and deductive reasoning [18].

3. **Identifying and avoiding reasoning errors** is essential for developing clear and CT. Reasoning errors, known as logical fallacies, can undermine the validity of an argument and lead to incorrect conclusions. Fallacies are types of

faulty reasoning that may seem valid at first glance, but upon closer analysis turn out to be logically flawed. For example, instead of responding to a point about climate change, we might say "We shouldn't listen to X because he is not an expert in climate science," thereby ignoring the argument being made. This fallacy does not attack the logic of the argument, but only targets the proponent, which is not a valid way of refuting an opinion [14].

An example of this highlights the importance of CT and metacognition in preventing diagnostic errors, a significant problem in the medical field, which is responsible for up to 70% of errors in this field. One of the root causes of these errors is the influence of cognitive biases, which can affect physicians' clinical decisions. Therefore, recognizing and managing these cognitive biases is essential to prevent errors. In this regard, medical education should include training in metacognition, which helps professionals to reflect on their own thinking and identify potential errors in reasoning. This process would lead to greater accuracy in diagnostic formulation and improved patient safety. Although studies on the effectiveness of training in CT and metacognition are mixed and have methodological limitations, this educational approach has significant potential for improving diagnostic accuracy. Training in metacognition not only helps to combat cognitive biases but also contributes to the development of self-assessment and self-corrective decision-making skills. By adopting such an educational approach, physicians could learn to improve their thinking, recognize errors in their reasoning, and make better informed decisions. Thus, these strategies could significantly contribute to reduce diagnostic errors and improve the quality of medical care, with a direct impact on patient safety [19].

In conclusion, CT plays an essential role in developing the ability to analyze, evaluate, and construct well-founded arguments. Elements such as logical and deductive reasoning, argument analysis, and identifying errors in reasoning are essential for reaching rational conclusions and avoiding cognitive pitfalls. By applying these principles, individuals can recognize logical fallacies, formulate clear arguments, and make more informed decisions. Thus, CT not only improves the ability to analyze information but also contributes to a more rational and constructive dialog across various areas.

16.3 Critical thinking techniques

CT techniques are fundamental tools for analyzing information, evaluating arguments, and making informed decisions. These techniques are essential in a variety of contexts: in problem-solving, in academic and professional debate, and in everyday life. They involve skills such as identifying the premises and conclusions of an argument, checking its coherence, and recognizing errors in reasoning. By applying CT techniques, individuals can avoid cognitive pitfalls, think more clearly, and make better informed decisions.

Basic critical thinking techniques

1. ***Analyzing arguments:*** Argument analysis involves identifying premises and conclusions to assess the coherence and validity of an argument. It helps to uncover possible contradictions or to check that conclusions are adequately supported by premises.
2. ***Clarifying questions:*** Asking clarifying questions is an essential method to better understand information and to determine the validity of an argument or claim. These questions can help to clarify terms or to further an idea.
3. ***Evaluating sources of information:*** An important technique in critical thinking is evaluating sources of information. It is essential to check the credibility and reliability of the sources of information to prevent manipulation or misinformation.
4. ***Recognizing errors in reasoning***: Critical thinking involves identifying and avoiding logical fallacies such as ad hominem arguments, false dilemmas, or hasty generalizations. Recognizing these helps to construct sound and correct reasoning.
5. ***Convergent and divergent thinking:*** Convergent thinking involves applying logical reasoning to arrive at a single valid conclusion, while divergent thinking involves exploring multiple possibilities. Both types of thinking are essential for effective problem-solving.
6. ***Compare and contrast:*** Comparing different perspectives or solutions allows you to assess the advantages and disadvantages of each option. These techniques are useful when there are several alternatives and the most appropriate has to be selected.
7. ***Looking for opposing arguments:*** Examining counterarguments helps to test the validity of your own beliefs. This not only improves your own arguments but also allows a deeper understanding of the topic under discussion.
8. ***Checking assumptions:*** It is essential to identify and test the assumptions implicit in an argument or decision. Testing them helps reduce the risk of error and ensures that the argument is based on facts and not unfounded assumptions.
9. ***Clarifying terms:*** Clearly defining ambiguous or vaguely defined terms in an argument is crucial to prevent confusion and to ensure a correct understanding of the topic under discussion.
10. ***Breaking down complex issues:*** Breaking down a large problem into smaller and more manageable parts helps for more effective understanding and resolution. This technique is essential for complex problems that require a structured.

Although the importance of developing this skill is widely recognized, current educational approaches often prioritize subject content over developing CT. This gap between educational goals and their actual implementation underlines the need to explore innovative methods and incorporate new technologies in teaching and learning CT [20,21].

Today, there are numerous CT techniques that can be applied across different contexts to support effective decision-making, regardless of the field.

16.3.1 Strategies for critical thinking development

One of the most effective ways to develop CT is by integrating metacognitive strategies and problem-based learning (PBL). These techniques encourage individuals to develop a reflective and analytical approach, helping them to avoid the uncritical acceptance of information. They challenge individuals to think more deeply and evaluate sources of information from multiple angles.

1. **Problem-based learning** is an active learning method that focuses on authentic problem-solving. In this process, students are actively involved in identifying and solving complex dilemmas, stimulating their CT and problem-solving abilities. Through PBL, students are encouraged to ask questions, develop hypotheses, analyze information, and propose reasoned solutions. A successful example of the application of PBL is the ARDESOS-DIAPROVE program, which uses this method to encourage students to develop CT skills within a structured educational framework [22,23].
2. **Debate and role-playing** is another effective technique for developing CT. These methods stimulate analytical thinking by actively involving students in the process of presenting, arguing, and analyzing opposing points of view. Participants are encouraged to adopt different perspectives, support their views with evidence, and argue rationally. In this way, they learn not only to express their opinions but also enhance their ability to communicate effectively and analyze arguments in a logical and coherent way.
3. **Integrating technology into learning** can significantly amplify the effectiveness of CT techniques. Digital platforms and online resources provide access to vast amounts of information, making it easier to analyze and discuss arguments. For example, data analytics has become a growing practice in education, particularly in the context of Big Data and AI. This allows students to critically examine information and make decisions based on solid evidence.
4. **Using AI to develop critical thinking**. Using AI in education can improve students' language skills and stimulate CT through intelligent interactions. For example, AI-based learning platforms can adapt content according to the needs and level of each student, facilitating personalized and deep learning. AI can also help increase self-confidence, open-mindedness, and cognitive maturity, helping students become more aware of their own thoughts and more deeply analyze their thought process [24,25].
5. **Collaborative learning** is another important aspect in developing CT. Working in small groups, such as through PBL or debate, allows individuals to share ideas, receive feedback, and refine their thinking in constructive dialog. This process not only helps to strengthen CT skills but also develops communication and negotiation skills, which are essential in professional settings [26].

However, there are challenges in implementing CT techniques in education. Although many educational institutions are beginning to adopt these methods, their systematic integration into curriculum is not yet universal. Some institutions focus more on the transmission of knowledge over the development of CT skills, and this may limit the potential of students to become independent thinkers capable of analyzing information from multiple perspectives.

From another perspective, we also identify the importance of an integrated approach to CT in higher education that recognizes the influence of affectivity on cognitive processes. This more nuanced perspective suggests that CT is not limited to logical reasoning alone; it also includes an emotional dimension, which can profoundly influence how students interact with complex and sensitive topics.

Encouraging open and reflective dialog that combines reasoning with empathy can facilitate a deeper understanding of social and political issues. Thus, integrating affectivity into CT has the potential to transform higher education by stimulating both intellectual and emotional growth. This holistic approach is essential for cultivating constructive dialog and managing diversity of perspectives in educational settings [27,28].

In conclusion, CT techniques are a fundamental element of modern education, having a significant impact not only on academic performance but also on the ability of individuals to navigate the challenges of professional and personal life. The effective implementation of these techniques into educational curricula, in particular through interactive methods and educational technologies, can contribute to the development of a society better prepare to address the rapid changes and complexities of the contemporary world.

16.4 Applications of critical thinking

16.4.1 Critical thinking in education

CT is an essential element of education, contributing significantly to the development of pupils' and students' cognitive skills. It helps them to understand subjects in a deeper way and to approach information not only from the perspective of memorizing but also of analyzing and evaluating it.

In an educational context, CT involves the ability to identify, analyze, and evaluate arguments and ideas, as well as to make well-informed decisions based on the information available.

Educational activities that develop CT include debates, case studies, analysis of complex texts, and interdisciplinary projects, all of which stimulate reflection, creativity, and innovative problem-solving. Project-based learning, in particular, contributes not only to the acquisition of knowledge but also to the development of essential skills for academic and professional success [29,30].

The benefits of CT in education are manifold, having a direct impact on academic success. It equips students to become not only receptive to information but also able to interpret, question, and apply it in different contexts. This preparation empowers them to navigate future challenges in both their careers and personal life.

Examples of activities to stimulate critical thinking in education:

1. ***Risk analysis and impact assessment:*** An example of an activity to develop critical thinking could be to organize a workshop in which the team identifies and analyzes the possible risks of a large project such as launching a new product on the market. Each team member will have to assess the risks and propose solutions to address them. Critical thinking is essential in accurately assessing possible risks and developing effective strategies to mitigate them.
2. ***Case study on change management:*** Another example could be the analysis of a case study in which a large company is going through significant changes such as digitizing processes or restructuring teams. They would identify how the changes affect different parts of the project, anticipate possible resistance to change, and develop plans to support the implementation of change efficiently and with minimal risk.
3. ***Creating a stakeholder interest management plan:*** In a construction project, the project team may be asked to analyze and create a detailed stakeholder interest management plan. Critical thinking is necessary to assess their needs and expectations and to find compromise solutions between parties that may have divergent interests. The team will need to assess the impact of decisions on the different groups and create a balanced plan to manage relationships and interests. effectively.
4. ***Crisis simulation and team decision-making:*** As part of a project management course, the team may be challenged to participate in a simulation where a crisis occurs (e.g., a major delivery delay or product quality problem). Each team member adopts a critical approach to identify solutions and assess the impact of their decisions on the success of the project. In this activity, critical thinking helps team to analyze different options and make informed decisions, even under time pressure.
5. ***Evaluating the performance and learning process in a project:*** After completing a project, the project team evaluates the process and its performance, using feedback and self-assessment techniques. Each member analyzes successes, identifies challenges, and suggests improvements for future projects. Critical thinking is essential to make an objective analysis of the whole process and to learn from mistakes, thus having a direct impact on long-term success.
6. ***Resource planning in a complex project:*** Another example of an activity that promotes critical thinking is simulating resource planning for a large-scale project involving multiple teams and departments. Participants analyze resource requirements (e.g., time, budget, staff) and establish priorities, taking into account resource constraints and the long-term impact of their decisions. In this process, critical thinking helps participants to make informed decisions and optimize the use of available resources.

16.4.2 Critical thinking in professional careers

CT plays a fundamental role in professional careers, facilitating effective decision-making, tackling complex problems, and generating innovative solutions. In a dynamic and constantly evolving professional environment, CT becomes essential for both individual success and the collective performance of teams. CT enables professionals to analyze situations in a logical and objective way, thus preventing impulsive or biased decisions.

Examples of activities to stimulate critical thinking in professional careers:

1. ***Financial risk assessment:*** A financial analyst evaluates investment options by analyzing financial data and market trends to make well-informed decisions about capital investments.
2. ***Lead multinational teams:*** A manager working with teams from different cultures backgrounds analyzes communication styles and team preferences to create effective collaboration strategies.
3. ***Evaluating a business partnership:*** An entrepreneur assesses potential business partners by analyzing their backgrounds, values, and goals to decide whether a long-term partnership is beneficial.
4. ***Managing work-life balance:*** A professional effectively plans their time, identifying key activities that contribute to career success and maintaining a healthy lifestyle.
5. ***Work process optimization:*** A production manager analyzes workflows and manufacturing processes to identify inefficiencies and propose solutions to increase productivity.
6. ***Selecting the most suitable candidates for a job:*** A recruiter reviews resumes and interviews candidates, asking questions that highlight their critical skills and ability to solve complex problems.
7. ***Managing an organizational change project:*** A consultant helps a company adopt new technology or implement restructuring, analyzing the impact on employees and proposing solutions for smooth transitions.
8. ***Monitoring teams' performance:***A manager monitors the progress of teams and analyzes the results to identify any bottlenecks, and implement adjustments to achieve the set objectives.
9. ***Negotiation strategy in a competitive environment:***A sales manager negotiates prices and payment terms with large customers, analyzing their position and identifying win-win solutions that benefit both parties.
10. ***Critical thinking in communicating with customers***
    A customer support agent resolves complaints by carefully analyzing customer concerns and offering appropriate solutions, while maintaining positive long-term relationships.

Through this approach, individuals can evaluate multiple alternatives and select the optimal option, based on well-reasoned judgment. Additionally, CT helps professionals to identify risks and opportunities, enabling them to develop coherent and well-argued strategies [31].

16.4.3 Critical thinking in daily life

Today, both online and offline life seem to be linked to the democratization of information. While this brings clear benefits, it raises the question: does fast and free access to vast amounts of information necessarily result in individuals being better informed, more rational, and better equipped to make decisions or solve problems? In an era dominated by fraudulent actors, outdated media, fake news, and chaotic streams of information, the ability to think critically is more important than ever.

Examples of activities to stimulate critical thinking in daily life:

1. ***Evaluating a news story:*** Read a news story and check the sources, analyze if the information is supported by facts or if there are possible errors or biases.
2. ***Logic games:*** Engage in logic games as Sudoku or chess to develop analytical thinking and strategic planning.
3. ***Argued discussion:*** Participate in a discussion on topical issues, arguing a logical and reasoned point of view.
4. ***Planning a budget:*** Create a monthly financial plan, identifying necessary expenses and setting priorities.
5. ***Personal Reflection:*** After making an important decision, analyze what factors influenced you and how you could improve your thought process in the future.
6. ***Product comparison:*** Analyze two similar products or services and compare prices, quality, and user reviews to make an informed choice.
7. ***Socratic Questions:*** Ask clarifying questions during a conversation, such as "Why do you think that?" or "What evidence support this opinion?" to encourage deeper thinking.
8. ***Conflict resolution:*** When conflict arises, identify the points of view of both sides and seek solutions that satisfy everyone's needs.
9. ***Reading a book critically:*** Read a book on a topic you are unfamiliar with and reflect on new concepts, considering how they apply to your life.
10. ***Food Decisions:*** Choose a balanced diet, analyzing the nutritional value of food and understanding their long-term health benefits.

CT is vital not only in education but also in professional and personal spheres, as well as in fulfilling social roles. However, CT is neither innate nor effortless; it must be deliberately cultivated and systematically developed [32].

In essence, CT is more than an intellectual skill; it is a mindset and a way of life that positively influences decisions, relationships, and success in everyday life. It is a vital tool for navigating the complexities of the modern world, helping individuals to actively engage in their professional and personal lives, make informed decisions, and contribute to building a more knowledgeable and responsible society [33].

So we identify the need to promote CT in everyday life by implementing simple and effective strategies. For instance, debates can be organized on current issues, where individuals can defend their points of view and argue rationally. In the workplace, brainstorming meetings can be used to analyze different solutions to problems, stimulating CT within team. In personal life, adopting habits such as checking sources of information before making important decisions, whether for online shopping or job selection, can enhance CT. Additionally, reflecting on past decisions and evaluating their results can help develop a more analytical and conscious thought process.

16.5 The role of educational technologies in supporting critical thinking

Educational technologies play an essential role in developing CT by integrating resources and tools that support analysis, evaluation, and argumentation. In today´s rapidly expanding digital age, where information is readily accessible to everyone, it is essential for students learn not only to acquire information but also to analyze and interpret it critically. CT is a key skill in education, and it can be cultivated through educational technologies.

Digital technologies are essential in supporting the development of CT, as they allow quick access to information, facilitate interaction between students and teachers, and encourage analytical thinking through interactive applications and simulations. Additionally, the use of educational technologies aligns with the goals of the 2030 Agenda for Sustainable Development, which promotes quality, inclusive, and equitable education for all [34].

16.5.1 Online platforms for interactive and collaborative learning

Online educational platforms are fundamental tools that support active, collaborative, and interactive learning, making them highly effective for developing CT. Popular platforms such as Moodle, Google Classroom, and edX allow students to participate in interactive activities, collaborate in groups, and express their perspectives. In this way, educational technologies foster a learning environment based on CT and the exchange of ideas:

1. **Online discussions and forums** – these are excellent platforms for debating complex topics and presenting rational arguments. Active participation in online discussions stimulates reflective thinking and helps students develop skills in argumentation and critically evaluating information.
2. **Collaborative projects and activities** – online platforms allow the creation of team projects and activities, where students apply theoretical knowledge to real-world contexts. Through collaboration, they learn to analyze and synthesize information from various sources, thus developing CT and problem-solving skills.
3. **Reflective questions and self-assessment** – online platforms often include open-ended questions that challenge students to critically analyze their own learning and reconsider their approaches. This is an effective way to encourage self-reflection, which is essential to developing CT skills.

16.5.2 Simulation and educational games

Simulations and educational games are valuable tools for stimulating CT, providing a safe and controlled environment where students can experiment different scenarios and make informed decisions. These methods are widely used across various fields, including engineering, business management, medicine, and financial education:

1. **Business simulations** – educational games such as "SimCity" or financial education apps challenge users to make long-term strategic decisions and analyze their impact on the business environment. These activities stimulate CT by evaluating various options and anticipating their potential effects.
2. **Medical simulations** – in the medical field, surgical or diagnostic simulations allow students to choose treatment options and analyze their outcomes. These simulations help develop CT and complex reasoning skills that are essential in medical decision-making.

16.5.3 Educational applications for logical thinking and argumentation

Educational apps for mobile devices are another important tool for developing CT, offering interactive activities that challenge users to solve problems, argue and develop logical reasoning:

1. **Logic and problem-solving apps** – apps like Lumosity or Elevate are designed to improve cognitive skills, including logical reasoning and problem-solving. These apps help users analyze information, make logical inferences, and approach complex problems in a structured way.
2. **Argumentation apps** – apps like "Argument Wars" challenge users to construct strong arguments and support logical points of view, thus developing essential CT skills such as formulating coherent reasoning and justifying their conclusions.

16.5.4 Automated evaluation and feedback

Another key aspect of educational technologies is continuous assessment and automated feedback. These tools allow students to receive immediate feedback on the quality of their arguments and solutions, contributing to self-reflection and continuous improvement of CT skills.

Educational platforms can provide instant feedback on student responses, allowing students to quickly review their reasoning and adjust their approaches, thereby stimulating active and continuous learning [35].

Educational technologies are essential tools for developing CT, offering multiple opportunities for active, collaborative, and interactive learning. Online platforms, educational simulations, and apps for logical thinking all contribute to the development of analytical and argumentative skills, which are essential for academic and professional success. While the integration of educational technologies presents challenges, such as unequal access to technology and the risks of dependency, their benefits are clear. When used appropriately, these tools can make education more efficient and accessible.

References

[1] Barta, A., Fodor, L.A., Tamas, B., and Szamoskozi, I. The development of students critical thinking abilities and dispositions through the concept mapping learning method – A meta-analysis. *Educational Research Review*, 2022, 37, 100481.

[2] Chen, S.L., Liang, T., Lee, M.L., and Liao, I.C. Effects of concept map teaching on students' critical thinking and approach to learning and studying. *Journal of Nursing Education*, 2011, 50(8), 466–469.

[3] Obidovna, D.Z. The art of questioning: Enhancing critical thinking through effective pedagogical techniques. *International Journal of Literature and Languages*, 2023, 3(11), 54–60.

[4] Sellars, M., Fakirmohammad, R., Bui, L., *et al.* Conversations on critical thinking: Can critical thinking find its way forward as the skill set and mindset of the century? *Education Sciences*, 2018, 8, 205.

[5] Paul, R., and Elder, L. *The miniature guide to critical thinking concepts and tools*, Foundation for Critical Thinking, 2014. Available online: https://www.criticalthinking.org/files/Concepts_Tools.pdf [Accessed 11 Dec 2024].

[6] Dewey, J. *How we think*, D C Heath, 1910.

[7] Paul, R., and Elder, L. *Critical thinking: Tools for taking charge of your learning and your life* (3rd ed.), Pearson, 2014.

[8] Facione, P.A. Critical thinking: A statement of expert consensus for purposes of educational assessment and instruction, 2011. Available online: https://eric.ed.gov/?id=ED315423 [Accessed 11 Dec 2024].

[9] Halpern, D.F. *Thought and knowledge: An introduction to critical thinking* (5th ed.), Psychology Press, 2014.

[10] Ennis, R.H. *Critical thinking: A streamlined conception*, Pearson, 2011.

[11] Holguin, B. Thinking, guessing, and believing. *Philosophers' Imprint*, 2022, 22(1), 1–34.

[12] Ho, Y.R., Chen, B.Y., and Li, C.M. Thinking more wisely: Using the Socratic method to develop critical thinking skills amongst healthcare students. *BMC Medical Education*, 2023, 23(1), 173.

[13] Copi, I.M. *Introduction to logic* (13th ed.), Pearson, 2011.

[14] Walton, D. *Informal logic: A pragmatic approach*. Cambridge University Press, 2008.

[15] Tato, A., Nkambou, R., Brisson, J., and Robert, S. Predicting learner's deductive reasoning skills using a Bayesian network. In *Artificial Intelligence in Education: 18th International Conference, AIED 2017*, Wuhan, China, June 28–July 1, 2017, 381–392. Springer International Publishing.

[16] Toulmin, S. *The Uses of Argument*. Cambridge: Cambridge University Press, 1958, p. 11.

[17] Johnson, R.H., and Blair, J.A. *Logical self-defense*. McGraw-Hill, 2006.

[18] Zambak, V.S., and Magiera, M.T. Supporting grades 1–8 pre-service teachers' argumentation skills: Constructing mathematical arguments in situations that facilitate analyzing cases. *International Journal of Mathematical Education in Science and Technology*, 2020, 51(8), 1196–1223.

[19] Royce, C.S., Hayes, M.M., and Schwartzstein, R.M. Teaching critical thinking: A case for instruction in cognitive biases to reduce diagnostic errors and improve patient safety. *Academic Medicine*, 2019, 94(2), 187–194.

[20] Alsaleh, N.J. Teaching critical thinking skills: literature review. *Turkish Online Journal of Educational Technology – TOJET*, 2020, 19(1), 21–39.

[21] Tiruneh, D.T., Verburgh, A., and Elen, J. Effectiveness of critical thinking instruction in higher education: A systematic review of intervention studies. *Higher Education Studies*, 2014, 4(1), 1–17.

[22] Liu, O.L., Frankel, L., and Roohr, K.C. Assessing critical thinking in higher education: Current state and directions for next-generation assessment. *ETS Research Report Series*, 2014, 2014(1), 1–23.

[23] Rivas, S.F., Saiz, C., and Ossa, C. Metacognitive strategies and development of critical thinking in higher education. *Frontiers in Psychology*, 2022, 13, 913219.

[24] Muthmainnah, Ibna Seraj, P.M., and Oteir, I. Playing with AI to investigate human-computer interaction technology and improving critical thinking skills to pursue 21st century age. *Education Research International*, 2022, 2022(1), 6468995.

[25] Solovyeva, N., Tapalova, O., and Smirnov, S. Specifics of the students' critical thinking formation within active learning space. *Frontiers in Education*, 2023, 8, 1132525.

[26] Xu, E., Wang, W., and Wang, Q. The effectiveness of collaborative problem solving in promoting students' critical thinking: A meta-analysis based on empirical literature. *Humanities and Social Sciences Communications*, 2023, 10(1), 1–11.

[27] Zembylas, M. Revisiting the notion of critical thinking in higher education: Theorizing the thinking-feeling entanglement using affect theory. *Teaching in Higher Education*, 2024, 29(6), 1606–1620.
[28] Lincoln, D., and Kearney, M.L. Promoting critical thinking in higher education. *Studies in Higher Education*, 2019, 44(5), 799–800.
[29] Zulyusri, Z., Elfira, I., Lufri, L., and Santosa, T.A. Literature study: Utilization of the PjBL model in science education to improve creativity and critical thinking skills. *Jurnal Penelitian Pendidikan IPA*, 2023, 9(1), 133–143.
[30] Aycicek, B. Integration of critical thinking into curriculum: Perspectives of prospective teachers. *Thinking Skills and Creativity*, 2021, 41, 100895.
[31] Hyslop-Margison, E.J., and Armstrong, J.L. Critical thinking in career education: The democratic importance of foundational rationality. *Journal of Career and Technical Education*, 2004, 21(1), 39–49.
[32] Franco, A., Marques Vieira, R., and Tenreiro-Vieira, C. Educating for critical thinking in university: The criticality of critical thinking in education and everyday life. *ESSACHESS – Journal for Communication Studies*, 2018, 11(2), 22.
[33] Kuhn, D. A role for reasoning in a dialogic approach to critical thinking. *Topoi*, 2018, 37, 121–128.
[34] Haleem, A., Javaid, M., Qadri, M.A., and Suman, R. Understanding the role of digital technologies in education: A review, *Sustainable Operations and Computers*, 2022, 3, 275–285.
[35] Hattie, J., and Timperley, H. The power of feedback. *Review of Educational Research*, 2007, 77(1), 81–112.

Chapter 17
Reflective thinking

Darius Andriukaitis[1], Neringa Dubauskienė[1], Jurgita Barynienė[2], Asta Daunorienė[3], Jurga Vitkuvienė[4], Algimantas Valinevičius[1], Mindaugas Žilys[1] and Dangirutis Navikas[1]

Reflective thinking is recognized as a powerful tool for the successful learning process, allowing students to deepen what they have already learned, relate existing knowledge to newly acquired information, and identify gaps in their knowledge and abilities. Reflective practice plays an essential role in enabling students to develop their professional and individual capabilities. Reflection allows students to evaluate their experience of the design and/or learning process more deeply, answering the questions of what was done well and what could be improved next time. This way, students identify areas for improvement and rethink their actions. Reflection can help them to avoid potential mistakes in the future when making innovative decisions that should be sustainable in the long term. Incorporating reflection enables students to focus not only on the end result but also on the value created throughout the entire learning journey. This is especially significant since the disciplines of engineering require deep knowledge of fundamental subjects, which students often perceive as purely theoretical and detached from their practical skills which they usually acquire in the later years of their studies. Even subjects directly related to their specialties contain broad theoretical components that support the narrower practical abilities that students gain. The perception of certain parts of their studies as theoretical or detached from their personal goals hinders students' involvement in deep learning by diminishing their intrinsic motivation. The usual method for presenting the importance of the knowledge to be learned is through lectures, which are delivered from the tutor's perspective without necessarily changing the student's viewpoint.

[1]Faculty of Electrical and Electronics Engineering, Kaunas University of Technology, Lithuania
[2]Faculty of Social Sciences, Arts and Humanities, Kaunas University of Technology, Lithuania
[3]School of Economics and Business, Kaunas University of Technology, Lithuania
[4]Faculty of Civil Engineering and Architecture, Kaunas University of Technology, Lithuania

Properly conducted reflection practices help students construct their own personal knowledge systems, engage in deeper learning, and increase their intrinsic motivation rather than merely trying to meet the minimal requirements for passing the course. Reflection not only helps students in the context of their studies but also enables them to view their work as professionals with impact to the real-world problem-solving, and to evaluate the consequences of their decisions across different dimensions.

However, it must be acknowledged that the practice of reflection is not commonly observed among engineering students, as they often lack a clear understanding of their own thinking processes. This is where design-based thinking methodology offers an excellent opportunity to encourage student reflection. Nonetheless, the choice of a reflection tool becomes a pertinent question. It is crucial to select reflection tools that facilitate a multi-perspective approach to engineering studies, fostering critical thinking and problem-solving skills. This chapter aims to analyze different reflection tools and their value for engineering students. The authors present descriptions of reflection tools, assess their effectiveness, and provide student feedback, aiming to offer the most effective reflection tools for engineering students utilizing a design thinking approach in their study process.

17.1 Definitions of reflection

There are several definitions of reflection in the context of higher education, and its practices also vary. Generally, reflection is understood as a thinking process that aids understanding, requires active engagement from the participant, and can be applied both to the current situation and in a broader social context [1]. Reflection plays an important role in bringing about quality change in the learning process. The power of reflective thinking in achieving learning outcomes has been demonstrated in the works of many scholars [2]. Research emphasizes that the learning process and reflection are closely interlinked, as students today are often confronted with unfamiliar, uncontrollable situations during the learning experiences [3]. Moreover, it can be argued that reflection is a necessary prerequisite for deep learning which is particularly relevant in the context of engineering studies.

Although engineering specialties and their studies are usually associated with deep expertise, hard skills, and methodologies, the elements and experiences of reflective learning could significantly enrich the social competencies, critical thinking, and creative thinking of modern engineers' skills. These skills are sorely needed for the rapidly changing labor market and the complex challenges of today's world [4]. Typically, engineering education programs are dominated by the view that the most important aspect of training students is to educate them as specialized experts with sufficient technical knowledge to solve the questions and problems of their field. Much less is said about the complexity, ambiguity, and uncertainty of engineering tasks that engineers face in real-life situations [5]. "Engineering students have to learn that people, policies, environmental aspects, politics, economics or cultural values often override disciplinary expertise" [6]. In this sense, reflection, as a set of activities aimed at reviewing, reflecting on, summarizing and evaluating the results and process of the actions carried out, and formulating conclusions and insights for the future from this information can serve different purposes:

- Get to better know your opportunities and potential;
- Gain additional motivation for activities and decisions;
- Evaluate the efficiency and success of the task performance process;
- Manage and summarize the new competencies and skills acquired during the process;
- Pursue new, more efficient results for the future;
- Recognize failures, analyze their causes and learn from them;
- Use reflection as a means of understanding the professional context and identity, and of forming a system of value principles [7,8].

In the study process of engineering fields, reflective learning is still applied rather reservedly and often receives mixed reactions and discussions from teachers and students, it has not yet received much empirical attention [8,9]. Resistance to learning "soft skills" has many causes, the most common being a lack of time, motivation, and perceived benefits. Pre-developed and structured reflection tools and forms that are more acceptable to engineering students could help in this situation. Structured reflection is considered to be a regular and systematic purposeful activity that contributes to the achievement of reflective learning goals [10]. In addition, reflection opens up avenues for understanding the effectiveness of the learning and teaching process and enables preventive solutions to problematic situations, as well as timely feedback and support. For example, a common element of engineering curricula is practical work in the laboratory, where students are expected to deepen their understanding of the subject by engaging in hands-on activities, which are commonly structured as following a set of instructions provided by the lecturer. These activities intend to promote the construction of student knowledge, but without adequate opportunities for self-learning, they become mindless and disconnected from the goals of the course. Integrating reflection is one of the ways to remedy this situation [11].

Reflection activities could be graded or ungraded, depending on the time allocated to the activity, the goals of the course, and other factors. While ungraded activities might be perceived by students as more authentic, grades provide them an additional incentive to complete the reflection activity. Figure 17.1 illustrates the structure and components of reflective thinking, highlighting the interplay between knowledge, skills, and attitudes over time [12].

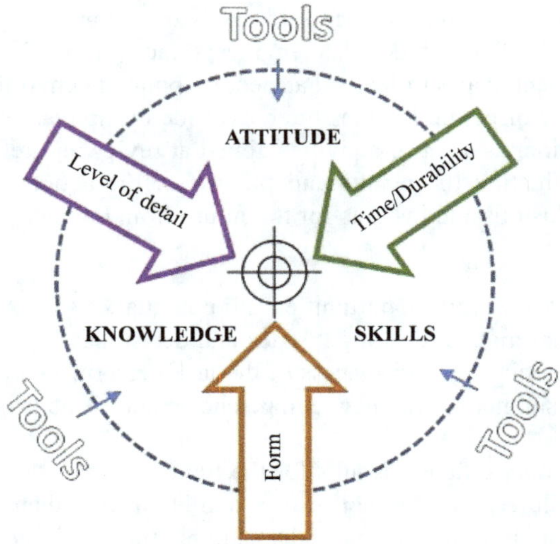

Figure 17.1 Structure of reflective thinking (based on [13])

17.2 Fundamentals of design thinking approach

Pedagogical approaches should focus on the ability to adapt learning to real-life scenarios and challenges, creating the conditions for developing new ways of thinking and mindset [14]. With the recent rethinking of teaching and learning strategies in engineering studies to focus as much as possible on student engagement and active involvement, the design thinking approach is emerging as one that meets this objective [15–17]. Design thinking has entered engineering studies as a result of the search for a new pedagogical approach that allows the development of original and innovative solutions that meet the needs of users. The design thinking approach enables students to deepen their knowledge of their discipline and, at the same time, acquire skills in solving multi-dimensional challenges and integrating cross-disciplinary experiences, developing effective problem-solving and empathy skills, becoming effective collaborators, and gaining intrinsic motivation for learning – all based on hands-on work, activity, and sharing of experiences.

The design thinking approach typically consists of six phases: understanding, observation, synthesis, ideation, prototyping, and testing [18]. The understanding phase focuses on considering the challenge within the team. The main objective of this phase is to operationalize the concepts and comprehend the challenge. The observation phase involves exploring the challenge. Its primary objective is to analyze the challenge from different perspectives, get to know the user, identify their needs, and gain empathy. Once a significant amount of data has been gathered, the synthesis phase begins, where students need to organize the data, highlight what is important, and draw inspiration for solving the challenge. The ideation phase is about generating original ideas to address the challenge. The prototyping phase involves turning the solution into something tangible that the user can see and use. Finally, the testing phase consists of evaluating the developed solution in a real environment and gathering feedback to improve the solution to meet the user's needs.

It should be emphasized that reflection plays a crucial role in design thinking. Since this is experiential learning, students need to be given the opportunity to reflect on their experiences both individually and in teams.

Why design thinking in engineering studies?

The application of the design thinking approach in the study process helps not only to deepen student´s knowledge of the discipline but also to develop their creativity, problem-solving, analytical thinking, teamwork, process management, entrepreneurship, openness to change, and innovation. This method allows theoretical knowledge to align more closely with real-world challenges and practical results, creating clear connections between existing and newly acquired knowledge. One of the key components of this approach is empathy (the ability to understand and relate to others, a clear grasp of the context, and the recognition of user's needs), which is interwoven with out-of-the-box thinking.

Before applying this method in the study process, the teacher must clearly answer one key question: Does this method help achieve the course's learning outcomes and if so, how?

17.3 Reflection practices in the teaching and learning process

The process of reflective learning and the tools used can be described and identified both by their content and by the form of information collection and presentation. Reflective learning tools are instrumental in helping students critically evaluate their experiences, identifying areas for improvement, and developing a deeper understanding of the subject matter. In the context of design thinking, these tools are essential for ensuring continuous learning, improvement, and alignment with user expectations.

Reflective learning tools can focus on various aspects, including:

- **Content analysis**: Reflecting on the material learned, its relevance, and how it can be applied in different contexts.
- **Process reflection**: Evaluating the steps taken during a project or activity, identifying what worked well and what did not.
- **Personal reflection**: Considering one's own experiences, feelings, and growth throughout the learning or project process.
- **Collaborative reflection**: Engaging with peers to discuss insights, challenges, and solutions, enhancing collective understanding.

The design of teaching and learning activities needs to explicitly consider the role of reflective learning. Reflective learning should not be employed merely for the sake of using it; reflections must have a clear purpose and create significant value for students. Typically, educators need to think about first, providing a unique experience for students to reflect on later in the process and second, methods, identifying the strengths and areas for improvement of that experience, and considering how it can be applied in subsequent activities.

Given the specificity of engineering studies, several models can be identified before implementing reflective learning in theoretical lectures, hands-on classes, and project-based activities.

The example in Figure 17.2 shows how reflection can be integrated into theoretical lectures. It should be noted that theoretical lectures in engineering sciences are characterized by a substantial amount of conceptual knowledge, making reflection an important tool for students to connect their previous experiences with new knowledge and skills. Figure 17.2 presents an example of a reflection following a theoretical input. During this reflection, students are invited to consider what new knowledge they have acquired and how they can relate it to their existing knowledge and other experiences.

The second example (see Figure 17.3) is similar to the first, but it adds an element of reflection at the beginning of the lecture regarding what the students have learned in previous meetings and how this relates to the material from earlier lectures.

The third example (Figure 17.4) illustrates the use of reflection in laboratory work. It should be noted that laboratory work poses challenges for engineering students, and educators face the issue of ensuring that students arrive at the hands-on sessions prepared. This is where reflection is particularly important, as it

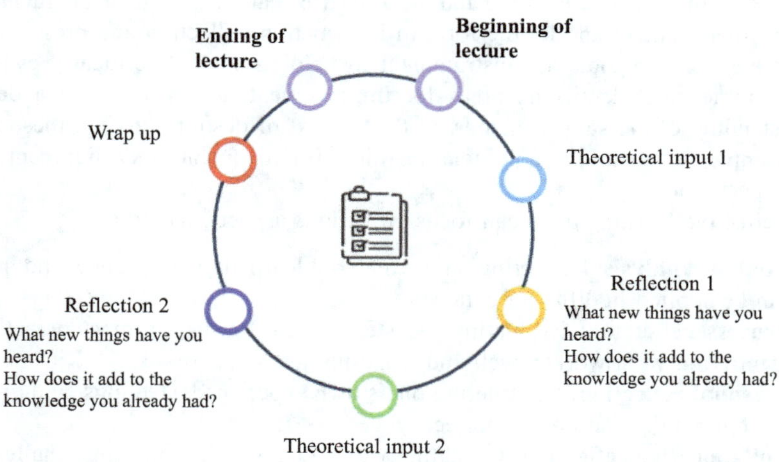

Figure 17.2 Example 1 of reflective learning in a theoretical lecture

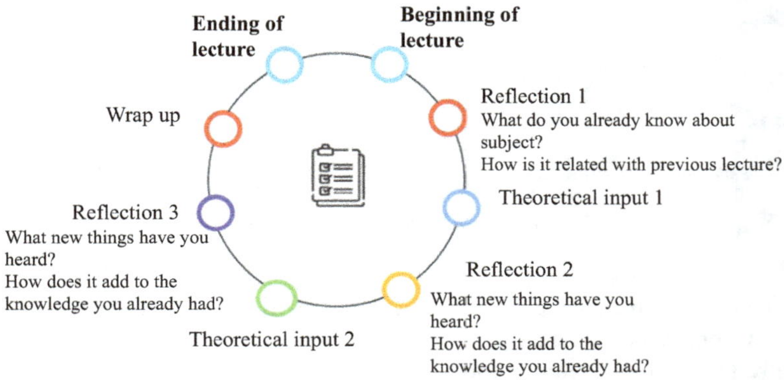

Figure 17.3 Example 2 of reflective learning in a theoretical lecture

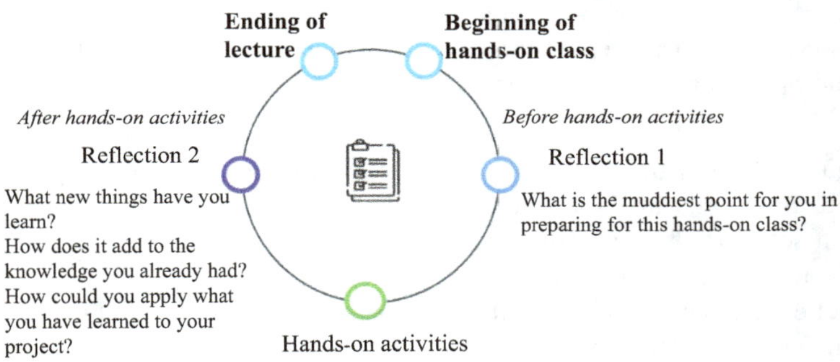

Figure 17.4 Example of reflective learning in hands-on classes

encourages students to reflect on their experience, ask themselves questions about the process, and consider areas for improvement.

The example in Figure 17.4 proposes two reflections with very clear, but distinct objectives:

- **Before hands-on activities**: The purpose of this reflection is to encourage students to think about their preparation for this meeting, including what they did not understand, or what was difficult for them.
- **After hands-on activities**: The purpose of this reflection is to evaluate the experience that has just taken place by asking questions such as: what new things they have learned, what they understood, how this relates to the knowledge they already possess, and how they can use the knowledge they have gained to develop their projects.

Reflection tools provide structured methods to analyze work, gather insights, and make informed decisions both for individual students and teams. These tools

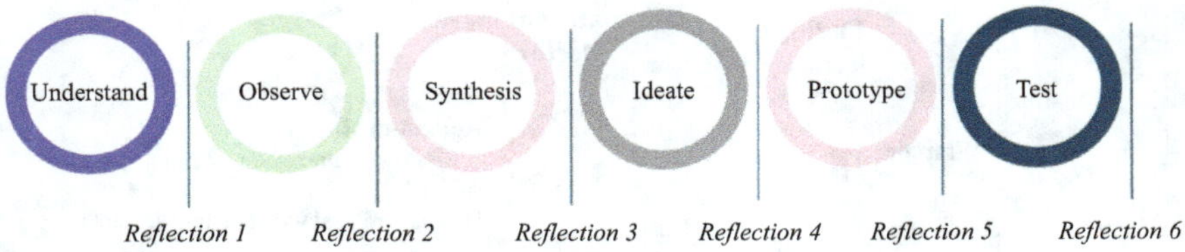

Figure 17.5 Reflections during design thinking process

help bridge the gap between theory and practice, enabling teams to refine their ideas and approaches based on empirical evidence and user feedback. By integrating reflection tools, teams can:

- **Enhance learning**: Facilitates deeper understanding and retention of knowledge by encouraging critical thinking about experiences and outcomes.
- **Improve solutions**: Allows teams to identify and address issues early, leading to more robust and user-centered solutions.
- **Foster collaboration**: Promotes open communication and shared understanding among team members, enhancing collaboration and collective problem-solving.

Figure 17.5 illustrates the use of reflection in applying the design thinking approach. Since this method involves experiential learning, it is logical to hold reflection sessions after each phase or to integrate reflection tools into each phase of the design thinking process.

If reflection is used as a tool after each phase of the design thinking process, it could encourage students to discuss a set of questions:

- What activities did you carry out during this phase?
- What challenges did you face, and how did you overcome them?
- What new insights have you gained during this phase?
- What is the most important outcome of this phase?
- Are you satisfied with the results you achieved?
- What have you learned so far from your work as a team?

If reflection is used as a tool during each phase of the design thinking process, teachers could utilize some of the reflection tools employed at different stages to support various objectives.

17.3.1 Understand phase

Mind maps: A visual tool that helps teams to organize complex information around a central theme and explore relationships between different aspects of a problem. Write the main problem in the center, create branches for major categories, add detailed information as sub-branches, use short phrases and visuals [19].

Concept maps: A visual tool that represents relationships between different concepts, helping teams understand and organize knowledge about a problem.

Write the main problem or topic at the top, list related key concepts below the central concept, draw lines to connect related concepts, labeling the lines to explain the relationships, and arrange concepts in a hierarchical structure from general to specific [20].

17.3.2 Observe phase

Empathy maps: A visual tool that helps teams understand the user's needs and experiences. Divide a sheet into four quadrants labeled "Says," "Thinks," "Does," and "Feels." Fill in each section based on user research and observations [21].

Journey mapping: A visual representation of the user's journey, highlighting pain points and opportunities. Plot the user's experience from start to finish, noting their actions, thoughts, and emotions at each step [22].

User persona reflection: Detailed profiles of different types of users. Regularly update and review personas based on new insights, discussing how these insights impact design decisions [23].

17.3.3 Synthesis phase

Problem statement analysis: A critical examination of the problem statement to ensure it accurately reflects the user's needs. Ask questions such as "Does this problem statement address the root cause?" and "Is it specific and actionable?" [24].

SWOT analysis tool: A tool to identify Strengths, Weaknesses, Opportunities, and Threats related to the problem. Fill out each quadrant with insights from user research and team discussions [25].

17.3.4 Ideation phase

Brainstorming reflection: A review and evaluation of brainstorming sessions. After a brainstorming session, categorize ideas, discuss the most promising ones, and reflect on the effectiveness of the ideation process [26].

SCAMPER technique: A creative thinking tool that prompts reflection through questions about Substituting, Combining, Adapting, Modifying, Putting to another use, Eliminating, and Rearranging. Use SCAMPER questions to reflect on existing ideas and generate new ones [27].

Dot voting: A method to prioritize ideas democratically. Team members use dots to vote on their favorite ideas. Reflect on why certain ideas received more votes and discuss their potential impact [28].

17.3.5 Prototype phase

Prototype feedback grid: A structured method to collect and reflect on feedback. Create a grid with sections for positives, negatives, questions, and ideas. Gather feedback from users and team members and reflect on how to iterate the prototype [29].

Storyboarding: A visual tool to outline how users will interact with the prototype. Create a storyboard of the user's interaction with the prototype, then reflect on potential improvements and overlooked details [30].

17.3.6 Test phase

Usability testing reflection: An evaluation of the prototype based on user testing sessions. Conduct usability tests, record observations, and discuss what worked and what did not. Reflect on changes needed before the next iteration [31].

Feedback loop analysis: A continuous process of collecting and reflecting on user feedback. Implement regular feedback loops with users, reflect on the insights gained, and integrate them into the design process [32].

This structured approach to reflection in design thinking not only enhances the effectiveness of the methodology but also fosters a culture of continuous improvement and innovation in engineering studies and practice.

17.4 Research methodology

To evaluate the effectiveness of reflection tools, a study was conducted in the engineering study programs of Kaunas University of Technology, which integrated the design thinking method. In this study, 35 programs were selected that use design thinking in at least one study module. During the selection stage, study modules incorporating reflection tools within the design thinking framework were analyzed. From the selected study programs, the design thinking approach utilizing reflection tools was applied in the following study modules: *Construction Logistics and Management, Sustainable Development, Basics of Fashion Design, Fashion Collections, Basics of Design, Experimental Mechanics, Innovations in Fashion Design, Building Information Modeling and Model Management, Low Energy and Modernized Buildings, Circular Economy and Sustainable Use of Resources in Construction, Dynamics of Nonlinear Systems, Mathematical Materials Science, Product Development Project, Fundamentals of Business Management, Movement of Biomechanical Systems, Technical Creation and Intellectual Property, Sketching Technique, Image Design and Communication, Prototyping Methods, Technologies for Industrial for Design, Bionics and Biomimicry in Design, Technical Creativity and Intellectual Property, Product Design, Creative Design Workshops, Basics of Mobile App Development, User Experience Design, Thermodynamics and Statistical Physics, Physics of Surfaces and Surface Phenomena*. In total, 27 study modules were included from a pool of 109 study modules. Some study modules were taught in different study programs. All study modules were divided into semesters and taught from the first year to the last year of undergraduate studies. Study modules of the 2022/2023 academic year were analyzed during the research.

The research utilized secondary data sources to conduct the study. The analysis involved examining the syllabus outlined in official course descriptions, as well as the study module descriptions within the Moodle system, to identify the reflection tools applied in the design thinking approach throughout the course. Additionally, feedback from students collected post-semester served as another secondary data source for analyzing the effectiveness of the reflection tools utilized during the design thinking process.

17.5 Findings and discussion

The research study revealed variations in the tasks and reflection requirements across different study modules (see Table 17.1). These differences were attributed to the diverse adaptation of design thinking approach to reflection tools within each module.

A variety of reflection tools were utilized across engineering study modules, each customized for a specific phase of the design thinking process. Not all the tools were necessarily labeled exactly as they were selected from the literature analysis. All reflection tools identified in the study modules were analyzed to determine which phase they corresponded to and to identify similarities with the proposed tools. Examples of student feedback are provided in italic font, separated by quote marks and brackets.

17.5.1 Understanding phase

Analysis has shown that there are several reflection tools useful in the understanding phase such as mind and concept maps. It should be noted that in the understanding phase, there is a lot of information and disconnected facts about the object of study and its context. These tools can significantly help students to reflect on what they already know and combine newly found information, knowledge, and context.

Mind maps can be characterized as a user-friendly tool that are especially suitable for handling complex engineering issues by breaking them down into manageable pieces. Students emphasize the importance of this tool in seeking

Table 17.1 Reflection tools used in the study modules

Design thinking phase	Reflection tool	Number of study modules that use the tool
Understand phase	Mind maps	7
	Concept maps	20
Observe phase	Empathy maps	6
	Journey mapping	13
	User persona reflection	9
Synthesis phase	Problem statement analysis	24
	SWOT analysis	4
Ideation phase	Brainstorming reflection,	19
	SCAMPER technique	4
	Dot voting	5
Prototype phase	Prototype feedback grid	23
	Storyboarding	3
Test phase	Usability testing reflection	11
	Feedback loop analysis	10

deep learning ("The mind maps have strong relevance to my learning context" and "The use of mind maps helped me studying the course content"). Furthermore, students see it as important to incorporate additional features like notes, resource links, and graphics ("Mind maps could have been more effective if they had additional features enabling in-depth analysis within challenge"). Mind maps offer a well-defined framework for integrating feedback and refining ideas, thereby enhancing the overall quality of design thinking projects. They are particularly effective for iterative learning. Integrating a pre-existing feature that allows peers or instructors to provide comments directly improves the effectiveness of feedback ("It would be beneficial to have constant feedback from teachers when creating mind maps").

In cases where a more hierarchical structure is needed or where more interrelationships occur, concept maps can be used. Concept maps serve as a reflective tool that enables students to precisely identify and establish connections between different project components ("The use of concept maps helped me to think critically about the interconnections among various team members' ideas" and "The concept map helped me understand the relationships between the different parts of the challenge, which at first glance seemed completely unrelated"). Concept maps offer a graphical framework that aids understanding of interconnections between concepts and the overarching objectives of the course ("Concept maps aided me in precisely defining and linking diverse design thinking elements"). This technique aids in maintaining project alignment with course learning objectives ("It ensured that our design thinking project remained aligned with the course outcomes").

What needs to be taken into account?

These tools effectively help learners to immerse themselves in the broader context of the topic and develop the ability to assess the impact of external factors on the issue at hand. However, it is important to note that the optimal balance between this phase and the time allocated to the task must be maintained to keep students' attention and focus on the core topic. Additionally, it is essential to assess:

- In what way do students prefer to express their thoughts?
- Is there a need for the development of new skills and techniques in the course?

17.5.2 Observation phase

In the observation phase, it is very important to get to know your user well and understand his/her perception of the world. For this reason, several reflective tools can be used such as empathy map, journey map, and user persona picture.

- **Empathy maps** are used in the observation phase. This tool encourages students to empathize with users and their experiences, helping them gain a deep understanding of the user's needs and perspectives. This is also evidenced by the students' statements ("Empathy maps helped me understand the user's needs and perspectives" and "Empathy maps helped me think about what I know about my user, and it also made me reflect on what I still need to know about him in order to meet the challenge"). Furthermore, empathy maps helped students organize user insights, facilitating a clear and manageable process of comprehending user experiences.
- **Journey maps** are another reflection tool which can be efficiently used in the observation phase. They are a valuable technique in engineering projects, as they help identify user interactions and pinpoint areas of difficulty in engineering solutions ("Journey mapping helped me see the user's experience"). The sequential arrangement of journey maps documents the user's journey. It is important to note that journey maps improve social competencies, critical thinking, and creative thinking by fostering collaboration and thorough examination of the user's experiences ("It helped me to collaborate with my team and conduct an in-depth examination of the user's journey"). Students emphasize the importance of having clear instructions on how to use this tool ("I needed more elaborate instructions and illustrations throughout the process of journey mapping").
- **User persona** reflection is another tool for aligning solutions with user needs, aiding in the achievement of learning objectives ("The creation of user personas helped us to create solutions understanding concrete user needs"). This tool can be very useful for comprehending target users in engineering projects, offering important insights into user behavior and preferences ("I found it highly relevant for understanding users in engineering projects"; "Creating a persona helped me to empathize with my user, allowing me to understand not only their needs but also the reasons behind them"; "Persona description helped me invent solutions focusing on individual user needs"). To enhance the clarity and ease of use for this tool, it is beneficial to provide additional templates and examples for developing comprehensive personas ("Teacher could offer more templates and examples for personas analysis").

What is important

This phase requires close, direct contact with the user, which can sometimes present practical challenges in organizing the process. It is also important to assess the following:

- How and in what way will the contact between the learner and the user occur? Will it be direct or indirect?
- How will personal data protection issues be managed?

Students' comments indicate that mixed forms of reflection are highly useful during this phase ("It is very beneficial to incorporate different tools, such as journey map and user personas"; "Using different tools seemed less useful at the beginning, but in a later phase, I realized that it allowed me to dig deeper into my user's needs and understand them better to propose the best engineering solution").

17.5.3 Synthesis phase

In the synthesis phase, students continue to analyze the problem, but it becomes essential for them to organize and interpret the available information. Various tools can be used for reflection during this phase such as problem statement analysis and SWOT analysis:

- During the synthesis phase, analyzing the problem statement serves as a reflective tool, helping students accurately describe the problem. Providing more explicit instructions and illustrative examples for constructing problem statements can be beneficial ("The teacher could assist by providing clearer guidelines and examples for formulating problem statements"). To increase the tool's relevance, it is advisable to focus solely on engineering-specific issues ("It would be ideal to concentrate on engineering-specific issues").
- SWOT analysis is another tool for reflection in this phase. It helps students thoroughly review projects and ensure alignment with design thinking challenge statement. This analysis evaluates both internal and external factors ("SWOT analysis helped me reflect on the solution from different points of view").

17.5.4 Ideation phase

In the ideation phase, students advance to the process of developing a solution, making various reflection tools valuable in this stage such as reflection on brainstorming sessions, SCAMPER, and Dot voting.

- Reflecting on brainstorming sessions serves as a tool to ensure that generated ideas align with the learning objectives ("Brainstorming sessions helped met generate ideas"; "The tool helped gather ideas from each team member"). This tool enables students to explore a broad range of possibilities ("At the beginning, I didn't know where to start. The tool and the guidelines encourage us to talk and share"). It is beneficial to provide methods specifically designed for capturing and organizing brainstorming ideas. Using engineering-specific brainstorming approaches can enhance the relevance of this tool ("It would be interesting to learn and experience engineering-specific brainstorming techniques if available").

How to implement it?

Synthesis is a crucial point in the process that determines the direction of the future solution, and therefore requires special attention and sufficient time. Key considerations include:

- Do students have enough material to summarize?
- How do they justify their choices?
- How does the outcome of the students' summaries align with the overall context of the topic defined in the previous phases of the process?

- SCAMPER reflection tool is used to systematically innovate and improve existing ideas ("SCAMPER was very useful in my engineering project as it provides a clear structure for modifying and improving solutions"). SCAMPER also fosters the development of social skills ("SCAMPER facilitated teamwork and creative thinking by helping us collectively explore different solution possibilities").
- Dot voting is another reflection tool used in ideation phase to evaluate ideas, prioritize them, and make decisions. It is particularly useful for achieving learning objectives related to efficient problem-solving. The dot voting process is highly effective in quickly identifying the most popular ideas ("It helped us easily vote on the ideas").

What is important?

In this phase, it is essential to strike a balance between convergent and divergent thinking by prioritizing the quantity of ideas and subsequently evaluating their quality and relevance. Key considerations include:

How can creativity be boosted?
How do students justify their choices?
How do the selected ideas address the topic?

17.5.5 Prototype phase

In the prototyping phase, students have an initial idea of the solution and further develop it to meet the user's needs. Reflection tools such as prototype feedback grid and storyboard are used during this stage

- The prototype feedback grid serves as a systematic tool for gathering feedback. This tool is essential for engineering projects that require prototyping and user input, making it practically applicable in engineering contexts ("This tool has helped me gain a deeper understanding of what my user is expressing"; "In my opinion, the prototype feedback grid is essential in the prototyping phase because it allowed me to anticipate what I could improve in the next phase"). The prototype feedback grid improves social skills by organizing the process of collecting and analyzing feedback. However, students note that this tool could be more technology-enriched ("We used it in a rather traditional way. Teachers could consider providing digital templates").

- Storyboarding is another reflection tool used in the prototyping phase. Storyboarding enhances social skills and imaginative problem-solving through the visual representation of concepts and solutions ("Visual tools have helped us understand complex problems more clearly and generate innovative solutions").

What could be relevant?

As this is one of the most complex phases of the process and involves the risk of errors, reflective skills require significant attention. Questions to consider include:

- In what way do students prefer to express their thoughts?
- Do we have enough time for this?
- What are the pros and cons of paper versus digital tools?

17.5.6 Test phase

The testing phase is crucial in engineering projects because it allows the improvement of provided solutions. Several tools can be used at this phase such as feedback loop analysis and usability testing reflection.

- Feedback loop analysis is a vital tool for reflection during the testing phase. It helps in comprehending iterative improvement processes and aligns well with learning objectives that emphasize continuous enhancement ("It was useful to have the possibility to test and repeat steps of the process and achieve better result"; "It would be beneficial to have more time for testing solutions").
- Usability testing reflection aids students in assessing the usability of engineering solutions, ensuring that they fulfill user requirements and function as intended. Reflecting on usability testing not only improves social competencies but also fosters critical and creative thinking through the collective examination of user feedback and design enhancements ("Improving our critical thinking and teamwork skills by working together helped us to make solution improvements").

What is important?

This phase serves as a summary of the entire process, providing an overview of the road ahead. Important questions to consider include:

- What are the testing parameters and the criteria for their selection?
- What alternative methods can be employed?

17.6 Conclusions

Recent educational research shows the unquestionable importance of reflection in the teaching and learning process. When used appropriately, reflection can help learners to understand theoretical knowledge, to combine theoretical knowledge with practical approach, to acquire skills relevant to their professional life, and to anticipate future improvement steps. Knowing the pros and cons of reflection tools and their application challenges, teachers can select the necessary reflection tools that are relevant to the desired outcome and help learners to fill learning gaps. When designing a learning strategy, the place of reflection, which can vary in the teaching and learning process, must also be considered in relation to the learning outcomes.

In order to assess the effectiveness of reflection tools, a study was carried out in the engineering courses at the Kaunas University of Technology, which apply the design thinking method. The empirical research findings highlighted the significance of aligning reflective practices with learning outcomes to achieve success in engineering education. Students emphasized various favorable features, observing that theoretical knowledge, engaging design thinking assignments, and reflection tools greatly contribute to attaining tangible educational results. The inclusion of a well-defined design thinking plan and the provision of opportunities for teamwork, as indicated in the feedback forms, are highly valued.

Reorganizing study process of the design thinking process curriculum in the course materials is another crucial factor. Students expressed their desire for more comprehensive and organized explanations within the learning management system. Restructuring the study modules to establish a more logical framework can improve students' understanding of the significance and applicability of reflecting on their studies.

When it comes to design thinking in engineering education, clarity and ease of use are essential factors to consider for effective reflection. The students' positive feedback suggested that the materials and instructions were generally perceived as valuable and comprehensible. The students appreciated the clear work assignments and chances for collaboration, as well as the teacher's feedback on their reflections, which demonstrated a high level of user-friendliness.

Students proposed that the study modules should commence with explicit indications of the objectives, significance, and assignment descriptions. They also emphasized the lack of clarity in the reflection procedure. Difficulties arose due to a dearth of examples and real-world relevance. Moreover, students expressed a desire for a more cohesive framework within the study modules. Observers notes discrepancies in the topics addressed during the lectures, such as transitioning from presentations to design thinking and subsequently to scientific readings and comments, without a discernible link or coherence.

The importance of context in engineering education cannot be overstated. It is crucial that instructional techniques and materials are closely aligned with the current needs and realities of the profession. The successful incorporation of engineering projects utilizing the design thinking technique in the curriculum was underscored by positive feedback. Students valued the clear work plans and opportunities for collaboration, which demonstrated organized instruction and effective task coordination.

Several students expressed the opinion that the reflection methods, such as mind mapping or persona analysis, were not well-suited for course subjects like data analytics and visualization. There were also concerns regarding the applicability of reflection activities to real-world engineering applications. Students suggested improving the reflection forms by including more precise information and minimizing broad interpretations. Students require more explicit instructions on how to effectively compose reflection reports and stressed the importance of receiving prompt and clear feedback from their lecturers.

Facilitating comprehension is vital in engineering education to guarantee that students not only acquire fundamental principles but also proficiently apply them in practical situations. The students' positive feedback emphasized the importance of reflecting on teamwork and collaboration. The presence of teamwork opportunities was highly valued, as these enabled effective participation in practical tasks that improve social skills and critical thinking.

References

[1] M. Tight, "Reflection: An assessment and critique of a pervasive trend in higher education," *European Journal of Higher Education*, vol. 14, no. 2, pp. 324–342, 2023, doi:10.1080/21568235.2023.2193345.

[2] B. Gibbs and G. C. Wood, "Reflection for Learning and Practice in Developing Engineers," in *Excellence in Engineering Education for the 21st Century: The Role of Engineering Education Research*, 2019, p. 118.

[3] S. Veine, A.-L. Haapasaari, S. Lindblom-Ylänne, *et al.*, "Reflection as a core student learning activity in higher education – Insights from nearly two decades of academic development," *International Journal for Academic Development*, vol. 25, no. 2, pp. 147–161, 2020, doi:10.1080/1360144X.2019.1659797.

[4] S. von Solms and H. Nel, "Reflective learning in engineering education: A case study of shell ECO-marathon," in *2017 IEEE International Conference on Industrial Engineering and Engineering Management (IEEM)*, IEEE, 2017, pp. 274–278. doi:10.1109/IEEM.2017.8289895.

[5] D. A. Schön, *The Reflective Practitioner*. Routledge, 2017. doi:10.4324/9781315237473.

[6] A. Kamp, *Navigating the Landscape of Higher Engineering Education: Coping with decades of accelerating change ahead*. Delft: TU Delft Open Publishing, 2023. doi: 10.59490/mg.72.

[7] A. Young, L. Dawes, and B. Senadji, "Using journey maps as a holistic, reflective approach to capture student engineering identity experiences," *European Journal of Engineering Education*, vol. 49, no. 1, pp. 22–44, 2024, doi:10.1080/03043797.2023.2268023.

[8] R. Thomas, S. Appelhans, S. Thompson, *et al.*, "Work in progress: Cultivating reflective engineers: Does providing a reflective ePortfolio experience in a first-year design course lead students to be more reflective in later courses?," in *2023 ASEE Annual Conference & Exposition Proceedings, ASEE Conference Proceedings*, 2023. doi:10.18260/1-2–44195.

[9] J.-H. Kim, N. T. T. Nguyen, R. C. Campbell, S. Yoo, R. Taraban, and D. D. Reible, "Developing reflective engineers through an arts-incorporated graduate course: A curriculum inquiry," *Thinking Skills and Creativity*, vol. 42, p. 100909, 2021, doi:10.1016/j.tsc.2021.100909.

[10] T. Ammersdörfer, D. Tartler, S. Kauffeld, and D. Inkermann, "Evaluating reflective behavior in engineering design retrospectives: An initial coding scheme," *Proceedings of the Design Society*, vol. 3, pp. 1685–1694, 2023, doi:10.1017/pds.2023.169.

[11] S. M. C. Atchia, "Integration of 'design thinking' in a reflection model to enhance the teaching of biology," *Journal of Biological Education*, vol. 57, no. 2, pp. 386–400, 2023, doi:10.1080/00219266.2021.1909642.

[12] A. Feest and K. Iwugo, "Making reflection count," *Engineering Education*, vol. 1, no. 1, pp. 25–31, 2006, doi:10.11120/ened.2006.01010025.

[13] N. H. Hoa and N. A. Tuấn, "Stimulating the students' reflection," *IOSR Journal of Research & Method in Education*, vol. 9, no. 6, pp. 18–21, 2019, doi: 10.9790/7388-0906041821.

[14] OECD, *Education at a Glance 2019*. OECD, 2019. doi: 10.1787/f8d7880d-en.

[15] E. Coleman, T. Shealy, J. Grohs, and A. Godwin, "Design thinking among first-year and senior engineering students: A cross-sectional, national study measuring perceived ability," *Journal of Engineering Education*, vol. 109, no. 1, pp. 72–87, 2020, doi:10.1002/jee.20298.

[16] N. Fila, S. McKIlligan, and K. Guerin, "Design Thinking in Engineering Course Design," in *2018 ASEE Annual Conference & Exposition Proceedings, ASEE Conferences*, 2018. doi:10.18260/1-2–30271.

[17] D. I. Levine, M. A. Lesniewski, and A. M. Agogino, "Design thinking in development engineering," *International Journal of Engineering Education*, vol. 32, no. 3, pp. 1396–1406, 2016.

[18] H. Plattner, C. Meinel, and U. Weinberg, *Design Thinking*. Landsberg am Lech: Mi-Fachverlag, 2009.

[19] L. A. Zampetakis, L. Tsironis, and V. Moustakis, "Creativity development in engineering education: The case of mind mapping," *Journal of Management*

Development, vol. 26, no. 4, pp. 370–380, 2007, doi:10.1108/026217107 10740110.

[20] G. W. Ellis, A. Rudnitsky, and B. Silverstein, "Using concept maps to enhance understanding in engineering education," *International Journal of Engineering Education*, vol. 20, no. 6, pp. 1012–1021, 2004.

[21] R. F. Dam and Y. S. Teo, *Empathy Map – Why and How to Use It*, Interaction Design Foundation – IxDF. 2025. [Online]. Available: https://www.interaction-design.org/literature/article/empathy-map-why-and-how-to-use-it [Accessed 20 Nov 2024].

[22] M. Stickdorn, M. E. Hormess, A. Lawrence, and J. Schneider, *This Is Service Design Doing*. O'Reilly Media, Inc., 2018. [Online]. Available: https://www.oreilly.com/library/view/this-is-service/9781491927175/ [Accessed 20 Nov 2024].

[23] D. Cronin, A. Cooper, C. Noessel, and R. Reimann, *About Face: The Essentials of Interaction Design*, 4th ed. Wiley.

[24] K. Ulrich, S. Eppinger, and M. C. Yang, *Product Design and Development*, 7th ed. McGraw Hill Education, 2020. [Online]. Available: https://www.mheducation.com/highered/product/product-design-development-ulrich-eppinger/M9781260043655.html [Accessed 20 Nov 2024].

[25] K. Goffin and R. Mitchell, *Innovation Management: Effective Strategy and Implementation*. Bloomsbury Academic, 2016. [Online]. Available: https://books.google.lt/books/about/Innovation_Management.html?id=Q3IQkAEACAAJ&redir_esc=y [Accessed 20 Nov 2024].

[26] M. Michalko, *Thinkertoys: A Handbook of Creative-Thinking Techniques*. Ten Speed Press, 2006. [Online]. Available: https://archive.org/details/thinkertoyshandb0000mich [Accessed 20 Nov 2024].

[27] M. Michalko, *Creative Thinkering: Putting Your Imagination to Work*. New World Library, 2011. [Online]. Available: https://archive.org/details/creativethinkeri0000mich [Accessed 20 Nov 2024].

[28] T. Brown, *Change by Design, How Design Thinking Transforms Organizations and Inspires Innovation*. HarperCollins Publishers, 2019. [Online]. Available: https://www.harperacademic.com/book/9780062856623/change-by-design-revised-and-updated/ [Accessed 20 Nov 2024].

[29] J. Gothelf and J. Seiden, Lean UX. 2013. [Online]. Available: https://www.oreilly.com/library/view/lean-ux/9781449366834/ [Accessed 20 Nov 2024].

[30] K. Goodwin and A. Cooper, *Designing for the Digital Age: How to Create Human-Centered Products and Services*. Wiley, 2009. [Online]. Available: https://www.wiley.com/en-gb/Designing+for+the+Digital+Age%3A+How+to+Create+Human-Centered+Products+and+Services-p-9780470229101 [Accessed 20 Nov 2024].

[31] C. M. Barnum, *Usability Testing Essentials: Ready, Set ...Test!* 2020.

[32] E. Ries, *The Lean Startup: How Today's Entrepreneurs Use Continuous Innovation to Create Radically Successful Businesses*. Crown, 2011.

Part V

Soft skills in academic activity

Chapter 18

A course sheet for training soft skills

Iñigo Cuiñas[1] *and Ioannis Tsampoulatidis*[2]

This course proposal outlines the contents of a core course on soft skills, which can be implemented as an independent program or integrated into a broader curriculum covering related topics such as project management, ethics, or leadership. The primary aim of this course sheet is to assist lecturers in incorporating soft skills into university curricula, thereby equipping students with essential non-technical competencies required for both professional and personal development.

The course is designed for undergraduate students from diverse disciplinary backgrounds. While these students may possess foundational technical knowledge in their respective fields, they often lack critical soft skills. This diverse and generalized definition of the student body is not viewed as a limitation; rather, it presents an opportunity for lecturers to customize lessons and activities to meet the specific needs of their audience. Flexibility is a key feature of the course design, allowing instructors to select from a selection of soft skills highlighted in this proposal and adapt the content to align with the specific backgrounds and career aspirations of the students.

The central objective of the course is to raise students' awareness of the importance of holistic development in their professional and personal lives, emphasizing the interplay between technical skills relevant to their fields and soft skills that govern interpersonal interactions. This objective is underpinned by two fundamental ideas:

1. Societal Problem-Solving: Society, composed of the individuals who constitute it, faces various challenges that professionals can address using their technical expertise. However, the ability to identify and respond effectively to these challenges relies heavily on soft skills such as communication, empathy, and critical thinking.
2. Ethical Responsibility: The professional activities carried out by individuals have a direct impact on society, influencing how people live and interact. This impact underscores the need for a strong ethical foundation, ensuring that professionals are not only technically competent but also morally responsible in their decisions and actions.

[1]Departamento de Teoría do Sinal e Comunicacións, Universidade de Vigo, atlanTTic, Spain
[2]Infalia PC Pilea, Thessaloniki, Greece

The course proposes an experiential learning approach, immersing students in scenarios that simulate real-world challenges. These scenarios are designed to enhance students' psychological resilience and provide them with methodological tools to navigate complex situations effectively. Each scenario is carefully aligned with one or more soft skills as defined in the European Training 2020 Peer Learning Activity (ET 2020 PLA) report on "Developing Future Skills in Education," published by the Directorate-General for Education and Culture in February 2016 [1].

The selection of specific soft skills to be included in the course may be informed by several sources. These include the comprehensive international survey of employers and university students detailed in Chapter 3 of this book, which identifies the most valued skills in the contemporary job market, as well as the practical insights of lecturers, who may tailor the course content to address the specific needs of their students. This adaptability ensures that the course remains both relevant and effective in equipping students with the skills essential for their future professional roles.

18.1 Course structure and credit allocation

The European Credit Transfer and Accumulation System (ECTS) credits allocated to this course may vary depending on the specific university where it is implemented. While the course is designed with a foundational structure equivalent to 5 ECTS, it can be expanded by incorporating more ambitious projects and additional learning activities. This flexibility allows the course to be adapted to institutional requirements and student needs.

For the standard 5 ECTS configuration, which corresponds to approximately 125 total hours of student engagement, the following time distribution is proposed:

1. **Face-to-Face Instruction (20 h):** These hours are dedicated to direct interaction between students and instructors, focusing on lectures, workshops, and in-class activities that introduce and practice key soft skills. This component is essential for fostering an interactive learning environment where students can receive immediate feedback and participate in collaborative exercises.
2. **E-Learning (36 h):** A significant portion of the course is designed to be delivered through e-learning platforms. This mode of delivery allows for a flexible and self-paced learning experience, where students can engage with multimedia content, participate in online discussions, and complete assignments that reinforce the concepts introduced in face-to-face sessions.
3. **Autonomous Work (69 h):** The remaining hours are allocated to autonomous work, which includes individual study but emphasizes teamwork. This autonomous work component is particularly conducive to the application of the soft skills developed during the course. It is recommended that students engage in case study analyses during these hours, as this method provides a practical framework for applying soft skills such as problem-solving, communication, and teamwork in real-world scenarios.

The integration of case studies into autonomous work not only reinforces the theoretical knowledge acquired during the course but also enables students to practice and refine their soft skills in contexts that reflect real-world professional challenges. This approach ensures that students can translate classroom learning into practical competencies, thereby enhancing their preparedness for future professional roles.

18.2 Mode of delivering

The course is designed to be delivered using a blended learning approach, combining face-to-face instruction, e-learning, and autonomous work. This hybrid delivery method leverages the strengths of each mode to provide a comprehensive learning experience. However, if the course is adapted for a fully online format, it is strongly recommended to utilize e-learning platforms that support collaborative work. Such platforms are essential for facilitating interaction among participants and enabling the development of essential teamwork skills.

In a fully online environment, the use of digital tools that allow real-time collaboration, group discussions, and interactive project management is crucial. These platforms should offer features such as virtual breakout rooms, shared workspaces, and communication channels to simulate the collaborative dynamics of face-to-face learning. This ensures that students can effectively engage with one another, even without physical classroom interactions.

By employing these technologies, the course can maintain its emphasis on soft skills development, particularly in areas like communication, collaboration, and teamwork. The collaborative work fostered through these platforms allows students to practice these skills in a digital environment, which is increasingly relevant in today's remote and hybrid work contexts. Thus, whether delivered in a traditional classroom, online, or through a combination of both, the course remains aligned with its core objectives of enhancing students' soft skills to prepare them for professional careers.

18.3 Objectives of the course unit

The course unit is designed with three primary objectives, each aimed at enhancing students' professional competencies and preparing them for successful integration into the workforce and broader society:

1. **Transfer of Foundational Knowledge on Soft Skills:** The first objective is to equip students with essential knowledge about a range of soft skills, highlighting their importance as vital tools for improving students' professional performance and societal engagement. At this stage, it is critical for students not only to recognize and define each soft skill but also to engage in activities and utilize tools that foster the integration of these skills into their personal and professional identities. This foundational understanding serves as the bedrock upon which students can build their soft skills throughout their careers.

2. **Application and Testing of Soft Skills in Team Settings:** The second objective focuses on familiarizing students with the practical effects of soft skills by encouraging their application in small, collaborative teams. While the training of soft skills is inherently valuable, students are more likely to commit to personal development when they observe the tangible benefits these skills offer to both personal and professional contexts. By participating in team-based exercises, students can experience firsthand the utility and impact of soft skills, thereby enhancing their motivation to further cultivate these competencies.
3. **Cross-Disciplinary Collaboration and Problem-Solving:** The third objective challenge students to work outside their primary areas of expertise, collaborate in diverse teams, and develop strategies to overcome various difficulties. In these team settings, students are expected to actively apply their soft skills while simultaneously working on specific tasks assigned by the lecturers. This objective not only reinforces the importance of teamwork but also highlights the relevance of soft skills in addressing complex, interdisciplinary challenges.

Lecturers are encouraged to incorporate additional objectives tailored to the specific workshop or course design they implement. These supplementary goals should align with the overall aim of soft skills development, allowing instructors to adapt the course content to better meet the needs of their students and the unique demands of their academic or professional environment.

18.4 Skills or competencies to be acquired

The following list comprises a selection of key skills and competencies identified from the comprehensive analysis presented in this book. These skills represent the core soft skills that students are expected to develop through participation in the course. While this set of skills serves as a foundational framework, it is crucial that lecturers tailor the emphasis on specific skills according to the context of the program in which the course is integrated. The selection and prioritization of these competencies should be aligned with the unique goals and needs of the student body and the academic environment.

For this proposal, eight core soft skills have been identified as universally essential, forming the baseline competencies that all students should acquire through the course. These skills were selected based on the results of the survey discussed in Chapter 3, which highlighted their critical importance in both professional and academic contexts. The core soft skills proposed for the basic course are

- Teamwork: The ability to collaborate effectively with others, contributing to group objectives while valuing diverse perspectives.
- Time Management: The capacity to plan and organize tasks efficiently, ensuring the timely completion of responsibilities.
- Problem-Solving: The aptitude for identifying issues and implementing effective solutions through a systematic approach.

- Analytical Thinking: The skill to assess information critically, breaking down complex problems into manageable components.
- Critical Thinking: The ability to evaluate arguments, identify biases, and make reasoned judgments based on evidence.
- Decision Making: The competence to make informed choices by weighing alternatives and considering potential outcomes.
- Creative Thinking: The capacity to generate innovative ideas and approaches, fostering originality and adaptability.
- Interpersonal Communication: The skill to convey information clearly and effectively, while also listening and responding appropriately in social interactions.

In addition to these core skills, lecturers are encouraged to incorporate additional competencies that align with the specific objectives of their local programs. This customization allows the course to more effectively address the particular needs and goals of the student population, ensuring that the training provided is both relevant and impactful.

18.5 Learning outcomes

The following collection of learning outcomes is designed to be adaptable to the specific program in which the course is implemented. Upon successful completion of the course, students should be able to:

- Apply Teamwork Principles: Demonstrate the ability to collaborate effectively with others, contributing to group objectives while fostering a cooperative and inclusive working environment.
- Manage Time Effectively: Organize and allocate time efficiently to perform various tasks within a project, ensuring timely completion and adherence to deadlines.
- Implement Problem-Solving Techniques: Utilize appropriate strategies and techniques to identify, analyze, and resolve problems in a systematic and effective manner.
- Conduct Methodical Analysis: Employ structured, methodical approaches to dissect and understand complex problems, leading to well-founded solutions.
- Engage in Critical Reflection: Critically evaluate situations by considering multiple perspectives, questioning underlying assumptions, and assessing the validity of various approaches.
- Make Informed Decisions: Make well-reasoned decisions by weighing the advantages and disadvantages of different options, considering both short-term and long-term consequences.
- Generate Creative Solutions: Produce innovative ideas and solutions by applying creative thinking techniques, demonstrating adaptability and originality in problem-solving.
- Communicate Effectively: Engage in clear and effective communication with others, facilitating productive group meetings and ensuring that ideas are conveyed and understood efficiently.

18.6 Contents

The basic course is organized into four modules, each focused on developing a specific set of soft skills categorized under cognitive, methodological, and social competencies as classified by the ET 2020 PLA report on "Developing Future Skills in Education," released by the Directorate-General for Education and Culture in February 2016 [1]. The configuration of these modules is as follows:

- Module 1: Organizational Skills
  - Teamwork: Developing the ability to collaborate effectively within a group, contributing to collective goals while respecting diverse perspectives.
  - Time Management: Equipping students with strategies to manage their time efficiently, ensuring that tasks are completed within set deadlines.
- Module 2: Creativity Skills
  - Interpersonal Communication: Enhancing students' capacity to communicate clearly and effectively, facilitating productive interactions in both personal and professional settings.
  - Creative Thinking: Encouraging the generation of innovative ideas and solutions, fostering adaptability and originality in problem-solving.
- Module 3: Problem Analysis Skills
  - Critical Thinking: Training students to critically assess situations, identify underlying assumptions, and evaluate arguments based on logical reasoning.
  - Analytical Thinking: Focusing on the ability to systematically break down complex problems and understand their components through structured analysis.
- Module 4: Solution Skills
  - Decision Making: Teaching students to make informed and effective decisions by evaluating different options and considering potential outcomes.
  - Problem-Solving: Providing students with practical tools and methodologies to identify problems, explore solutions, and implement effective strategies.

Each module combines two skills from different categories, ensuring a balanced development of cognitive (three skills), methodological (three skills), and social (two skills) competencies across the four basic modules. This integrative approach allows students to apply and reinforce their skills in diverse contexts, preparing them for real-world challenges.

Additional modules can be incorporated into the course based on the specific needs of the local program or the expertise of the lecturers. This flexibility enables the course to be tailored to the unique goals and academic environments of different institutions, ensuring that students receive the most relevant and impactful training.

The course will be structured around the following key activities, designed to enhance the students' understanding and application of selected soft skills:

1. Case-Forum Discussions:
   Each module will begin with a case forum centered on a specific situation related to the targeted soft skills. The primary objective of this activity is to immerse students in real-world scenarios – preferably drawn from their own experiences – that challenge them to develop solutions, suggestions, or proposals. This approach aims to make students aware of the relevance and application of soft skills even before they formally recognize or understand them as such. The process encourages experiential learning, where the recognition and importance of these skills emerge organically through problem-solving. Pre-prepared examples of cases will be provided to guide the implementation of this activity.
2. Theoretical Framework for Soft Skills:
   This section of the course will deliver foundational theoretical knowledge on each of the selected soft skills. Students will explore historical contexts, key theories, and various conceptual frameworks related to the skills under study. These theoretical insights will be disseminated primarily through e-learning platforms, allowing students to engage with the material at their own pace while ensuring a comprehensive understanding of the underlying principles.
3. Team Formation:
   Effective team construction is crucial for the collaborative activities in this course. Lecturers will organize students into groups of four to seven members. Several methods for group formation may be employed, including:
   Allowing students to self-select their groups based on personal preference.
   Randomly assigning students to groups to encourage diverse interactions.
   Using the results of a psychological assessment administered during the first session to create balanced teams, ensuring a mix of personalities and strengths.
   Any other suitable method that aligns with the course objectives and student needs.
4. Team-Based Autonomous Work:
   The teamwork component will be conducted as autonomous work, where each group is tasked with preparing an essay or an oral presentation that proposes a solution to a given case. This task will require students to identify the soft skills necessary for addressing the problem and to articulate the significance of these skills in their proposed solutions. A collection of case study ideas, provided at the end of this chapter and also in Chapter 19, will serve as a resource for initiating this work.

   To facilitate collaboration, especially for groups unable to meet in person, an online platform will be provided. This platform will function as a repository for resources, a meeting space for sharing ideas, and a tool for collaboratively creating documents. Each group will document their findings and process through reports or contributions to an online forum or wiki. Additionally, they will prepare and deliver a presentation to the entire class. The outcomes of these activities will be assessed using agreed-upon rubrics that evaluate both content and collaboration.

Interaction with lecturers will occur both face-to-face and online. Forums will be available for ongoing discussions during the research phase, while email will be used for the exchange of ideas. Scheduled meetings will be arranged to monitor progress and provide guidance. Additionally, students will have the option to request short calls or video meetings with lecturers to address specific questions or challenges, ensuring timely and effective support throughout the course.

18.7 Planning

The course is generally designed to be worth 5 ECTS credits. However, depending on the specific requirements of local programs, additional ECTS credits may be assigned. This flexibility allows for the expansion of the course content and methodology, enabling a deeper exploration of the material or more ambitious project work.

The basic course structure would be as follows:

- In-Class Instruction: Students will engage with lecturers for a total of 20 h of face-to-face instruction. This time will be dedicated to interactive lectures, workshops, and seminars where foundational concepts and practical applications of soft skills are introduced and discussed.
- Theoretical Content: The theoretical component of the course will be delivered through 32 h of e-learning activities. This includes the study of core concepts, historical contexts, and theoretical frameworks related to the soft skills covered in the course.
- Autonomous Work: The remaining course hours will be allocated to autonomous work, which encompasses individual study and team-based projects outside of formal classroom sessions. The total hours dedicated to autonomous work will vary depending on the local program's total credit load. Students are expected to invest additional time in completing these tasks, which will involve analyzing cases, developing solutions, and preparing presentations. The depth and quality of the solutions are anticipated to correspond with the number of hours dedicated to this component.

In summary, the total time commitment for the course includes 20 h of direct instruction, 32 h of e-learning, and additional autonomous work hours as specified by the local program (as summarized in Table 18.1). The flexibility in credit allocation and time commitment ensures that the course can be tailored to meet diverse academic requirements, providing students with a comprehensive and adaptable learning experience.

Table 18.1 Planning

Activity	**Face to face**	**e-learning**	**Autonomous work**
Introductory activities	2	–	-
Classroom work: cases related to soft skills (2 h each skill)	16	–	–
Tutorials on soft skills: history, theories, applications (4 h each skill)	–	32	–
Team projects	–	–	L.P. (min. 73)
Project presentations	2	–	–
Total hours	20	32	73+

*L.P. indicates that the number or hours depends on local programs.

18.8 Methodologies

To facilitate the acquisition of the proposed competencies, the course will employ a variety of methodologies designed to actively engage students actively and deepen their understanding of the soft skills covered. The key methodologies are outlined as follows:

- Introductory Activities:
  Lecturers will commence the course by outlining the objectives and providing practical insights into soft skills and teamwork. This initial phase will also involve the formation of student teams, setting the stage for collaborative work throughout the course.
- Classroom Sessions:
  These sessions will combine lectures with interactive components, such as group work and debates, to explore and understand each soft skill in depth. The course will consist of eight two-hour sessions, each dedicated to a specific soft skill included in the program. The pedagogical approach will emphasize active learning, encouraging students to engage in various activities rather than passively receiving information.
- Soft Skills Training:
  Students will receive targeted training on the soft skills addressed in the course. This training will involve detailed explanations provided by the lecturers, including content from specialized authors, precise definitions, historical applications, and contemporary case studies. The training will be delivered through e-learning methods, allowing for flexible and comprehensive engagement with the material.
- Projects and Meetings with Tutors:
  A central component of the course is a major project undertaken by small students' team. This project will require students to apply their acquired knowledge to a practical case study, which may be proposed either by the students themselves

or by faculty members. Throughout the course, teams will work closely together to meet project objectives, with regular supervision provided through weekly meetings via email, forums, or face-to-face interactions with lecturers. At the conclusion of the course, each team will present their project in oral and/or poster sessions. Alternatively, students may be required to record a video presentation of their project, which will be assessed by lecturers and potentially by their peers.

- Presentations and Exhibitions:
  Teams will present their projects to their peers, with presentations potentially supplemented by video recordings. The oral presentations can be delivered by one or more team members and must include evidence demonstrating the work completed and the results achieved. Following the presentations, all team members must be available for questions and assessment. Whenever possible, these presentations could be organized as part of a workshop or trade show, with invitations extended to industry professionals and members of the university community to provide feedback and insights on the projects.

18.9 Assessment

Although Chapter 20 of this book is dedicated to the assessment of soft skills, this section outlines the specific methods for evaluating student performance based on the course structure described herein. The assessment will ideally involve both lecturers and peers, with a proposed distribution of 50% of the grade assigned by lecturers and 50% by peers. In cases where local regulations preclude peer assessment, the entire evaluation will be conducted by the lecturers, constituting 100% of the final grade.

As part of their autonomous work, students will collaboratively develop a solution to a case study. Each team will be required to record a short video presenting their insights and proposed solutions. This video will serve as a medium for both lecturers and peers to evaluate the team's performance.

The evaluation of the project will be based on pre-published rubrics, which will detail the criteria for assessing the application of soft skills in solving the case study. These rubrics will be introduced to students within the first two months of the course to provide clear expectations for their work. Both lecturers and, where permitted, peers will use these rubrics to assess the video presentations. The dual assessment approach ensures a comprehensive evaluation of the students' performance, incorporating diverse perspectives on their application of soft skills.

The rubrics for evaluating the project will be detailed and made available to students at the outset of the course. These rubrics will provide specific criteria for assessing various aspects of the project, including the effectiveness of the proposed solutions and the integration of soft skills. The use of standardized rubrics aims to ensure consistency and fairness in the evaluation process.

In addition to the project evaluation, students will be asked to complete a satisfaction survey regarding the course content and its delivery. This feedback will help assess the effectiveness of the course and identify areas for improvement.

18.9.1 Evaluation rubric

Each lecturer and student will evaluate the solutions presented by the participants based on the integration and demonstration of the various soft skills emphasized throughout the course. The evaluation will focus on how effectively each soft skill has been incorporated into the proposed solution to the case study.

Each soft skill will be assessed on a scale of three levels, as outlined below:

- 1 Point: The soft skill is not demonstrated or evident in the proposed solution.
- 2 Points: The soft skill is partially demonstrated but lacks depth or clear application within the solution.
- 3 Points: The soft skill is adequately demonstrated with clear application and relevance to the solution.

The rubrics for each soft skill will be detailed and will describe the criteria for awarding points at each level. These rubrics will ensure that the assessment is objective, consistent, and aligned with the course objectives. They will be provided to students at the beginning of the course to guide their understanding of the expectations and criteria for each soft skill. The assessment will follow a two-step process:

1. Lecturer Assessment: Lecturers will assess each solution based on the rubric, assigning points for each soft skill demonstrated in the proposed solution. The total score for each solution will reflect the combined effectiveness of all assessed soft skills. This assessment will provide a comprehensive evaluation of how well students have integrated the various competencies into their solution, ensuring alignment with the learning objectives of the course.
2. Peer Assessment: Where peer assessment is permitted, students will also evaluate their peers' solutions using the same rubric. This approach provides additional perspectives and enhances the evaluation process by promoting collaborative learning. Peer feedback encourages students to critically assess the work of others, strengthening their understanding of soft skills and how to apply them effectively in diverse contexts.

A proposal for an assessment rubric is deployed in Table 18.2. This rubric will guide both lecturer and peer assessments, ensuring consistency and clarity in evaluating the demonstration of each soft skill.

18.9.2 Satisfaction survey

In addition to peer assessment, which plays a crucial role in evaluating student performance, a satisfaction survey will be administered to measure the overall experience and satisfaction of each participant in the course. The survey will provide valuable insights into the effectiveness of the course content, delivery, and overall structure. The satisfaction survey will serve as an important tool for continuous improvement and ensuring the course aligns with the needs and expectations of the students.

Table 18.2 Assessment rubric

Soft skill	Acquisition level		
	1	2	3
1a. Team working	Students were able to work together in a group	Students constructed a team and coordinated the different roles of the individuals	Students worked in a team, coordinating roles, and they realized the usefulness of this way of working
1b. Time management	Students were able to analyse the way they use their time	Students were able to identify the different tools that can help them to manage their time	Students were able to use classic tools and digital tools to manage their time
2a. Interpersonal communication	Students are able to identify the different types of communication	Students can define the characteristics of communication and to understand the importance of effective communication	Students are able to communicate in an effective way
2b. Creative thinking	Students are able to understand the importance of group work in a spirit of openness and tolerance to the most diverse and original ideas	Students are able to appropriate the spirit and the method of the creativity techniques	Students are able to use creativity techniques, with confidence within a group, to provide original solutions to a given problem
3a. Critical thinking	Students were able to define and understand the concept of critical thinking	Students were able to identify and use critical thinking skills, processes, and techniques that will assist them in their personal lives and careers	Students develop their ability to investigate and evaluate thinking from different viewpoints; also, identifying, analyzing, and formulating solutions to problems as they arise
3b. Analytical thinking	Students were able to think in an orderly and analytical way	Students correctly identified the issue of the given task requiring a solution and hypothesized	Students collected information necessary to solve a task, analyzed it and drew logical conclusions
4a. Decision making	Students were able to define and understand the sources of the most common difficulties in the decision-making process	Students were able to define and understand the action of factors affecting the decision-making process	Students were able to use techniques that facilitate decision-making process
4b. Problem-solving	Students were able to solve a problem giving a reasonable solution	Students give a solution based on following the steps given. They have specified some of the steps followed	A whole definition of the problem-solving resolution is given by the student team. They give a good description of every approach followed in the different steps of the problem-solving process

The survey has several purposes:

- Assess Course Effectiveness: The survey will gauge participants' perceptions of how well the course met its objectives and how effectively the soft skills were taught and integrated into their learning experience.
- Identify Strengths and Areas for Improvement: Through the survey, feedback collected will help identify strengths in the course design and delivery, as well as areas that may require adjustments or enhancements. This feedback will be crucial for fine-tuning specific elements such as the course structure, instructional methods, and assessment approaches.
- Enhance Future Offerings: Insights gained from the survey will be used to refine and improve the course for future iterations, ensuring that it remains relevant, engaging, and effective, and aligned with contemporary educational needs. Adjustments may include modifications to the course content, changes in delivery methods, or enhancements to activities that strengthen the application of soft skills. By collecting and analyzing this feedback, instructors and program coordinators can ensure the course's continued success and relevance. The satisfaction survey will be administered at the end of the course to allow participants to reflect on their full experience.

The proposed survey questions cover a comprehensive range of aspects related to the course and will help gather valuable insights into the students' experience and satisfaction. It could be as follows:

Demographic questions: Please, answer the following questions for statistics analysis purposes:

1. Age
2. Gender: male/female/other/prefer not to answer
3. Program you are currently following
4. Employment situation: employed/unemployed/other
5. Country of residence/work

Training quality questionnaire: Rate the following aspects of the course using a score scale from 1 to 5 (1 lowest score and 5 highest score).

Course organization

6. The course has been well organized (information, meeting dates and times, material delivery)
7. The number of students in the group has been adequate for the development of the course

Course Contents

8. The contents of the course have responded to my training needs
9. There has been an adequate combination of theory and practical application

Duration and schedule

10. The duration of the course has been sufficient according to its objectives and contents
11. The schedule has favored course attendance

Trainers/Tutors

12. The way of teaching or tutoring the course has facilitated learning
13. The trainers/tutors know the topics taught in depth

Didactic means (guides, manuals, files, ...)

14. The documentation and materials delivered are compressible and adequate
15. The didactic methodology is updated

Facilities and technical means (whiteboard, screen, projector, TV, video, computer, programs, machines, tools, ...)

16. The classroom, workshop, or facilities have been appropriate for the development of the course
17. The technical means have been adequate to develop the course content (computers, whiteboard, projector, TV, machines)

Only when the course has been carried out in the distance, e-learning or mixed mode

18. Tutorial guides and teaching materials have made it easy to complete the printed course (e-learning applications, printed materials, etc.)
19. There has been sufficient support (individual tutorials, mail and distribution lists, teleconference, virtual library, search engines, etc.)

Mechanisms for learning assessment

20. Evaluation and self-assessment tests have been available that allow me to know the level of learning achieved (the answer is yes or no)

Course overall rating

21. The course can contribute to my incorporation into the labor market
22. The course has allowed me to acquire new skills/abilities that I can apply to a job position
23. The course has improved my possibilities to change jobs in or outside the company
24. The course has expanded knowledge to progress in my professional career
25. The course has favoured my personal development
26. Degree of overall satisfaction with the course

Optional: Open-Ended Feedback

27. If you wish to make any suggestion or observation, please use the space reserved below

18.10 Examples of use cases

Soft skills play a pivotal role in fostering a productive work environment, ensuring interpersonal relationships, and enhancing employee performance. Below, there are several key scenarios illustrating the importance of soft skills in addressing common workplace challenges and driving positive outcomes for individuals and organizations.

Soft skills such as communication, time management, conflict resolution, and leadership are essential for addressing a range of challenges within the workplace. By equipping employees and managers with these skills, organizations can create a more productive, positive, and resilient work environment. Understanding these use cases emphasizes the importance of ongoing training and development programs to ensure these skills are continuously fostered and applied effectively.

18.10.1 Low motivation and job satisfaction

Employees generally strive to perform well and contribute positively to their organization. Their motivation often stems from a desire to be valued by both managers and colleagues. However, when managers fail to acknowledge employees' contributions, whether intentionally or unintentionally, it can lead to disengagement and dissatisfaction. A lack of recognition may leave employees feeling underappreciated, ultimately reducing their enthusiasm for their work.

Roles Involved

- Employees
- Managers

Requirements

To combat low motivation, it is crucial to implement strategies that enhance employee morale and recognition systems are needed.

Solution

Employees can take proactive steps to keep their supervisors informed of their contributions through *interpersonal communication*. For instance, an employee who successfully created an ad campaign that pleased a key client might remind their manager of this achievement, using *verbal communication* to highlight how it benefited the team and the organization. Such communication fosters a transparent relationship, enabling managers to be more aware of individual contributions.

Alternatives

Managers may consider implementing new *teamwork* guidelines or recognition programs to promote a more positive and engaging environment.

Issues

Employees should balance the need for recognition with a realistic understanding of their role, avoiding expectations of acknowledgment for routine tasks (e.g., submitting reports on time).

Soft Skills Involved

- Interpersonal communication
- Verbal communication
- Teamwork

18.10.2 Interpersonal conflict

Despite best efforts, conflicts with colleagues, superiors, or subordinates are sometimes unavoidable. Differences in professional opinions or communication styles can cause friction, hindering productivity and collaboration.

Roles Involved

- Employees
- Teams

Requirements

For teams to function effectively, collaboration and mutual respect are critical, even during disagreements.

Solution

To address interpersonal conflict, individuals should adopt a neutral, non-confrontational approach. Using *positive attitude*, employees can start by acknowledging the hard work of their counterpart and suggest a constructive conversation to resolve differences. This process can include *accepting feedback* and working together to find common ground that supports productivity.

Alternatives

In unresolved cases, managers may need to step in with formal *conflict management* strategies.

Issues

Effective teamwork requires negotiation, an appreciation of diverse perspectives, and the ability to both give and receive feedback constructively.

Soft Skills Involved

- Positive attitude
- Accepting feedback
- Conflict management

18.10.3 Time management

Modern workplaces often impose heavy workloads and tight deadlines, which can overwhelm employees. The 2018 Employee Experience Report by Udemy highlighted that many employees and managers struggle with *time management*, often exacerbated by productivity tools that are not always as helpful as intended.

Roles Involved

- Employees
- Managers

Requirements

Both employees and managers need effective strategies to improve time management and reduce work-related stress.

Solution

Organizations can provide training on time management through Massive Open Online Courses (MOOCs) such as "Productivity and Time Management for the Overwhelmed." These courses teach employees to identify factors that lead to procrastination, develop productive habits, and use productivity-enhancing tricks. Such structured approaches help employees manage their workload more efficiently.

Alternatives

Self-driven learning, where employees seek out additional resources independently, can also be effective.

Issues

The broad scope of *time management* issues requires ongoing training and reinforcement, often necessitating a series of courses or workshops.

Soft Skills Involved

- Productivity
- Time management
- Self-driven decision-making

18.10.4 Lack of training

In some cases, employees feel underprepared for their roles due to insufficient or inadequate training. This gap in preparation can hinder their performance and overall confidence in their work.

Roles Involved

- Employees
- Training Department

Requirements

There is a clear need for systematic and continuous training programs to equip employees with the necessary skills for their roles.

Solution

Employees should take initiative by communicating with their managers about their training needs. *Communication skills* are essential here, as employees must articulate the specific areas where they require additional instruction. Additionally, *critical thinking* and *design learning* skills can help them evaluate which training options would be most beneficial for their career development.

Alternatives

Employees may also choose to pursue self-driven learning opportunities, such as online courses or professional certifications, if formal training programs are unavailable.

Issues

Organizations need to establish life-long learning programs that continually address the evolving skills required in the workplace.

Soft Skills Involved

- Communication skills
- Design learning
- Critical thinking

18.10.5 Lack of leadership

Leadership skills, including *conflict resolution*, are valuable in various job sectors because unresolved conflicts can harm productivity, morale, and employee retention. Leaders with strong conflict resolution abilities can mediate disputes effectively, ensuring a positive work environment.

Roles Involved

- Managers

Requirements

Successful collaboration between group members often depends on strong leadership that encourages open communication and swift problem-solving.

Solution

While not every role requires formal leadership, most employers expect their staff to demonstrate decision-making abilities during challenging situations. Employees who can step up during conflicts, mediate, and manage situations will stand out. Managers need to cultivate these skills within their teams, fostering *leadership* and *delegation* methods.

Alternatives

In some cases, delegation of leadership tasks to specific team members or creating leadership development programs can help address this gap.

Issues

It takes time for employees and managers to adapt to new leadership methods and strategies, particularly in high-pressure situations.

Soft Skills Involved

- Conflict resolution
- Leadership
- Managing situations
- Delegation methods

Reference

[1] ET2020 Working Group on Modernization of Higher Education, "Developing future skills in higher education", *Peer Learning Activity (PLA)*, European Commission, Directorate-General for Education and Culture, Brussels, February 25–26, 2026.

Chapter 19

A collection of cases of use for understanding soft skills

Susana Pimentel[1]

The engineering profession demands a distinct set of skills for a successful career. While technical proficiency is crucial, engineers often underestimate the importance of "soft skills," such as communication, leadership, adaptability, and teamwork. These skills are equally vital in a continually evolving field.

The World Economic Forum's Future of the Jobs Report 2023 [1], identifies key skills required by today's workforce. Analytical thinking is commonly chosen as a core skill, with creative thinking, resilience, flexibility, agility, empathy, lifelong learning, leadership, and influencing skills also emphasized.

This chapter explores the practical applications of key soft skills, offering valuable tools for training students and professionals. It highlights real-world use cases that demonstrate the significance of these skills. Additionally, it provides strategies to develop soft skills both on an individual and team level. By emphasizing the importance of these skills and providing actionable approaches, this chapter aims to equip readers with the knowledge and resources needed to enhance their soft skills effectively.

19.1 Methodological skill – time management

Time management skills give us the ability to use our time wisely. Effective time management helps us accomplish more, feel less stressed, and maintain a healthier balance of free time.

Some examples of time management skills include prioritizing tasks and setting deadlines. Managing time effectively is considered a soft skill because it cannot be taught in a formal classroom setting; instead, it requires practice and experience to develop. Here is a dynamic activity that will help educators teach this skill to their students in the classroom.

[1]Associação Empresarial de Portugal, Portugal

19.1.1 Eisenhower Matrix [2]

Eisenhower Matrix helps individuals differentiate between the urgent and the important, allowing them to prioritize in both business and daily routine. The Eisenhower Matrix was named in honor of the 34th president of the USA (1953–1961), Dwight D. Eisenhower. As president, general in the US Army, and in other leadership positions, Eisenhower developed a system that helped him prioritize his list of daily tasks, categorizing them into urgent and important. This effective prioritization made him a better decision-maker in both military and governmental contexts.

The matrix divides tasks into four quadrants. In this sense, he categorized each demand according to how urgent and important it was, and assigned it to one of the four quadrants:

Do: important and urgent.
Schedule: important, but not urgent.
Delegate: not important, but urgent.
Delete: neither important nor urgent.

Use this in the classroom

Provide the group a list of pending tasks and ask them to divide them into the four quadrants, creating different combinations between them. This visualization, shown in Figure 19.1, helps participants better understand how to prioritize tasks based on their urgency and importance.

19.1.2 Benefits of using the Eisenhower Matrix in companies

Using the Eisenhower Matrix offers several benefits for companies, including the following [3]:

1. *Helps you prioritize your tasks*

   By considering the importance and urgency of tasks, you can prioritize your to-do list focusing on what truly matters.

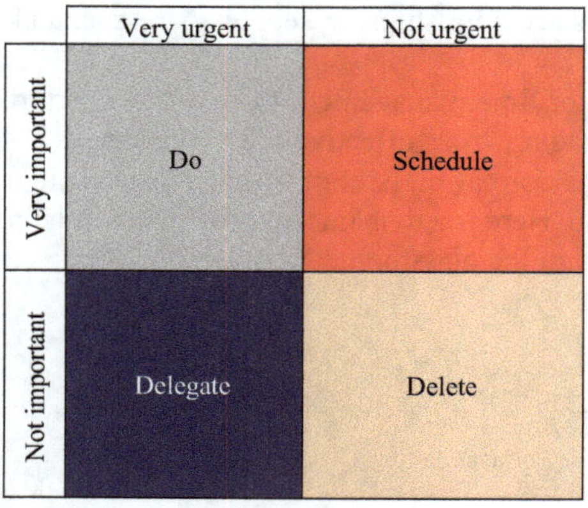

Figure 19.1 Eisenhower Matrix: four quadrants

2. *Helps you to be more efficient*
   By focusing on the tasks in the top left quadrant, you can tackle the most important and urgent tasks first, rather than wasting time on tasks that are less important or can be put off.
3. *Helps you delegate or postpone tasks*
   By identifying tasks that are either urgent, but not important or pressing, you can delegate them to someone else or postponed.
4. *Helps you eliminate unnecessary tasks*
   By identifying tasks that are neither important nor urgent, you can remove them from your to-do list, freeing up time and energy for higher-priority tasks.

19.2 Methodological skill – decision making

Decision-making skills encompass the ability to select between two or more alternatives, effectively solving problems and determining the most suitable course of action. These skills are fundamental to cognitive and social development, as they allow individuals to navigate the complexities of daily life and interpersonal relationships.

While decision-making is an unavoidable part of everyday life, it can often be challenging. We tend to think that there are only a few difficult decisions in life, such as "should I take this job or that job?" or "should I buy a house or rent?" Although these are significant decisions, their difficulty often lies in the perceived similarity of the available options. When choices appear equally viable, making a decision becomes more complex. However, there are ways to make the decision less difficult by weighing up the options, analyzing potential long-term results, and reflecting on the decisions made.

So, for good decision making, you must answer those key questions:

- What are the different types of decisions?
- What is a decision-making model?
- What factors contribute to making a decision difficult?

Decision-making games in the classroom

Interactive activities such as mock trials, stock market simulations, or role-playing scenarios provide students with opportunities to analyze information, anticipate outcomes, and make decisions within a controlled but dynamic environment. These games hone analytical skills and prepare students for real-world challenges [4].

19.2.1 Decision-making model in rugby

A complex system in nature is composed of many different interacting components. Sport teams, as examples of complex systems, are highly integrated and composed of multiple components (e.g., individual players). This integration affords the emergence of rich behavioral patterns within dynamically changing environments. During competitive games, the decisions and actions of each player are constrained by multiple factors that produce multiple effects [5].

The context-dependent constraints applied by the neighboring components of the system lead to an interdependency of decisions and actions. This context dependence of behavior leads to non-linearity that characterizes actions in team sports, i.e., outputs of the interactions are not deterministic (entirely predictable), nor are they completely random (entirely variable). This dynamic relationship between perception and action, known as circular causality, is illustrated in Figure 19.2.

To understand the practical implications of these key concepts from the complexity sciences in team games, let us consider an example of emergent behavior in rugby during the situation of two attackers against one defender. For the attacker who carries the ball, the decision of when and where to perform a pass is constrained by several factors, such as the position of teammates and the nearest opponents, the approaching speed of adjacent opponents, the running line speed, the proximity of their nearest teammate, and key boundary markers, such as the try line and the sideline. In turn, the relative position of each player (i.e., teammate and opponent) also depends on the ball carrier's behavior. From this viewpoint, the neighboring components of the team game as a complex system are constraints that shape the interdependence of players' decisions and actions. This interdependence leads to the emergence of each player's behavior. The decisions and actions performed by each component of the subsystem (i.e., players) are no longer independent; they are now context-dependent.

Clearly, this theoretical description of decision making and action involves developing an understanding of the concept of constraints. Perhaps the most insightful categorization of constraints is Newell's (1986) model, which explains how coordination and control in human motor behavior emerge. According to Newell (1986), players' decisions and actions are bounded by the interactions of

(i) their own **individual characteristics** such as feelings, emotions, thoughts (i.e., psychological states); physiological responses, technical and tactical skills;

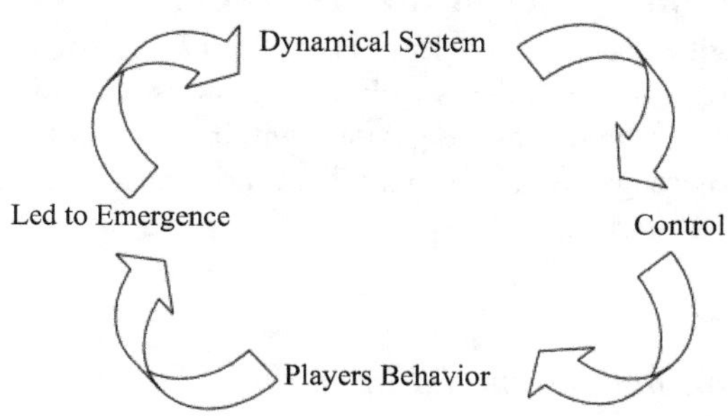

Figure 19.2 Circular causality of perception and action

(ii) the **characteristics of the specific task**, such as the rules, the specific goals to be achieved, boundary lines, the length of the performance field, the number of opponents involved, the number of teammates involved, situational characteristics of opponents (e.g., relative position; approach speed);
(iii) the **characteristics of the environment**, including physical characteristics such as the weather, surface conditions, and altitude, as well as social factors like societal expectations, the presence of the media, and the presence and characteristics of an audience.

The constraints that bind players' decisions and actions at every moment of a Rugby Union match are the multiple causes that shape the behavior of the whole dynamical system (i.e., the game) created by the interactions of adjacent players and the multiple outcomes from those ongoing interactions (e.g., a ruck, a tackle, a grubber kick).

19.3 Methodological skill – problem-solving

Surprisingly, some individuals who are excellent at solving problems in their personal lives may not necessarily be expert problem solvers in the workplace. Those with strong problem-solving skills, in professional settings, can handle challenges and adapt effectively to unforeseen circumstances by calmly evaluating situations. These skills not only enhance team performance and morale but also help companies stay on track during uncertain times, making expert problem solvers highly sought after by employers [6].

However, problem-solving is not just a single skill. Like many soft skills, problem-solving is a broad competency created by several smaller skills, including

- Active listening
- Analytical mindset
- Strong research abilities
- Creative problem-solving approaches
- Effective communication
- Capacity to make quick and effective decisions

19.3.1 The goat in the living room

The parable of the goat in the living room emerges as an old folk tale, symbolizing the way we deal with complex problems in our lives [7].

At the heart of this story is a farmer, overwhelmed by family and professional problems, who receives an unusual piece of advice from a priest: to introduce a goat into his living room. Following the advice, the farmer puts the goat in the room, creating an uncomfortable situation that is impossible to ignore.

The goat's unpleasant presence symbolizes unresolved problems that, although uncomfortable, need attention. The physical presence of the goat forces the farmer and his family to confront the difficulties, transforming their perception of the existing problems.

Take the goat into the classroom

Start the dynamic by telling the parable of the goat in the room, emphasizing how the story symbolizes the importance of facing problems directly. This introduction is crucial for contextualizing the problem-solving dynamic, creating a parallel between the parable and the challenges faced in the business environment. Reinforce the idea that recognizing and addressing problems directly is essential to finding effective solutions.

Divide the participants into multidisciplinary groups, considering the diversity of experiences and skills. This diversity enriches the process by bringing different perspectives and ideas, which are fundamental to a comprehensive analysis of the problems. Encourage the inclusion and active collaboration of all group members to maximize the effectiveness of the dynamic.

Conduct a brainstorming session where each group uses post-its to record their certainties, assumptions, and doubts about the problems identified. This stage of the problem-solving dynamic encourages active participation and the expression of thoughts and ideas, providing a solid basis for identifying both common and unique problems within the organization.

At this stage of the problem-solving dynamic, organize the post-its into categories of "Certainties," "Assumptions," and "Doubts," making it easier to see and understand the different perspectives on the problems. This categorization process allows for a more structured and focused discussion, helping the group to better understand the challenges they face and devise viable solutions.

After the discussion, guide the groups in creating an action plan that is SMART (i.e., specific, measurable, achievable, relevant, and time-bound), addressing the problems identified in a concrete and realistic way.

Finally, assign responsibility for each objective in the action plan and set up a regular monitoring system. This follow-up is essential to ensure the effective implementation of solutions and to prevent problems, or the "goat," from returning to the classroom.

The problem-solving dynamic of the goat in the room transcends simple problem-solving, acting as a catalyst for the development of a more integrated and innovative work environment. By promoting active individual participation, this approach strengthens critical thinking and creativity, key elements of innovation. Intense collaboration during the problem-solving dynamic not only solves problems, but also consolidates interpersonal relationships, creating a solid foundation for effective internal communication and a stronger team spirit.

During the week with the goat in the living room, the farmer's family begins to reflect on the issues that plague them. The metaphor of the goat helps them realize that problems, no matter how unpleasant, do not go away on their own.

This moment of introspection is crucial, highlighting the importance of actively recognizing problems before seeking solutions within the problem-solving dynamic.

After a week, the farmer removes the goat from the room. This act symbolizes the resolution of problems and subsequent relief. The family realizes that although the goat was a disturbing presence, it was the key to raising awareness and a stimulus for change. Removing the goat brings a new sense of harmony and understanding, reflecting the idea that confronting problems directly can restore balance.

The story encourages a proactive approach to problem-solving, promoting the idea that transparency, open communication, and collaboration are key to creating a healthy and productive working environment. By tackling challenges head-on, teams can overcome obstacles together, strengthening the collaborative spirit and organizational efficiency.

19.3.2 Rever use-case

Rever is an easy-to-use platform that helps companies expose hidden losses, activate the right people to solve them, and amplify the impact across organizations. In most organizations today, problems are addressed in an inconsistent, superficial way that manages symptoms without resolving the root causes [8].

The Rever platform allows all company employees to drive a consistent DMAIC process to contain symptoms and address the root causes of problems. This structured approach is visually represented in Figure 19.3, which outlines the five key phases of the DMAIC cycle: Define, Measure, Analyze, Improve, and Control.

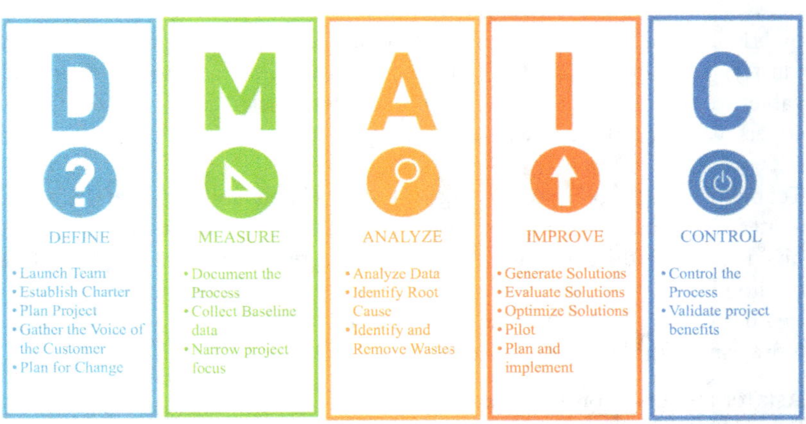

Figure 19.3 DMAIC process model by Six Sigma Development Solutions

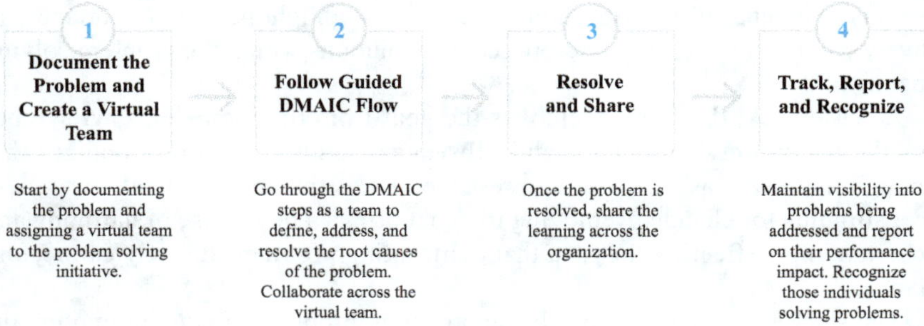

Figure 19.4 How Rever helps

With this method, Rever provides a set of results, including the following:

- A standardized process, available to everyone, to address problems in an effective manner.
- The ability to build virtual teams to ensure the right skills and expertise are included as a part of the team to solve the problem.
- A system to maintaining and sharing learnings to eliminate the need to solve the same problem multiple times.
- Dashboards for visibility into the activity and performance impact of solving operational problems.
- Recognition of individuals' driving performance.

Figure 19.4 illustrates how the platform guides users through each step—from documenting the problem and forming a virtual team to resolving the issue and tracking results—ensuring a collaborative and transparent problem-solving process.

19.4 Social skill – teamwork

Teamwork skills are essential for success in any workplace, regardless of job title or industry. Working well with colleagues, management, and clients creates an enjoyable work environment and helps complete tasks quickly and efficiently. Employers value employees who are team players that can help the organization reach its goals.

Teamwork skills refer to a range of interrelated abilities that help you to cooperate with others in different situations, meetings, and projects, while being organized and empathic. Individuals who are mature and possess people skills tend to have good teamwork abilities, as these qualities allow them to work effectively with team members to achieve the organization's goals.

How can you improve your teamwork skills?

- Ask for and give constructive feedback
- Foster mutual trust
- Resolve conflict quickly

- Be a team player
- Define roles and responsibilities clearly
- Learn from others with strong teamwork skills
- Define clear goals for the team
- Conduct regular meetings
- Create an office space that promotes teamwork

Improve your class teamwork with a balloon

Inflate a balloon and explain that all participants must touch it. Naturally, each person will take the balloon and pass it to the person next to them until everyone has had a turn. Next, issue the challenge: they must all touch the balloon again, but this time in only 10% of the time it took initially.

The first reaction will likely be say that it is impossible. However, after some time, the participants will organize themselves in one of two ways: either everyone lines up in a row and holds out their hand, while one person runs around, touching the balloon in the hands of all the others, or they form a circle and each person simultaneously places their hand on the balloon at the center.

19.4.1 Badger maps cross-training

Sharing experience with your peers is important, but recognizing where they can best help you improve is even more crucial [9].

According to **Steven Benson**, founder, and CEO of Badger Maps, his company has benefited from one self-initiated cross-training session which resulted in an increase both teamwork and productivity.

"An example of successful teamwork at our company was when the customer relation department put together an initiative of cross-training and specializing team members for different roles. After deciding who will focus on what, the group sat down and taught one another what they would need to become the expert in their respective area.

Because people were cross-trained, they had a broader set of skills they could use to handle customer interaction - which resulted in fewer hand-offs. This not only enhanced teamwork and productivity, but also improved customer satisfaction. Everyone worked as a team and covered for one another, which made everything move smoothly and quickly."

19.5 Social skill – conflict resolution

Conflict resolution can be defined as the informal or formal process through which two or more parties work together to find a peaceful solution to their dispute [10].

Several common cognitive and emotional traps, many of which are unconscious, can exacerbate conflict and contribute to the need for conflict resolution:

- Self-serving fairness interpretations
- Overconfidence
- Escalation of commitment
- Conflict avoidance
- Negotiation
- Mediation
- Arbitration
- Litigation

The Orange Negotiation

The "Orange Negotiation" is a conflict resolution activity designed for groups.

The group should be divided into two teams, with the facilitator acting as the keeper of a rare orange. The teams must understand that the orange is unique and the last of its kind. Both teams are supposed to buy the fruit, but they need to convince the facilitator that they are the right team that deserves the fruit.

At the beginning of the activity, Team A will be informed that the rind of the rare orange can be used to invent a new element, while Team B will be told that the pulp of the orange can cure a disease. The first speaker of Team A will present their argument in favor of their team's needs, followed by the first speaker of Team B, and so on, alternating turns.

The objective of the game is to encourage the two teams to negotiate and collaborate. Both teams need different parts of the fruit, so through discussion and negotiation, they will find a way to break the deadlock and resolve the conflict.

In general, it makes sense to start with less-expensive, less-formal conflict resolution procedures, before committing significant time and resources to more formal processes like arbitration and litigation. Conflict resolution training can further enhance your ability to negotiate satisfactory resolutions for your disputes.

19.5.1 Case study of conflict management

In the book "Difficult Conversations: How to Discuss What Matters Most," the authors explore how to engage in uncomfortable conversations, whether in our professional or personal lives, by examining a case study of conflict management. However, honest conversations are critical for managers, whether they need to change team culture, manage internal conflicts, deliver negative performance evaluation, disagree with others, or offer an apology [11].

To set the stage for a productive discussion, start a difficult conversation with the Third Story, an impartial observer's perspective, similar to that of a mediator. This approach presents a version of events that both sides can agree on, helping to facilitate open communication.

Suppose two regional sales reps share responsibility for sending weekly updates to their manager. Brad always submits them on time, but Frank often turns them in late. Saying, "Frank, you've turned in the sales reports late again" would only put Frank on the defensive. Instead, Brad opens the conversation this: "Frank, you and I seem to place different values on deadlines. I want to explain why meeting them is important to me, and then I'd like to hear your perspective."

Brad learns that when faced with the choice of possibly making a sale or compiling the report, Frank believes he should focus on the sale. With this insight, Brad proposes an alternative way to share responsibilities: Brad will complete the report when it is Frank's turn, as long as Frank gives Brad two hours' notice and shares a portion of any commission he earns from the sale, allowing him to continue pursuing the lead.

Conflict often triggers strong emotions and can lead to hurt feelings, disappointment, and discomfort. When handled in an unhealthy manner, it can cause irreparable rifts, resentments, and breakups. However, when conflict is resolved in a healthy way, it increases your understanding of the other person, builds trust, and strengthens your relationships.

19.6 Cognitive skill – analytical thinking

Analytical thinking involves breaking down complex information into smaller components and understanding their connections. It requires in-depth research, data analysis, problem-solving, and decision-making.

This higher-order thinking skill is important in problem-solving and decision-making by allowing individuals to systematically process information and identify key patterns, trends, and relationships.

To apply analytical thinking skills, you must develop proficiency in five key areas [12]:

- Data analysis
- Logical thinking
- Research
- Creativity
- Communication

You can assess and record your current level of analytical thinking by

- Evaluating your ability to break down complex information
- Identifying in-depth connections
- Analyzing data
- Solving problems
- Making decisions

To foster the development of analytical thinking skill, you can

- Network and engage in conversations with individuals from diverse backgrounds,
- Seek out mentors, and
- Attend workshops and seminars focused on enhancing analytical thinking.

Integrating analytical thinking into the curriculum

Developing analytical thinking skills should not be limited to specific subjects or activities. It is important to integrate analytical thinking across the curriculum, providing students with consistent opportunities to practice and apply these skills. Here are some ways to incorporate analytical thinking into different subjects [13]:

1. *Language Arts: Analyzing characters and plotlines*
   In language arts, encourage students to analyze characters' motivations, actions, and relationships in the literature. Discuss the development of plotlines and encourage students to make predictions and draw conclusions based on evidence from the text.
2. *Math: Solving complex problems and puzzles*
   In math, provide students with complex problems and puzzles that require analytical thinking. Encourage them to analyze the problem, identify relevant information, and develop a logical plan to solve it. This helps them develop critical thinking and problem-solving skills.
3. *Science: Conducting experiments and making observations*
   In science, engage students in hands-on experiments that require them to observe, analyze, and draw conclusions based on their findings. Encourage them to think critically about the scientific method and evaluate the validity of their results.

19.6.1 Leaders known for their analytical thinking

Famed and celebrated poet, **Maya Angelou**, had a very deliberate and disciplined approach to her writing. In an interview with George Plimpton for The Paris Review, she explained how she would rent a hotel room for a few months at a time, transforming it into her dedicated writing space. Rather than focusing on high output, Angelou exercised immense patient with her work. "It might take me two or three weeks just to describe what I'm seeing now," she revealed in the same interview. Needless to say, she possessed high levels of reflection and patience. One of her well-known quotes is, "Seek patience and passion in equal amounts. Patience alone will not build the temple. Passion alone will destroy its walls [14]."

It would be difficult to find someone less reactive than the Indian writer and social activist **Mahatma Gandhi**. He was known for his patient, kindness, and stood firmly behind his ideals of non-violence and non-retaliation. Gandhi remained steady even in situations that might provoke others, he was steady. He once said, "It is not that I do not get angry. I don't give vent to my anger. I cultivate the quality of patience and angerlessness, and generally speaking, I succeed."

Author and businesswoman, **Arianna Huffington**, built a thriving media empire, but her success was not solely driven by constant action. In 2007, she was working so hard that she reportedly collapsed from exhaustion. She considers that a turning point in her own career. It impacted her own approach to life, as well as her leadership style, emphasizing reflection and patience. Now, Huffington seeks employees who share her values. In an interview with Director Magazine, she remarked, “I’m also looking for people who aren’t too reactive and easily affected by the challenges the business faces every day.”

How to increase analytical thinking:

1. Pinpoint your triggers and emotions: Recognize situations or patterns that evoke strong reactions and understand why they affect you.
2. Pause before responding: Take a moment to process information and assess the situation before reacting.
3. Let history be your teacher: Reflect on past experiences to identify patterns, lessons, and strategies for making better decisions.
4. Be patient: Analytical thinking often requires time to fully evaluate and understand complex information.

19.7 Cognitive skill – critical thinking

Critical thinking is the ability to make informed decisions by objectively evaluating multiple sources of information. Critical thinkers possess a variety of essential skills, including analysis, creativity, problem-solving, and empathy.

Employers have always found critical thinking extremely valuable for its role in effective decision-making and problem-solving. However, it is often discussed only in theoretical terms, with little emphasis on how it translates into real-world applications. As a result, many individuals have never really understood the practical importance of critical thinking in business.

Critical thinking skills are crucial for organizations across all industries. In today’s constantly changing world, businesses require individuals who can adapt and apply their thinking to new challenges and situations. Moreover, these skills play a vital role when pursuing career advancement or promotions.

As examples of critical thinking, we can find problem-solving, risk assessment, data analysis, and talent hiring.

The Worst-Case Scenario

The Worst-Case Scenario is a classic team-building game that encourages participation from everyone involved [15].

To play, divide participants into small groups and ask them to imagine themselves in the worst-case scenario, such as surviving a plane crash or being stranded on a desert island. Each group must brainstorm and agree on 10 essential survival items they would need to overcome the situation.

Through this creative and out-of-the-box thought process, team members will discuss and justify their choices, explaining how each item would be useful in their survival strategy.

It is important to make all team participants collaborate in the activity. Active collaboration is key, as this activity enhances critical thinking, teamwork, and decision-making skills across the group.

19.7.1 XYZ Company's innovation challenge

XYZ Company, a technology start-up based in Bengaluru, faced declining market share and stagnating product innovation. To address these challenges, the company launched an Innovation Challenge, designed to foster critical thinking, creativity, and collaboration among employees [16].

During the Innovation Challenge, employees were organized into cross-functional teams, each assigned a specific problem to solve. The teams were encouraged to think outside the box, apply problem-solving frameworks, gather relevant data, and analyze market trends. By leveraging critical thinking techniques, they evaluated various perspectives and options and proposed innovative solutions.

Throughout this challenge, employees developed a growth mindset, embraced feedback, learned from failures, and refined their ideas through ideation and experimentation. The process ultimately led to several innovative solutions that addressed the company's core challenges, positioning XYZ Company for renewed growth and competitiveness.

References

[1] World Economic Forum, Future of the Jobs Report 2023 (https://www.weforum.org/reports/the-future-ofjobs-report-2023/) [Accessed 09 Dec 2024].

[2] G4 Educação (https://g4educacao.com/portal/matriz-de-eisenhower) [Accessed 09 Dec 2024].

[3] https://macroconsulting.pt/o-que-e-a-matriz-de-eisenhower/ [Accessed 09 Dec 2024].

[4] https://www.storyboardthat.com/pt/lesson-plans/habilidades-de-tomada-de-decis%C3%A3o#toc-6 [Accessed 09 Dec 2024].

[5] Dynamical decision making in rugby: identifying interpersonal coordination patterns – Passos, Pedro José Madaleno (http://hdl.handle.net/10400.5/509) [Accessed 09 Dec 2024].

[6] https://www.clevry.com/en/blog/problem-solving-skills-the-ultimate-guide/ [Accessed 09 Dec 2024].

[7] Keeps – Desenvolvendo e impulsionando conhecimento – Gustavo Formighieri (https://keeps.com.br/dinamica-de-resolucao-de-problemas-e-tomada-de-decisao-o-bode-na-sala-2/) [Accessed 09 Dec 2024].

[8] https://reverscore.com/2021-use-case-problem-solving/ [Accessed 09 Dec 2024].

[9] https://activecollab.com/blog/collaboration/real-world-examples-of-successful-teamwork [Accessed 09 Dec 2024].

[10] Program on negotiation – Harvard Law School – Daily Blog (https://www.pon.harvard.edu/daily/conflict-resolution/) [Accessed 09 Dec 2024].

[11] D. Stone, B. Patton, and S. Heen. *Difficult Conversations: How to Discuss What Matters Most*. New York: Viking Penguin; 2000.

[12] Vizologi – Examples of Analytical Thinking Made Easy (https://vizologi.com/examples-of-analytical-thinking-made-easy/#What_Is_Analytical_Thinking) [Accessed 09 Dec 2024].

[13] Everyday Speech – Building Strong Foundations: How to Teach Analytical Thinking to Elementary Students (https://everydayspeech.com/sel-implementation/building-strong-foundations-how-to-teach-analytical-thinking-to-elementary-students/) [Accessed 09 Dec 2024].

[14] Marlee – Fingerprint for Success – Analytical Thinking (https://getmarlee.com/traits/analytical-thinking) [Accessed 09 Dec 2024].

[15] Teamland – 10 Critical Thinking Team Building activities (https://www.teamland.com/post/critical-thinking-team-building-activities) [Accessed 09 Dec 2024].

[16] Kuril Founders B-School (https://www.linkedin.com/pulse/detailed-case-study-critical-thinking-business-aetrf/) [Accessed 09 Dec 2024].

Chapter 20

Assessing soft skills development

María del Pino Díaz-Pereira[1], Joseba Delgado-Parada[2], María Carmen Pérez-Llantada Rueda[3] and Andrés López de la Llave Rodríguez[3]

Experts widely recognise the importance of developing soft skills in the university context. Skills like problem-solving, teamwork, planning or creative thinking enhance performance, both in academic and later professional contexts.

Although soft skills are often integrated into training programmes and included as key objectives, educators frequently lack effective methods, procedures, instruments and techniques to assess how these skills are developed and impact the students.

Assessment is essential, as it allows for measuring the effectiveness and impact of any intervention. Through this process, one can determine if the objectives are being met and resources are effectively used to achieve the desired results. It also highlights areas for improvement, as analysing processes and outcomes reveal a programme's strengths and weaknesses. This allows for adjustments that can optimise its success.

Programme assessment can contribute to organisational learning. By reflecting on what does not work, organisations can apply these insights to future projects, continually improving their practices and outcomes – ultimately benefiting their target audiences. From a financial perspective, assessment is crucial for accountability. Funders, sponsors and stakeholders want to know how resources are used and what benefits are achieved. A well-executed assessment provides evidence of a programme's value.

Assessing programmes implies a methodological process similar to applied research, designed primarily to examine the effects of systematic, targeted actions that aim to meet specific objectives. A characteristic aspect of this process is that

[1]Departamento de Didácticas Especiais, Facultade de Educación e Traballo Social, Universidade de Vigo, Spain
[2]Departamento de Didáctica e Organización Escolar, Facultade de Educación e Traballo Social, Universidade de Vigo, Spain
[3]Departamento de Metodología de las Ciencias del Comportamiento, Facultad de Psicología, Universidad Nacional de Educación a Distancia (UNED), Spain

these actions are implemented and developed in a real-world setting and respond to identified needs.

In summary, programme assessment – especially for those aimed at soft skills development – is crucial for ensuring effectiveness, driving continuous improvement and promoting transparency in the use of resources.

This chapter will explore the importance of assessing soft skills development, the challenges involved and common assessment tools. It will also present a case study from an interuniversity postgraduate course in Galicia, Spain, designed to highlight the value of soft skills and their practical relevance to business operations.

20.1 Assessment in soft skills development

Two main reasons contribute to the traditional lack of attention on assessing soft skills in various contexts where interventions are implemented.

The first reason involves the challenges in defining soft skills operationally [1]. The abstract and subjective nature of these skills (e.g., teamwork, creative thinking) often makes it difficult to establish clear and measurable assessment criteria that can be consistently applied. Moreover, soft skills can manifest differently depending on the context. For example, effective communication in one environment might look quite different in another. Therefore, this variability makes it difficult to measure.

A second challenge arises from the interconnected nature of soft skills. Many of these skills influence each other (e.g., empathy can influence effective communication), which complicates defining them independently. Moreover, there is no universal consensus on which soft skills are most important or how they should be assessed. Different sectors and organisations prioritise dissimilar skills, which complicates the creation of a standard definition and the design and implementation of useful and efficient assessment tools and techniques. Unlike technical skills, which can be measured through objective tests, soft skills often require qualitative methods, such as interviews [1], which may provide subjective results.

It is necessary to recognise that soft skills can evolve and adapt to new work and social contexts. This adaptability means that any operational definition can quickly become obsolete if it is not regularly updated. Additionally, regarding the relative importance of 'soft' and 'hard' skills, previous literature does not reach a consensus [2]. While recent studies highlight the crucial role of soft skills in the workplace [3], some authors argue that technical or 'hard' skills are crucial for successful work, particularly in fields like engineering or within multidisciplinary teams [4]. While technical skills are vital for performing specific tasks and meeting job requirements, skills like problem-solving, planning or creative thinking are essential for effective interaction, teamwork and adaptability across diverse work environments.

Soft skills often become the differentiating factor between candidates with similar technical skills. For example, a professional with strong communication and teamwork skills can have a far greater impact on their organisation than someone who possesses only technical expertise. Furthermore, in an ever-changing world of work, the ability to learn, adapt and collaborate becomes increasingly valuable. Therefore, although some authors view soft skills as secondary competencies, their importance in professional and organisational success should not be underestimated.

Undoubtedly, to address the difficulties in assessing soft skills effectively, a collaborative and open approach is necessary. Since these difficulties can be solved, the assessment of interventions aimed at improving soft skills should be prioritised, not overlooked. Thus, the results of these assessment processes will contribute to improving intervention programmes.

20.2 Programme assessment

20.2.1 Phases in programme assessment

Programme assessment is a process similar to that followed in any research. However, there are significant differences between the two:

- Inclusion of intermediate assessments: Programme assessment often incorporates intermediate steps, which are not typically found in traditional research methods.
- Target population: In programme assessment, there is no sampling of the population. Instead, the intervention is directed at a specific population with distinct needs that the programme is intended to address.
- Methods and objectives: The methods used in programme assessment, unlike those with generalising or inferential purposes, focus on determining the effectiveness of the intervention rather than on generalising the results. This last issue is related to the characteristics of the participants, who are not part of a probabilistic sample. As a result, inferential techniques cannot be used, preventing the results from being generalised to the population.

Considering the peculiar characteristics of the programme assessment or other types of interventions (e.g., educational), the following are the stages or phases of the programme assessment process [5]:

1. The first phase to consider is **planning**. This activity is carried out before implementation. Here, you define the goals of the programme and establish indicators of success, as well as the measurement tools and statistical processes that will be applied to the data collected.
2. During implementation, the programme's **progress** is assessed in real time. This allows for adjustments to improve the effectiveness and efficiency of the programme. In relation to the assessment of results or utility, three basic concepts must be considered:
   - Effectiveness: Refers to whether the programme has achieved an impact related to its objectives, regardless of specific goals.

- Efficacy: Related to the empirical verification that the intervention has caused the desired changes in the intended direction and to the expected degree.
- Efficiency: Examines the relationship between the costs and the benefits obtained.

3. Once the programme has been completed, a final assessment is carried out to measure the results or outcomes and assess the extent to which the objectives were achieved.
4. In addition, periodic **follow-up** assessments are valuable over time to monitor the programme's impact and guide improvements for future interventions.

20.2.2 Strategies and instruments for data collection in programme assessment

Assessment examines the extent to which the programme objectives have been achieved. This requires taking measurements with data collection instruments at various points in time: before implementing the programme (pre-assessment, pre-test or initial assessment), during the intervention (mid-term assessment or process assessment), at the end of the intervention (post-assessment, post-test or final assessment) and time later (follow-up).

The most commonly used instruments for data collection are surveys and questionnaires, interviews, qualitative rubrics and portfolios. The characteristics of these instruments are described below.

Surveys and **questionnaires** are versatile tools that allow the collection of quantitative data. They can include both closed or open-ended questions, gathering numerical (quantitative) data as well as qualitative feedback. The design of questionnaires is a fundamental step in research and programme assessment, as it allows data to be collected in a structured way. Developing a questionnaire suitable for assessment involves specific steps and considerations.

Soft skills questionnaires are useful tools for identifying and assessing interpersonal and communication skills. Some examples of the elements assessed with this type of instrument include:

- Communication (e.g., How do you feel when expressing your ideas in a group? How do you handle disagreements within a team? What role do you usually assume in a work group and why?).
- Problem-solving (e.g., Do you prefer to work independently or collaboratively to find solutions? Why do you prefer this approach?).
- Adaptability (e.g., What do you do to remain flexible in a changing work environment?).
- Empathy (e.g., How do you show an understanding of others' feelings?).
- Leadership (e.g., What qualities are essential in a good leader?).

Steps and considerations for creating an effective questionnaire

1. Define the objective, such as the information needed and the research questions.
2. Identify the target population, considering the number of respondents and their demographic and psychographic characteristics.
3. Design the questions, including question types, e.g., open questions, closed questions, and response scales. The 11-point Likert scale, ranging from 0 to 10, is recommended [6].
4. Structure the questionnaire in a way that explains clearly the purpose of the questionnaire and assures respondents of confidentiality. In addition, group the questions into logical sections and order them appropriately, e.g., starting with easy questions to capture interest and then progressing to more complex topics.
5. Conduct a pilot test with a small group before launching the full questionnaire to gather feedback on the clarity, length and relevance of the questions.
6. Revise the questions as needed based on the results of the pilot test.
7. Choose an appropriate distribution method (e.g., online, paper, interviews) according to your target population.
8. Analyse the data. Once responses are collected, analyse the data using appropriate statistical tools and present the results clearly and understandably.

Interviews are characterised as interaction between the assessor and the participant. They are often useful in pre-assessment to establish objectives and identify needs or in post-assessment to discover participants' perceptions of a programme. This interaction must be carefully planned to ensure that the content of the interview is relevant to the intended purpose. In general, structured or semi-structured interviews are recommended, with a predefined script to guide the content being sought. In addition, creating interview sheets that allow for categorised coding of responses can enhance objectivity and simplify the process of analysing the results.

Qualitative rubrics are assessment tools that allow a more detailed and specific assessment of the performance or quality of a piece of work, project, or activity (Table 20.1). Unlike quantitative rubrics, which assign numerical scores, qualitative rubrics focus on descriptions and criteria that provide a clearer understanding of achievement levels. The components of a qualitative rubric include items such as:

- Criteria are specific aspects to be assessed (e.g., in an essay, these might include clarity of argumentation, organisation of content, appropriate use of sources and creativity).

Table 20.1 Example of a qualitative rubric

Criteria	Excellent (4)	Good (3)	Satisfactory (2)	Insufficient (1)
Clarity	Ideas expressed with great clarity are easy to understand	Clear ideas but with some minor confusion	Some ideas are confusing and difficult to follow the discourse	Very confusing and difficult to understand
Organisation	Logical and coherent structure, fluency of presentation	Good organisation and some logical jumps	Basic organisation and lack of fluency	Disorganised, no clear structure
Sources	Relevant and well-integrated sources	Adequate sources but limited integration	Limited use of sources, not always relevant	Absence of sources or irrelevant
Creativity	Original and innovative approach, multiple ideas of different nature	Original and innovative approach, some creative ideas	Combination of conventional and creative ideas	Lack of original and innovative ideas

- Levels of performance referred to as quality are described for each criterion, often categorised with verbal labels.
- Clear and detailed descriptions of each combination of criteria and level of performance to help assessors understand what is expected at each level.

A **portfolio** allows for assessing learning and skills development through a collection of work, projects and reflections from programme participants. This approach is particularly useful in educational contexts when the aim is to assess not only the final product but also the learning process.

Some key components of portfolio assessment are

- Content selection: Participants should carefully select the work to be included in their portfolio so that the content selected shows a variety of skills and knowledge acquired.
- Reflection: Each element of the portfolio should be accompanied by a written reflection.
- Organisation: It is useful to divide the portfolio into thematic or chronological sections.
- Assessment criteria: For a portfolio, it is essential.
- Self-assessment: Allows learners to reflect on their learning and development.
- Peer assessment: Promotes collaboration and peer learning.

20.2.3 Data analysis in programme assessment

Data analysis in programme assessment is a critical process that allows assessors to interpret the information collected to determine the effectiveness, efficiency and relevance of a programme. This analysis helps to make informed decision-making on whether to continue, improve or modify the programme being assessed. Both quantitative and qualitative analysis can be carried out and these approaches can complement each other, giving rise to mixed analyses.

In relation to quantitative analysis, it is necessary to consider that, in most cases, the implementation of programmes or interventions does not involve the selection of samples, but rather they are applied to specific 'target populations' with particular needs. In the case of programme assessment, calculating statistics to establish probabilistic relationships between the results obtained in a group of participants in an intervention and the general population is not appropriate. This approach does not meet most of the conditions required for inferential statistical techniques. For this reason, the analysis of results in the assessment of programmes focuses on descriptive analysis and emphasises the use of tables and graphs to compare the results obtained at the different moments of measurement (pre-test and post-test). In most cases, the assessment of a programme does not require the calculation of statistical indices, typically found in more rigorous studies and research, as the assumptions about the characteristics and probabilistic conditions, for example, extraction is not met in this context.

Qualitative analysis allows us to obtain a detailed information about participants' experiences, perceptions and opinions, providing insight into how and why a programme works or does not work in a specific context. In addition, qualitative methods encourage participation and allow you to adjust the questions or direction of the analysis as new areas of interest or problems are discovered during the assessment. Qualitative analysis also provides concrete examples derived from the application of a programme, which can be useful.

Steps and considerations in data analysis

1. **Data collection**: Data can be quantitative or qualitative. Sources may include surveys, interviews, focus groups, administrative records and observations.
2. **Data organisation and preparation**: It is important to clean and organise collected data. This may include coding qualitative responses, checking quantitative data and creating databases. Data must be prepared for analysis by eliminating errors and inconsistencies.
3. **Analysis**:
    - **Quantitative**: This analysis may include descriptive analysis (means, medians, modes), comparisons between groups, correlational analysis (correlations between variables) and regressions to understand more complex relationships.
    - **Qualitative**: Techniques may include thematic analysis (identifies patterns and recurring themes in responses), content analysis (examines textual content to extract meaning) and open coding (assigns codes to relevant text fragments and presents qualitative findings with textual quotations or thematic summaries).

20.3 Exemplification: intervention programme 'Soft skills. Human factor skills'

In the context of the interuniversity postgraduate course 'Entrepreneurship in the food sector' – offered by Universidade de Vigo, Universidade de Santiago de Compostela and Universidade da Coruña in Galicia, Spain – a training programme has been developed specifically for the subject 'Soft skills. Human factor skills'. This programme aims to equip participating entrepreneurs with tools for the systematic observation and assessment of soft skills, including mainly leadership, creativity, communication, teamwork and planning.

20.3.1 Context and participants

The training programme on 'Soft skills. Human factor skills' was designed for professionals from emerging companies in the food sector in Galicia (Spain). The aim was to support their training to facilitate the consolidation of their projects, as well as to promote innovation, viability and scalability within their companies.

The interuniversity postgraduate course 'Entrepreneurship in the food sector' was developed online over two sessions, with a duration of 4 h each, in May 2024. Eighteen participants (10 men and 8 women) aged between 19 and 52 ($M = 40.89$; $SD = 9.17$), with backgrounds in Mathematics, Biotechnology, Chemistry, Agricultural Engineering, Systems Engineering, Advertising or Physics, took part. In their companies, they held various positions, including CEO, management and deputy management, process, project, R&D, consultancy, design or commercial managers.

To develop the activities related to the training programme, the course teachers (three PhDs in psychology and one researcher in educational and socio-educational contexts) grouped the participants into two heterogeneous groups, considering their academic backgrounds and the positions in their companies.

20.3.2 Programme objectives

The purpose of this postgraduate course was to raise participants' awareness of the importance of soft skills and their applicability to various aspects related to business operations. To achieve this, the following specific objectives were established:

- To improve participants' understanding of soft skills and their relevance in work environments, particularly concerning group cohesion and leadership.
- To conceive soft skills as behaviours that can be systematically assessed.
- To encourage reflection on soft skills to design and implement an action plan aimed at improving soft skills in a real professional context.

20.3.3 Methodology

The development and assessment of soft skills, within this postgraduate course, were implemented through a gamification methodology and activities involving teamwork.

Gamification refers to the use of game elements in non-game contexts to make a product, service or application more engaging and motivating [7]. The fundamentals of gamification consist of dynamics, mechanics and components. Dynamics represent the implicit structure of the game through a narrative, emotions, constraints, etc. Mechanics are the processes that drive the game's development through challenges, rewards and competition. Components are the practical implementations of the dynamics and mechanics (e.g., avatars, badges, points, rankings, teams) [8].

Effective teamwork requires knowledge, skills and abilities in different areas. It involves the implementation of different skills and actions, such as leadership, communication, group decision-making, negotiation skills, conflict resolution, motivation, social skills and understanding of diversity and responsibility [9]. The integration of teamwork in online classes can increase students' teamwork skills and competencies, as well as their satisfaction with online learning experiences [10].

20.3.4 Activities and associated soft skills

The course lecturers designed and implemented four gamified activities that were intended to simulate situations or challenges whose resolution required the use of different soft skills: teamwork, flexibility, planning and leadership. These activities involved competition between two teams of nine participants each. Each activity had different rules and a scoring system based on observable indicators, which allowed the lecturers to objectively assess the performance of each team. At the end of the activities, the total scores obtained from each challenge determined the winning team. These gamified activities were carried out over two sessions of four hours each (Table 20.2).

Table 20.2 Gamified activities and associated soft skills

Session	Time	Title, Description, Rules and Scoring system	Soft skills
1	30′	Title: Our boat Description: Collaboratively, draw a company's boat using an online programme, screen-printing its logo and a name Rules: All team members must participate in a balanced way in the final drawing Scoring system: The teaching team assesses the productions considering the unity achieved, the originality of the proposal and the degree of elaboration	Teamwork: Participants' ability to work in a network [11]. The activity involves other skills and actions such as communication, participation, respect, performance and enjoyment

(Continues)

Table 20.2 (*Continued*)

Session	Time	Title, Description, Rules and Scoring system	Soft skills
	30′	Title: Classify as you can. Description: From pictures of different objects (e.g., a telephone, a pair of scissors, a sculpture), each team must classify them using different criteria Rules: One person from each team should record the different classifications together with the objects belonging to each classification. As many classifications as possible should be identified Scoring system: The teaching team assesses the classifications considering originality, flexibility and fluency	Flexibility: Adaptability, willingness to change and acceptance of new perspectives [12]. This activity requires the development of divergent thinking at individual and group levels. It involves other creative indicators, such as fluency or originality
2	45′	Title: The riddles Description: 14 riddles must be solved Rules: All team members must collaborate and no electronic devices are allowed Scoring system: The teaching team awards one point for each correct guess. For each mistake, one point is deducted. Six points are awarded to the team that submits the answers first	Planning: Mental simulations of the actions needed to achieve a goal [13]. The activity involves other skills, such as decision-making, problem-solving or organisation
	45′	Title: Our hymn Description: Each team must create their company hymn Rules: Each member must find a different musical instrument. The lyrics must be original and the final product must be presented in a video which must be at least one-minute long Scoring system: The teaching team assesses the quality of the proposal, departure from the model, inclusiveness, emotional intensity and quality	Leadership: Ability to direct, coordinate, train and supervise the behaviour of the members, delegating tasks [14]. The activity involves other actions, such as leader selection and performance

Once the course was over, a complementary task was proposed to facilitate the transfer of the experience to a real professional context. This task aimed to identify a weakness related to soft skills and design an intervention plan to optimise it. Specifically, the participants were asked to select a weakness they want to improve, define it operationally through a functional analysis and design and implement an action plan to develop it. They were also required to design and use a tool to measure the initial level of the behaviour and monitor changes over time. Once the changes achieved had been objectively assessed, they would present a final reflection on the impact of the change on their own and the working group's functioning in their company.

20.3.5 Assessment: procedure, instruments and techniques

During the assessment process, different instruments and techniques were applied depending on the objectives and variables involved at each stage. These moments, instruments and variables are summarised in Figure 20.1.

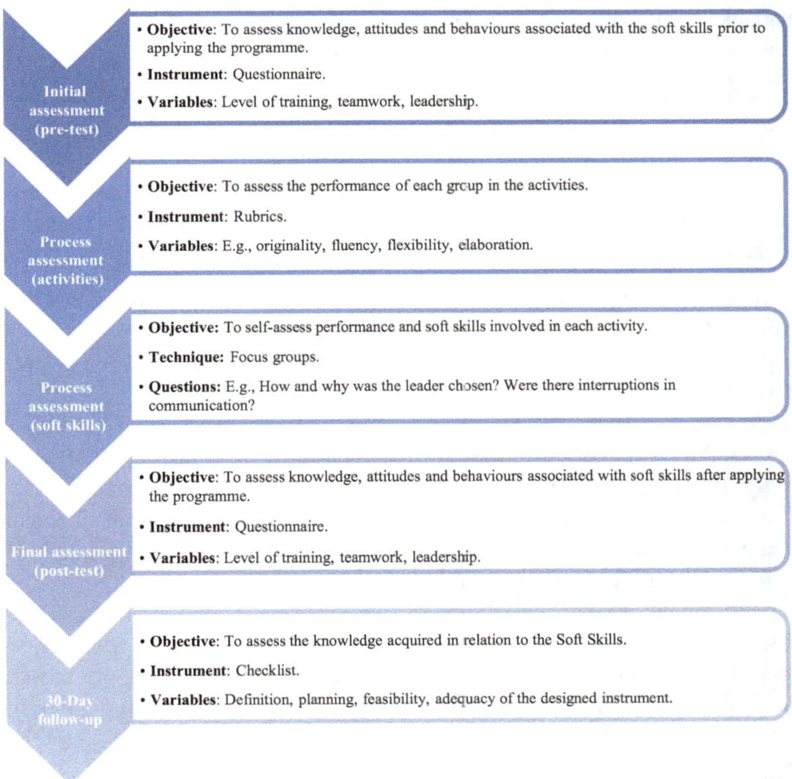

Figure 20.1 Assessment moments, objectives, instruments/techniques and variables/questions

An assessment system was implemented at different stages of the course

- **Initial**: Before carrying out the programme, a pre-test was conducted to assess participants' knowledge and the importance they attributed to soft skills in work environments.
- **Process**: Two different rubrics were proposed to assess each team's performance in the activities and to facilitate group reflection on the soft skills that were engaged during the course.
- **Final**: A post-test was suggested to observe the degree of knowledge and importance given by the participants to the soft skills in the work environments once the programme had finished.
- **30-day follow-up**: Finally, a complementary task was assigned to the participants, to solve autonomously. The aim was to assess the transfer of the knowledge acquired to a real professional context.

In terms of the agents involved, different types of assessment were used:

- **Hetero-assessment**, carried out by the course teachers to grade each activity and proclaim the winning team, as well as to assess the complementary task.
- **Self-assessment** and **co-assessment**, implemented during large group reflections on the soft skills developed at the end of each activity.

Regarding the instruments and techniques applied in the assessment process, questionnaires (pre-test and post-test), rubrics (assessment of the performance of each group in the gamified activities), discussion groups (reflection on the soft skills acquired), and checklists (to assess the complementary task) were applied.

In terms of the assessment instruments applied, a 42-item online **questionnaire** was designed for the initial and final assessment (pre-test and post-test):

- Soft skills training assessment: The first 11 items required participants to score from 0 (none) to 10 points (many), their level of training in soft skills, the resources used in their companies to develop these skills, and the importance given to them.
- Importance of soft skills: Following the same scale (from 0 to 10 points), participants were asked to assess the importance of 13 soft skills (e.g., creativity, planning, adaptability to change) in professional performance.
- Teamwork assessment: Eight items related to teamwork within their companies were included (e.g., 'My team members get along well with each other', 'I feel that I am part of my team') [15].
- Leadership assessment: Participants were asked to rate their agreement or disagreement with 10 statements regarding their leadership abilities within their companies (e.g., 'I ask for my colleagues' opinions on important issues at work', 'I allow my colleagues to participate in decision making'), based on a previously adapted scale [16].

During the process assessment, rubrics based on behavioural criteria and indicators were designed for each activity according to the soft skills assessed. Specifically, two types of **rubrics** were used:

- **Team performance assessment**: This rubric assesses the performance of each of the two teams in solving each of the four proposed activities, with hetero-assessment carried out by the course teachers. Before performing the activities, the students were informed of the assessment criteria. Once each activity was completed, the teachers consensually assigned a score to each criterion on a scale of 0–10. The final score for each team was the sum of the scores obtained for each criterion included in the activities (Table 20.3).
- **Behaviour assessment**: This rubric assesses behaviours specifically associated with the implementation of the soft skills developed in each activity through self-assessment carried out by the participants. At the end of each of the activities, four discussion groups were organised (each one associated with one of the activities carried out). In these groups, participants considered the soft skills they had applied based on various questions posed by the teaching team and then shared their reflections in a larger group. This approach encouraged reflection and debate (Table 20.4).
- To assess the extent to which the soft skills developed during the course were meaningfully understood and integrated into the participants' real professional

Table 20.3 Rubrics for assessing performance in the activities

Activity	**Criteria**	**Team score 0–10**
Our boat	Unity achieved: It seems to have been executed by a single person	
	Originality of the pictorial proposal: It departs from the model and offers a personalised proposal	
	Level of detail: Degree of elaboration	
	The originality of the title: Rupture, a departure from what is expected, from what is known	
	The best logo: Appropriateness, originality, aesthetics	
Classify as you can	Originality of criteria: Each unique criterion = 1 point	
	Fluidity: Each different classification = 1 point	
	Flexibility: Each different category = 1 point	
The riddles	Number of successes: Each success = +1 point	
	Number of errors: Each error = −1 point	
	Bonus: The first team to submit answers = +6 points	
Our hymn	Quality of the instrumental proposal: Variety and sound of the instruments	
	Deviation from the proposed song as a model	
	Integration between the participants in the constitution of a group	
	The emotional intensity of the message	
	Quality of the video presented	
	Total score (Σ Team scores)	

Table 20.4 Questions raised in the focus groups at the end of each activity

Activity	Soft skills	Questions
Our boat	Teamwork	How many team members were involved in the decisions? Were they always the same or did they vary?
		Did the team act to win or to have fun?
		Were there interruptions in communication between team members?
		Did team members respect decisions and execute agreements?
Classify as you can	Flexibility	How many different classification criteria have you been able to find?
		Were the criteria used divergent from each other or did they belong to the same functional category?
		Were the criteria used original, personalised and far from stereotypes?
		How was the team organised when looking for creative solutions to the problem?
The riddles	Planning	Did you act impulsively or was there a period of reflection and decision making beforehand?
		What were the strategies used to plan the work?
		In case of not being able to solve an activity, what strategy was adopted?
Our hymn	Leadership	How and why has the leader been chosen?
		Did the leader set specific objectives/challenges? Were they common or individualised?
		Did the leader organise and plan the tasks to be performed?
		Did the leader make decisions alone or in consultation?
		Did the leader delegate or take control?
		Did the leader minimise conflicts diplomatically?
		How did the leader respond to failures or errors?
		Did the leader reward or punish participation?

practice, the teaching team designed and applied a **checklist** to the complementary task (Table 20.5).

This complementary task presents several characteristics that contribute to its authenticity and validity. First, it involves the resolution of real problems that are inserted in relevant and significant contexts, allowing participants to observe and understand the origin and developments of the case under study. Second, the case presented is open-ended; it does not contain in its formulation all the relevant explanatory variables. As a result, the ability to define the problem operatively and delimit its scope becomes an assessable aspect of the task. Lastly, the

Table 20.5 Checklist for the assessment of the complementary task

	Criteria	No	Partially	Yes	Remarks
Problem definition	Adequately identifies the causes, antecedents and triggers of the behaviour				
	Provides objective data on the intensity and frequency of the behaviour to be modified				
	Adequately identifies the consequences derived from the behaviour				
Planning	It specifies and defines the objective of the intended change				
	The actions to be carried out are precisely detailed				
Feasibility	The implementation time is adjusted				
	The cost of implementation is appropriate				
	The plan foreseen is coherent and adequate to achieve the change				
Instrument	The instrument and measures designed are adequate/sensitive to verify the desired changes				

case allows for multiple possible solutions, provided they meet the conditions that guarantee their applicability. Its resolution is not a question of achieving a specific solution but of determining whether the proposed solution meets the criteria of efficiency and economy. In addition, the assessment should allow to check if the participant can apply techniques, principles and methods to analyse, identify and solve problems or weaknesses related to soft skills in their working environments.

20.3.6 Results

The results are presented according to the different assessment moments: initial and final, through the scores obtained by the participants in the questionnaires; process, in relation to the rubrics for the grading the activities and the discussion groups; and follow-up, after 30 days, related to the assessment of the complementary task by the teaching team using checklists.

20.3.6.1 Initial and final assessment

The mean scores obtained by the participants in the post-test, compared to the pre-test, showed significant differences, when applying the *t*-test, in item 1: 'What is your level of training in relation to soft skills?' (pre-test = 5.50, post-test = 7.11; $p = .008$); and item 11: 'To what degree do you think soft skills can be modified?' (pre-test = 6.89, post-test = 7.72; $p = .014$). In these items, the mean scores obtained in the pre-test were lower than those obtained in the post-test. For the remaining items, no significant differences were obtained. Moreover, for the items where the mean scores decreased in the post-test with respect to the pre-test, the differences

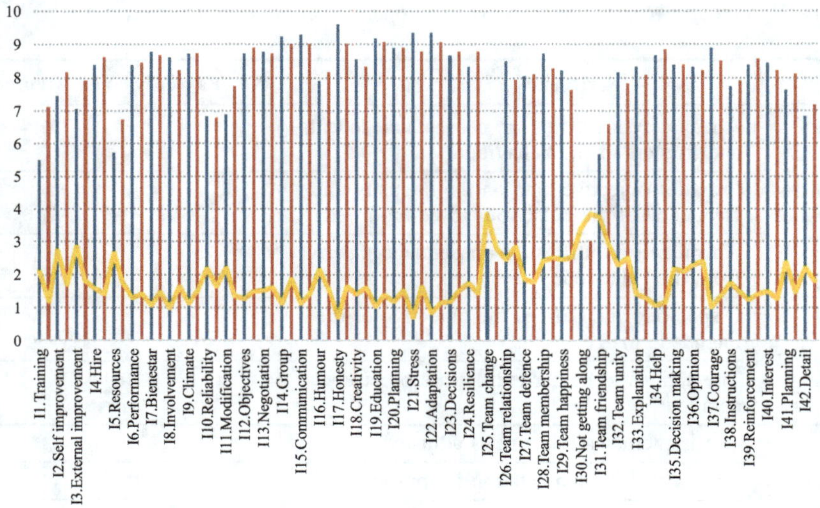

Figure 20.2 Mean scores on the items that made up the pre-test and post-test. The blue bars represent the mean scores on the pre-test for each item. The red bars represent the mean scores on the post-test for each item. The yellow line represents the standard deviation in the scores.

obtained were in all cases less than 1 point. These score variations across items are represented in Figure 20.2.

20.3.6.2 Process assessment

Each team created different products in the first activity, 'Our Ship', as shown in Figure 20.3.

In the second activity, 'Classify as you can', Team 1 identified seven classification types for the objects (e.g., based on where they can be purchased, their size and their shape) corresponding to five different typologies of categorisation (e.g., characteristics of the objects, locations where they are bought and sold). In contrast, Team 2 identified eight categorisation types (e.g., based on where they can be placed in a house and the age of the people who can use them) corresponding to six different categorisation categories (e.g., characteristics of the people who use them and characteristics of the objects). To assign a score for the originality of the classifications, points were awarded on a comparative basis, with one point given for each unique category that was not repeated between the two teams.

In the third activity, 'The riddles', Team 1 was able to solve 10 riddles, missing one, while Team 2 answered 9 correctly but had two incorrect answers. Despite this, Team 2 was the first to submit their answers and therefore earned the bonus associated with this activity. Below are two examples of the puzzles presented and their solution:

- Riddle 1: How can we make four nines result in a hundred? Resolution: 99 + 9/9 = 100.

Team 1	Team 2

Figure 20.3 Products related to the activity 'Our ship'

- Riddle 2: A person wakes up at night and discovers that there is no light in his room. They open the glove drawer, which contains 10 black gloves and 10 blue gloves. How many gloves must they take to ensure they get a pair of the same colour? Resolution: He must take 3 gloves. In the worst-case scenario, they could take out a black glove and a blue glove, one of each colour. When they draw a third glove, it must necessarily match the colour of one of the previous two because there are only two colours available.

For the last activity, 'Our hymn', results cannot be provided in textual format, as it was delivered in video format.

The rubrics to assess the team performance in the activities (team performance assessment) were scored in consensus by the teacher. For each team, the criteria were scored from 0 to 10. The sum of these scores (total score) proclaimed Team 2 as the winner of the competition with 113 points. Finally, this team received a diploma, as reflected in Figure 20.4.

In the reflection on the soft skills developed in each activity during the focus groups, a summary of some participant responses is presented as an example in Figure 20.5.

20.3.6.3 30-Day follow-up

The presentation of cases or difficulties related to soft skills, as well as the analysis and resolution process designed in a real professional context, has demonstrated that participants generally understood the course contents in a meaningful and contextualised way. Through the complementary task assigned 30 days after the end of the course, participants implemented some behavioural changes aimed at optimising their performance in different soft skills.

Participants selected different skills to improve, such as time management, active listening or communication skills as a leader. For example, one of the participants, who selected 'Procrastination and lack of planning in task management', identified the following background triggers: large or complex tasks without a proper breakdown, unclear priorities, a work environment with many distractions

Figure 20.4 Diploma for the team with the highest final score

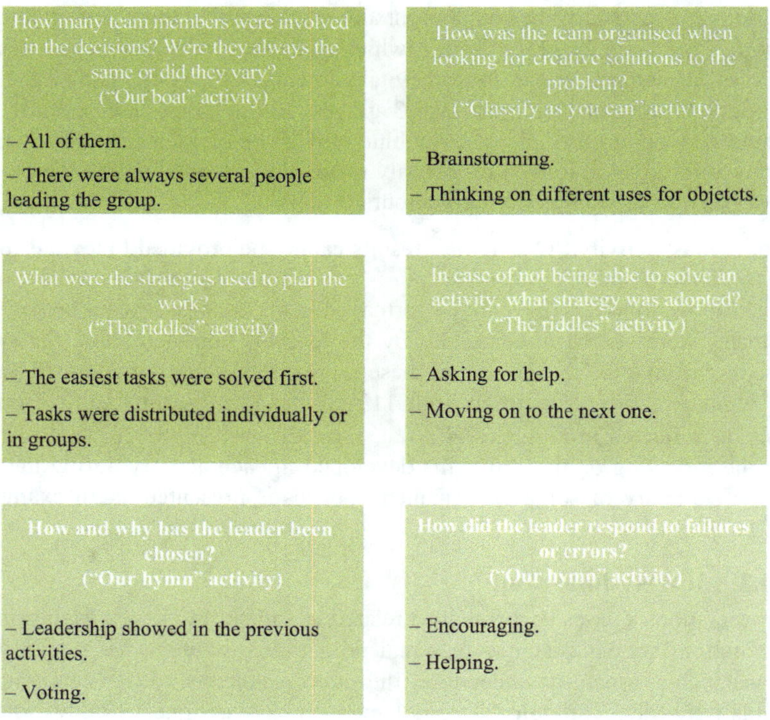

Figure 20.5 Examples of responses from the focus groups

and unrealistic or poorly defined deadlines. The consequences of this behaviour are stress and anxiety due to missed deadlines, low productivity and efficiency, negative impact on collaboration, general dissatisfaction and possible job burnout. The proposed actions to be taken have been setting specific and achievable daily and weekly goals, segmenting large tasks into smaller and more manageable steps, setting priorities, reducing distractions, creating a more focused work environment and organising weekly meetings to review progress.

The instruments proposed to verify the changes included an initial and final questionnaire to assess changes in the perception of time management, as well as a diary for participants to record the following: tasks planned and completed, time spent on each task, obstacles encountered and their sense of achievement at the end of the day.

The results derived from the implementation of the action plan showed that, after 30 days, significant changes in time management were observed. The implementation of the plan not only improved team efficiency and productivity, but also reduced stress and improved overall well-being. Team members reported greater clarity in their tasks and better organisation of their time. In summary, the experience demonstrated that a structured, collaborative approach can transform dysfunctional behaviours into productive habits, strengthening both individual and group performance.

By way of example, the assessment of this task, carried out by the teaching team, is presented (Table 20.6). Based on the general observations, the identified

Table 20.6 Checklist to assess the complementary task

Criteria		**No**	**Partially**	**Yes**	**Remarks**
Problem definition	Adequately identifies the causes, antecedents and triggers of the behaviour			X	It identifies them precisely
	Provides objective data on the intensity and frequency of the behaviour to be modified	X			No data is provided to determine the impact
	Adequately identifies the consequences derived from the behaviour			X	They are correctly described
Planning	It specifies and defines the objective of the intended change	X			They are not precisely and operationally defined
	The actions to be carried out are precisely detailed			X	They are precise and adequate
Feasibility	The implementation time is adjusted		X		Ambitious plan
	The cost of implementation is appropriate		X		High cost of implementation
	The plan foreseen is coherent and adequate to achieve the change			X	Correctly aligned
Instrument	The instrument and measures designed are adequate/sensitive to verify the desired changes		X		Prioritise indicators, records and evidence

problem is relevant and requires a more precise definition to better target efforts. The planning is well aligned, although its feasibility may be compromised by a high number of actions and instruments considering the timing. Changes are described at the end of 30 days; however, no objective data are provided.

20.4 Conclusions

This study describes a training experience aimed at raising awareness of the importance of soft skills. Specifically, it outlines the methodologies, activities, instruments and assessment techniques used during the programme, as well as the results obtained from the online implementation with professionals from emerging companies in the food sector in Galicia (Spain).

Given the results obtained from the programme through different types of assessment (initial, processual, final and a 30-day follow-up), both quantitative and qualitative information support the effectiveness of the designed programme.

First, the results obtained from the pre-test and post-test indicate significant positive changes in the participants' perception of their level of training in relation to soft skills, as well as their belief that these skills can be developed. This suggests that participants are likely to place greater emphasis on aspects such as planning, teamwork and creativity in their workplaces.

Regarding the assessment process, a competitive system allowed the assessment of the performance of each team in the activities. Furthermore, the discussion groups carried out also allowed each member to self-reflect on the soft skills developed. This task is the first step towards the completion of the complementary task.

Alongside the initial, process and final assessment, it is important to assess the transfer of the acquired knowledge about soft skills to the real working environment, specifically 30 days after the conclusion of the programme.

In summary, this work highlights the necessity of assessing experiences aimed at developing emphasising the importance of moving beyond mere playfulness, making participants aware of their relevance. Furthermore, the results obtained from the assessment process demonstrated the programme's effectiveness in shifting participants' perceptions of soft skills. Additionally, the instruments and techniques employed have proven sensitive in detecting such changes.

20.5 Limitations and practical implications

The first limitation of this study is the short duration of the programme, consisting of only two sessions. It would be beneficial to implement a similar programme over a more extended period, incorporating a greater variety of activities focused on additional soft skills and applying diverse assessment instruments. Additionally, while the online format of the programme offers multiple advantages, it also presented difficulties for participants. Some were unfamiliar with the virtual platform used and others had connection problems.

In terms of the practical implications of the study, trainers were provided with a detailed description of a programme designed to develop and raise awareness of the relevance of soft skills in the workplace. This includes concrete examples of assessment procedures and tools that could be useful in similar training initiatives.

References

[1] Ahmad, S., Ahmad, S., and Ameen, K. 'A qualitative study of soft skills development opportunities: Perceptions of university information professionals'. *Global Knowledge, Memory and Communication*. 2021; **70**(6/7): 489–503. https://doi.org/10.1108/GKMC-06-2020-0073

[2] Lyu, W., and Liu, J. 'Soft skills, hard skills: What matters most? Evidence from job postings'. *Applied Energy*. 2021; **300**: 117307. https://doi.org/10.1016/j.apenergy.2021.117307

[3] Poláková, M., Suleimanová, J. H., Madzík, P., Copuš, L., Molnárová, I., and Polednová, J. 'Soft skills and their importance in the labour market under the conditions of Industry 5.0'. *Heliyon*. 2023; **9**(8): e18670. https://doi.org/10.1016/j.heliyon.2023.e18670

[4] Martins, J., Duarte, M., Reis Cunha, S., Almada Lobo, B., Torres Marques, A., and Magalhães, B. 'The role of hard and soft skills on engineering education'. Proceedings of the International Conference on Engineering Education; Coimbra, Portugal, September 2007. ICEE; 2007 pp. 1–6.

[5] Pérez-Llantada, M. C., López de la Llave, A., and Gutiérrez, T. *Evaluación de Programas e Intervenciones en Psicología*. Madrid: Dykinson; 2009.

[6] Moral-Bofill L, López de la Llave A., Pérez-Llantada M. C., and Holgado-Tello F. P. 'Adaptation to Spanish and psychometric study of the Flow State Scale-2 in the field of musical performers'. *PLoS ONE*. 2020; **15**(4): e0231054. https://doi.org/10.1371/journal.pone.0231054

[7] Deterding, S., Khaled, R., Nacke, L., and Dixon, D. 'Gamification: Toward a Definition'. CHI 2011 Gamification workshop proceedings; Vancouver, Canada, May 2011. New York: ACM Press; 2011. pp. 12–15.

[8] Werbach, K. *For the Win: How Game Thinking Can Revolutionize Your Business*. Wharton: Wharton Digital Press; 2012.

[9] Konak, A., Kulturel-Konak, S., and Cheung, G. W. 'Teamwork attitudes, interest and self-efficacy between online and face-to-face information technology students'. *Team Performance Management: An International Journal*. 2019; **25**(5/6): 253–278. https://doi.org/10.1108/TPM-05-2018-0035

[10] Biasutti, M. 'The student experience of a collaborative e-learning university module'. *Computers & Education*. 2011; **57**(3): 1865–1875. https://doi.org/10.1016/j.compedu.2011.04.006

[11] Matook, S., Cummings, J., and Bala, H. 'Are you feeling lonely? The impact of relationship characteristics and online social network features on loneliness'. *Journal of Management Information Systems*. 2015; **31**(4): 278–310. https://doi.org/10.1080/07421222.2014.1001282

[12] Robles, M. 'Executive perception of the top 10 soft skills needed in todays workplace'. *Business Communication Quarterly*. 2012; **75**(4): 453–465. https://doi.org/10.1177/1080569912460400

[13] Mumford, M. D., Todd, E. M., Higgs, C., and McIntosh, T. 'Cognitive skills and leadership performance: The nine critical skills'. *The Leadership Quarterly*. 2017; **28**(1): 24–39. https://doi.org/10.1016/j.leaqua.2016.10.012

[14] Andrés, J. C., Aguado, D., and García-Izquierdo, A. L. 'Big Four LinkedIn dimensions: Signals of soft skills?' *Revista de Psicología del Trabajo y de las Organizaciones*. 2023; **39**(2): 75–88. https://doi.org/10.5093/jwop2023a9

[15] Dobbins, G. H., and Zaccaro, S. J. 'The effects of group cohesion and leader behavior on subordinate satisfaction'. *Group & Organization Studies*. 1986; **11**(3): 203–219. https://doi.org/10.1177/105960118601100305

[16] Marcén, C., Gimeno, F., and Gómez, C. 'Adaptación de la Escala de Liderazgo para el Deporte (LSS) para deportistas y entrenadores de un centro de tecnificación'. *Cuadernos de Psicología del Deporte*. 2016; **16**(3): 21–32.

Chapter 21

Soft skills in university classrooms: best practices and lessons learned

Bogdan Włodarczyk[1] and Manuel J. Fernández Iglesias[2]

The expectations of potential employees are changing faster than ever, and their training needs and learning preferences are diverging from traditional teaching initiatives [1]. In this context, time became a significant obstacle in the world of learning and personal development, as people are busy, coping with multi-tasking and working from anywhere at any time using mobile devices, among other reasons.

In fact, teaching soft skills face-to-face in a classroom presents a significant challenge in an era where technology enables people to work, learn, and collaborate anytime and anywhere. The traditional classroom setting competes with a digital world where students are accustomed to accessing knowledge on demand, engaging in virtual teamwork, and developing communication skills through online interactions – which are, incidentally, key soft skills. This shift in learning habits means that face-to-face instruction must offer something that technology alone cannot provide, namely real-time, dynamic human interaction, spontaneous problem-solving, and immediate feedback.

However, sustaining student engagement in a physical space can be difficult when digital tools provide seemingly more convenient and flexible alternatives. Educators must carefully design in-person learning experiences that justify the need for physical presence, incorporating hands-on activities, role-playing, and peer-to-peer interactions that cannot be fully replicated in an online environment. Moreover, balancing the use of technology within the classroom and leveraging its benefits without allowing it to overshadow the interpersonal nature of soft skills development adds another layer of complexity. As a result, teaching soft skills face-to-face requires a thoughtful, adaptive approach that integrates both digital resources and the unique benefits of in-person engagement.

Besides the challenges just described, teaching soft skills in university classrooms presents additional endeavors, but also opportunities. Unlike hard skills, which are often technical, objective, and measurable, soft skills encompass qualities that are

[1]Department of Finance, Faculty of Economic Sciences, University of Warmia and Mazury in Olsztyn, Poland
[2]atlanTTic Research Center, Departamento de Enxeñaría Telemática, University of Vigo, Spain

essential in today's dynamic professional landscape such as interpersonal abilities, problem-solving strategies, adaptability, and emotional intelligence [2]. Our experience in teaching these competencies highlights the importance of both transmitting knowledge and fostering an environment where students can practice and refine these skills in real-world contexts.

One of the key lessons that we learned is that traditional lecture-based approaches are insufficient for developing soft skills. Instead, interactive and experiential learning methods, such as group discussions, case studies, simulations, and project-based tasks, prove to be more effective because these approaches encourage students to engage actively with content, collaborate with peers, and reflect on their own learning processes. The role of the instructor also shifts from being a source of knowledge to acting as a facilitator, guiding students through self-discovery and skill development. With students able to access lectures, research papers, and expert insights at their fingertips, the real value of in-person education cannot lie in content delivery but in helping students critically engage with that content, apply it in real-world contexts, and develop essential soft skills like communication, teamwork, and problem-solving. The ability to filter, analyze, and synthesize information is more important than merely acquiring it, and this is where the instructor becomes a crucial guide, challenging students to think deeply, encouraging collaboration, and fostering the self-reflection needed for meaningful skill development. By creating interactive, student-centered learning environments, instructors prepare students to use acquired knowledge effectively in diverse and dynamic professional settings.

Additionally, our experience reveals that students often underestimate the importance of soft skills at the beginning of their learning journey, in most cases because they take for granted that abilities like teamwork, communication, and time management are either innate or secondary to technical expertise. However, structured exercises and real-life applications help them recognize the value of these skills, that they can be trained and improved, and the ways in which they can impact their professional careers. Furthermore, cultural and individual differences influence students' approaches to these skills, making it necessary to tailor teaching strategies to diverse learning styles and expectations.

In this chapter, we will explore the key insights gained from our experience in teaching soft skills. We will begin by examining best practices in instructional design and classroom strategies, followed by an analysis of the role of students in their own learning process. Next, we will discuss the impact of technology and innovative pedagogical approaches on the development of soft skills. We will also reflect on the broader lessons learned and provide recommendations for educators seeking to enhance soft skills instruction in higher education.

21.1 Importance of the leader

Lecturers should pay attention to didactic methods that activate students to find a solution together, require them to draw conclusions, and teach them to eventually make decisions. Monotonous lecturing is hardly enthralling. In turn, the presentation of practical aspects of the material taught, especially applications in industry, is highly appreciated.

The educator is a specialist in the subject matter of the class. Concepts and principles that are obvious to the lecturer may be difficult to understand for those who are encountering the lecture topic for the first time. Students do not yet have sufficient education and experience, and academics often assume that they have sufficient knowledge and skills to understand the material being taught. It is therefore necessary to make sure that content is understood and to encourage them to ask questions or raise concerns.

When addressing practical problems, breaking tasks into clear, manageable steps enhances understanding, efficiency, and problem-solving abilities. Instead of presenting a challenge as a whole, guiding students through a structured, step-by-step approach allows them to grasp each stage of the process, identify key decision points, and build confidence in their ability to solve complex problems. This approach also helps prevent cognitive overload by allowing students to focus on one aspect of the problem at a time before integrating their understanding into a complete solution. Additionally, explaining each step in detail, providing details about why it is necessary, how it connects to the next, and what potential pitfalls to expect, fosters critical thinking and adaptability, as students learn not just what to do but how to approach future challenges independently. Encouraging students to reflect on the process and explore alternative approaches further strengthens their analytical and problem-solving skills, making them more effective learners and professionals.

Furthermore, lecturers act as role models for the very skills that they teach. The way in which an instructor communicates, manages discussions, handles conflicts, and fosters collaboration within the classroom directly influences students' perception and internalization of soft skills. A lecturer who demonstrates active listening, adaptability, and constructive feedback in interactions with students provides a powerful, implicit lesson in professional behavior. Similarly, an instructor who encourages open dialogue, critical thinking, and respectful disagreement fosters an environment where students can safely practice and refine these skills themselves.

Another relevant aspect of the lecturer's role is adapting instructional methods to diverse learning styles. Some students prefer interactive discussions, while others find more inspiring structured exercises or reflective writing tasks. Incorporating a mix of methodologies – such as case studies, peer mentoring, debates, and role-playing – ensures that all students have opportunities to engage with the material in ways that suit their strengths and learning preferences. This requires lecturers to be both experts in their subject matter and skilled facilitators who can create inclusive, participatory learning environments [3].

Finally, the lecturer's ability to foster a growth mindset among students is essential for soft skills development. Unlike technical skills, which often follow a clear learning progression, soft skills require continuous practice and refinement, often involving trial and error. By normalizing mistakes as part of the learning process and encouraging students to reflect on their experiences, lecturers help build resilience and confidence. Encouraging students to step out of their comfort zones, whether by leading a group discussion, delivering a presentation, or resolving a simulated workplace conflict, strengthens their ability to navigate real-world professional challenges.

In sum, teaching soft skills in a face-to-face environment requires lecturers to go beyond traditional instructional methods and embrace the role of facilitators, mentors, and role models. Lecturers play a crucial role in shaping students' professional competencies by designing interactive learning experiences, demonstrating soft skills in practice, and adapting to the needs of diverse learners. The effectiveness of soft skills education depends both on the content delivered and on the way it is taught and embodied by those leading the learning process.

21.2 Importance of students

Students are ultimately responsible for their own learning, but the way in which knowledge is transmitted and the instructor's contact with students is instrumental in guiding their thinking and behavior. As discussed in the previous section, the lecturer takes on a variety of roles, depending on the objectives that they wish to achieve, such as facilitator, commentator, or critic. These roles should be aligned with the learning outcomes and support teaching activities. For example, if the aim is for students to analyze a case study, the role of the instructor is to moderate the discussion. If the aim is to teach students the art of argumentation and to defend the results that they have obtained or the choices that they have made when presenting their work, the role of the teacher will be that of a critical evaluator inspiring students to present arguments in defense of their ideas and to consider alternative possibilities.

However, while the teacher's guidance is essential, students must take an active role in their own skill development. Soft skills cannot be acquired passively; they require consistent practice, self-reflection, and a willingness to engage in activities that may initially feel uncomfortable or challenging. Students must recognize that these competencies, much like technical skills, improve through experience and deliberate effort. Those who embrace opportunities to lead discussions, work in teams, manage projects, and engage in critical debates, will develop these skills more effectively than those who remain passive observers in the learning process. The ability to step outside of one's comfort zone and take risks – whether by expressing an opinion in a debate, assuming leadership in a group project, or handling constructive criticism – plays a fundamental role in soft skills acquisition.

Moreover, self-awareness and reflection are especially relevant for students to develop a deeper understanding of their own strengths and areas for improvement. Soft skills, by nature, involve interpersonal interactions, self-regulation, and adaptability, which requires students willing to evaluate their own performance and seek constructive feedback. Encouraging students to engage in self-assessment exercises, maintain learning journals, or participate in peer reviews can enhance their ability to recognize progress and identify areas that need further development. Reflection is particularly important in skills such as emotional intelligence, conflict resolution, and decision-making, where understanding one's own biases and reactions is a necessary step to advance.

Another critical factor in the student's role in acquiring soft skills is engagement in collaborative learning experiences. Soft skills are inherently social, and they are best developed in contexts that require interaction, negotiation, and teamwork. Students who actively participate in group projects, case studies, and role-playing activities gain first-hand experience in problem-solving, leadership, and effective communication. However, successful collaboration requires more than simply working alongside others, as it demands active listening, adaptability, and a commitment to shared goals. Those who contribute meaningfully to group work, provide thoughtful feedback to peers, and take responsibility for their roles within a team will gain far more from the learning experience than those who approach it with a passive or transactional mindset.

Furthermore, students must understand that soft skills are not only relevant within the classroom but are also essential for their future professional and personal success. Many students initially underestimate the importance of these competencies, focusing primarily on acquiring technical expertise. However, employers consistently highlight the value of communication, teamwork, adaptability, and problem-solving abilities in professional settings. By recognizing the broader impact of soft skills early in their education, students can proactively seek out opportunities to refine these abilities, both inside and outside the classroom. Participation in extracurricular activities, internships, and volunteer work can further enhance soft skills development by exposing students to diverse real-world scenarios that challenge them to think critically, collaborate effectively, and navigate complex social dynamics.

Finally, motivation and mindset also play a role in the acquisition of soft skills. Students who approach learning with curiosity, a willingness to embrace challenges, and a growth mindset are more likely to develop and refine these competencies over time. Unlike technical knowledge, which can often be learned through structured study, soft skills require ongoing practice and adaptation to different contexts. Developing resilience, learning from failures, and maintaining a proactive attitude toward personal and professional growth are instrumental for long-term success.

In conclusion, while instructors play a guiding role in shaping the learning environment, students must take ownership of their soft skills development by actively engaging in learning activities, reflecting on their experiences, collaborating with others, and applying these skills beyond the classroom. The most effective learning happens when students recognize that soft skills are not just academic requirements but essential tools for navigating both their careers and personal lives.

21.3 Importance of learning content

Care should be taken to ensure that the information and materials provided to students are as attractive and updated as possible and persuade them to expand their knowledge of the subject on their own. Teaching requires constant adaptation of the content to the changing environment. When teaching the subject in subsequent years, you should be ready to make changes, for example, when the methods used so far have not yielded the expected results or new issues and problems have arisen in the scientific area. Feedback on the effectiveness of teaching methods can be obtained through student evaluations, course surveys, interviews, or post-teaching reflections.

The quality and relevance of learning content play a fundamental role in the development of soft skills, as these competencies are deeply contextual and shaped by real-world applications. Unlike technical knowledge, which often follows a structured curriculum, soft skills development requires exposure to authentic, dynamic, and situational learning experiences. Course materials should not only provide theoretical foundations but also illustrate how these skills apply in professional and social settings. Case studies, industry reports, and real-life scenarios can make learning more engaging and relatable, bridging the gap between abstract concepts and practical use. For example, a lesson on conflict resolution becomes far more impactful when students analyze real workplace disputes and propose solutions based on theoretical models.

Interactive and experiential learning content is particularly effective in developing soft skills when compared to traditional textbooks and lectures, which may not be sufficient to develop skills such as communication, teamwork, or emotional intelligence. Instead, incorporating simulations, role-playing exercises, problem-based learning, and collaborative projects enhances engagement and allows students to practice these skills in realistic contexts [4]. For instance, a course on leadership should include activities where students take on leadership roles, make decisions under pressure, and receive feedback on their performance. The more students can immerse themselves in practical applications, the more they will internalize and refine these competencies.

Keeping learning content aligned with industry trends and real-world demands is another essential aspect of soft skills education. The skills required for success in the workplace are constantly evolving, influenced by technological advancements, globalization, and shifts in workplace culture. For example, with the rise of remote and hybrid work environments consequence of the pandemics, digital communication and virtual collaboration became compulsory skills for professionals. Learning materials should reflect these changes, incorporating discussions on remote teamwork, cross-cultural communication, and digital etiquette to ensure that students are prepared for the modern workforce. Bringing in guest speakers, industry professionals, or alumni to share insights and experiences can further enhance the relevance of course content, providing students with practical perspectives and real-world applications.

Moreover, the accessibility and format of learning materials can significantly impact how effectively students engage with soft skills education. Given that today's learners often prefer digital and multimedia formats, integrating videos, podcasts, interactive modules, and online discussion forums can make the content more appealing and accessible. For example, instead of assigning a lengthy reading on negotiation tactics, providing a series of short video case studies featuring real negotiations can make the concepts more digestible and engaging. Similarly, incorporating gamified learning elements – such as challenges, leaderboards, or scenario-based quizzes – can enhance motivation and encourage active participation.

Feedback mechanisms are equally critical in ensuring that learning content remains effective and responsive to student needs. Continuous assessment of materials through student feedback and course evaluations allows instructors to refine their approaches and incorporate improvements based on actual learning experiences. Encouraging students to reflect on which learning activities were most valuable, which challenges they faced, and how they applied newly acquired skills can provide invaluable insights for course development. A cycle of testing, adapting, and refining ensures that the learning content evolves to meet both academic and professional demands.

In conclusion, well-designed, relevant, and engaging learning content is fundamental to the successful development of soft skills. By incorporating real-world applications, interactive learning methods, industry trends, and diverse content formats, educators can create an environment that fosters meaningful skill acquisition. Moreover, continuously updating and refining materials based on feedback ensures that students receive the most effective and applicable training possible. Developing soft skills is not just about what students learn but how they learn, and the content provided to them plays a critical role in shaping that journey.

21.4 Diversity and impartiality

The university is a place open to the exchange of views and neutral discussion on any topic concerning the world around it. The diversity of worldviews and the clarity with which they are conveyed is an added value desirable in an academic environment, as it shapes the future intellectual leaders of society. However, when teaching a class, the purpose of the lecture should be kept in mind, and both the discussions and debates should be directly related to it to fulfil the planned scope of work envisaged for the class. Where appropriate, the instructor should moderate the discussion by maintaining objectivity and neutrality in the judgements expressed by themselves and the students. The university should be unbiased in its views, seeking through staff and students to identify both the advantages and disadvantages of an issue, and the statements presented should respect the diversity of participants, their personal well-being, and their perception of the world.

Soft skills, by their very nature, involve interpersonal interactions, communication, and adaptability – all of which are enriched by diverse perspectives. A classroom that embraces diversity allows students to engage with viewpoints different from their own, fostering empathy, cultural awareness, and critical thinking. Exposure to a variety of perspectives enhances students' ability to navigate complex social and professional environments, equipping them with skills that are increasingly valued in globalized workplaces. Whether in teamwork, leadership, negotiation, or problem-solving, the ability to work effectively with individuals from diverse backgrounds is essential for professional and personal success.

Moreover, recognizing and valuing diversity encourages inclusive learning, ensuring that all students feel respected and listened to [5]. Instructors should actively create an environment where students feel comfortable expressing their points of view without fear of judgment or exclusion. This can be achieved through structured discussions, debate formats that encourage respectful disagreement, and activities that highlight the importance of diverse problem-solving approaches. By making inclusivity a core component of soft skills education, educators help students build the confidence and flexibility needed to interact successfully in diverse settings.

While diversity is essential in enriching classroom discussions, impartiality is equally important in maintaining a fair and balanced learning environment. When teaching soft skills, instructors must be conscious of their own biases and ensure that they do not unintentionally favor certain viewpoints over others. The role of the instructor is not to impose personal beliefs but to facilitate a space where students can critically analyze different perspectives and form their own reasoned conclusions.

To maintain impartiality, teachers should ensure that classroom discussions are grounded in evidence-based reasoning, encouraging students to support their arguments with logic, research, and well-founded evidence rather than relying on emotional or ideological positions. It is also important to foster balanced participation, creating an environment where all voices are heard, including those who may be hesitant to express dissenting opinions. Implementing structured dialogue formats, such as debate structures, role rotations, or assignments that require students to argue perspectives they may not personally hold, helps develop objective reasoning and respect for diverse viewpoints. Additionally, lecturers should be aware of language and framing, as the way a topic is introduced can subtly shape students' perceptions. Using neutral phrasing and presenting issues from multiple perspectives can help maintain fairness and encourage critical thinking.

The ability to engage in respectful discourse with individuals of differing perspectives is a skill that extends far beyond the classroom. In professional settings, employees frequently encounter colleagues, clients, and stakeholders with contrasting viewpoints, cultural backgrounds, and work styles. Developing the ability to listen actively, consider alternative perspectives, and communicate effectively without personal bias is a crucial component of leadership, negotiation, and conflict resolution. Universities that emphasize these competencies prepare students not only for academic success but for the complexities of globalized work environments.

Additionally, embracing diversity while maintaining impartiality enhances decision-making and problem-solving. Studies have consistently shown that diverse teams produce more innovative solutions, as differing perspectives challenge assumptions and lead to more comprehensive analyses of problems [6,7]. However, for this process to be effective, individuals must feel that their contributions are valued and that discussions remain fair and objective. By practicing these principles in the classroom, students learn to approach problem-solving with open-mindedness, adaptability, and a commitment to rational, informed decision-making.

Teaching soft skills in a way that embraces diversity while maintaining impartiality is essential for fostering an inclusive and intellectually stimulating learning environment. By ensuring that students are exposed to a range of perspectives while also learning to evaluate them critically and fairly, educators help develop essential professional competencies. The ability to engage in meaningful discussions, navigate different viewpoints, and make impartial, well-reasoned judgments is not only a vital soft skill but also a foundation for responsible citizenship and effective leadership in an increasingly complex world.

21.5 Design approach

The implementation of soft skills education in engineering curricula requires a carefully structured methodological framework that balances traditional classroom instruction with modern pedagogical approaches. Our teaching methodology was designed to create an environment where students could develop these skills through a combination of formal instruction and practical application. We placed particular emphasis on clear communication of expectations, structured deliverables, and technology-enhanced learning experiences. This approach acknowledges that while technical competency is obviously fundamental to engineering education, the development of soft skills requires a different pedagogical paradigm, and more specifically one that emphasizes continuous learning, social interaction, and workplace relevance.

On most occasions, courses on soft skills will require students to produce some kind of coursework (e.g., work output reports, project descriptions, interim work, reports). In this case, there are some aspects that should be carefully considered:

- Students should be provided with a clear description of the task, together with precise instructions on how to complete it. This applies especially to projects.
- The purpose of the assignment should be clearly defined.
- The editing requirements and presentation format should be clearly specified, as they directly develop professional communication competencies.
- A deadline for the delivery of the work was indicated, and the consequences of failing to meet the deadline were specified. Note that this is also associated with the development of soft skills, such as time management, accountability, and responsibility, or planning and organization.
- Assessment criteria and their contribution to the final grade of the course should be provided in advance.

The practice of providing example studies and model work serves as a powerful learning tool in soft skills education. By examining concrete examples, students gain insights into professional standards and best practices that might otherwise remain abstract concepts. These examples help bridge the gap between theory and practice, offering students tangible references for their own work while developing their ability to analyze and evaluate quality in professional communications. This approach is particularly valuable in engineering education, where students often benefit from seeing how complex technical information can be effectively presented and communicated in various professional contexts.

The design of coursework in soft skills education must serve multiple interconnected objectives that go beyond simple knowledge acquisition. Each assignment should function as a scaffold that helps students organize and internalize their understanding while simultaneously encouraging them to explore beyond the prescribed boundaries of the subject matter. This dual approach of structured learning and independent exploration mirrors the real-world engineering environment, where professionals must balance established methodologies with innovative problem-solving. Furthermore, the coursework should be deliberately crafted to develop both technical and soft skills in tandem, recognizing that these competencies are increasingly inseparable in modern engineering practice. This integrated approach helps students understand that soft skills are not merely supplementary abilities but are fundamental tools that enhance their technical expertise and professional effectiveness.

The establishment of clear assessment criteria and their weighted contribution to the final course grade serves multiple pedagogical purposes in soft skills development. When students understand exactly how their work will be evaluated, they learn to align their efforts with professional expectations and quality standards. This transparency in assessment helps students develop critical self-evaluation skills and an understanding of performance metrics, which in turn are competencies that directly translate to workplace scenarios, where engineers must meet specific project requirements and quality benchmarks. Furthermore, clear assessment criteria help students understand the relative importance of different soft skills components, allowing them to prioritize their development efforts effectively.

The implementation of specific editing requirements and presentation formats in course assignments serves a deeper purpose beyond mere academic structure. These requirements function as a practical training ground for professional communication skills, teaching students that effective communication in engineering extends beyond technical accuracy. When students must adhere to precise formatting guidelines and presentation standards, they develop critical workplace competencies such as attention to detail and audience awareness. This approach particularly benefits engineers who must frequently translate complex technical information for diverse stakeholders, where the manner of presentation can be as important as the technical content itself.

Similarly, the establishment of clear deadlines and consequences for assignments cultivates a range of essential professional capabilities. This practice develops students' time management skills and strengthens their sense of accountability, mirroring real-world project environments where timeline adherence is critical. Through deadline-driven assignments, students learn to handle pressure, prioritize tasks, and make strategic decisions about resource allocation.

Moreover, they develop the professional maturity to communicate proactively when challenges arise, a skill particularly valuable in engineering projects where delays can have significant ripple effects. These experiences prepare students for the realities of engineering practice, where project milestone management and deadline adherence are fundamental to success.

Finally, while traditional classroom learning will continue to be practiced, particularly in basic skills, the learning process will shift from formal education to an ongoing process where learning is continuous, social, informal, and embedded in the workplace.

Note that emphasis should be placed on the use of technology as a key element of future learning. Technology provides greater opportunities to deliver content on a "here and now" basis, combining recognition of the need for development with timely delivery of relevant content. This means that new approaches will need to be taken to structure learning based on changing realities [8].

Using technology requires a different way of thinking about learning, which is not necessarily about delivering content, but rather providing people with the means to find and share information themselves [9]. Understanding the psychology of relationships and networks will be fundamental to creating effective learning environments.

The teaching approach described above represents a significant departure from conventional engineering education methods. By establishing clear frameworks for coursework while simultaneously embracing technological innovation and informal learning opportunities, we created a learning environment that better reflects the real-world demands facing modern engineers. This methodology proved particularly effective in bridging the gap between academic requirements and workplace expectations. The emphasis on structured deliverables, combined with the flexibility to incorporate social and informal learning, allowed students to develop soft skills in a way that felt both authentic and purposeful. Moving forward, this approach demonstrates that effective soft skills development must be both systematic in its implementation and adaptive in its delivery, ensuring that students are well-prepared for the evolving demands of the engineering profession.

21.6 Training design

The design of soft skills training in engineering education requires a fundamental shift from traditional teaching paradigms to more dynamic and learner-centered approaches. As the professional landscape evolves, the methodology for delivering soft skills education must adapt to meet both current needs and future challenges. This section explores five core areas of training design that are essential for effective soft skills development, emphasizing the importance of contextual learning, accessibility, data-driven approaches, technological innovation, and environmental considerations. Each of these elements contributes to creating a comprehensive learning experience that prepares engineering students for the complexities of modern professional practice. The future is both about delivering the same content in a new way and about giving people the ability to find information themselves and learn independently.

Thus, we identified five core areas of training design that soft skills learning professionals should consider.

- **Contextual learning.** This is all about access to learning and experience for the learner [10]. Learning needs must be personalized to meet the specific requirements of each learner. We know that people learn better by doing. Therefore, the learning process should be practical and based on real-life situations. Learning should be project-integrated, with more emphasis on group work, project collaboration, social interaction (access and sharing), game inspiration, and embedding in digital media.
- **Learning must be micro and mobile friendly.** Learning must be student-centered, meaning that the learner must be able to access the desired content and how and when they want to access it. Learning must be easily accessible and based on micro-learning concepts [11]. Smaller is better. Educational initiatives such as videos, podcasts, and online short courses are good examples of this.
- **Learning should be data driven.** R&D professionals should learn to use new technologies themselves and should consider using data analytics platforms to predict trainee demand based on technology use, such as social media, just as advertising agencies predict the types of ads to send based on technology use and search engine analytics [12].
- **Virtual reality (VR) and augmented reality (AR).** Technology should build innovative learning tools that enhance the overall user experience. AR and VR tools enable immersive learning experiences. For example, learners can step into a 360-degree video shot or create a realistic 3D environment [13,14].
- **Learning environment.** A key element in the knowledge transfer process is the consideration of the learning environment [15]. In traditional environments, we are used to the school mode, and although we can set up the tables in the room in a U-shape, this mode still hinders true collaboration between trainees and participants.

In this context, to foster an environment that promotes innovative learning, we need to start doing things differently. Classroom design must stimulate creativity and creative thinking and enable collaborative learning. For example, the use of round tables instead of rows allows for natural group discussion and interaction.

In addition, the appropriate design of the learning environment can positively influence the learning experience by enabling additional collaboration and creative thinking, through devices such as walls on which to draw mind maps, record the results of brainstorming sessions and group reflections.

The multifaceted approach to training design described above represents a significant evolution in soft skills education for engineers. By integrating contextual learning, mobile accessibility, data analytics, immersive technologies, and thoughtfully designed learning environments, we created a robust framework that supports comprehensive skill development. This holistic approach acknowledges that effective soft skills training cannot be achieved through traditional methods alone but requires an ecosystem of learning opportunities that mirror the complexity of real-world professional interactions. The success of this training design lies in its ability to blend technological innovation with human-centered learning principles, creating an educational experience that is both engaging and effective in preparing engineering students for their future careers.

21.7 A case study: Project UNIFORS 2020

The Erasmus+ UNIFORS 2020 project is a good example of the power of international collaboration in advancing innovative educational practices. Bringing together a diverse consortium of European higher education institutions and industry partners – including the University of Warmia and Mazury (Poland) and the University of Vigo (Spain), among others – the initiative aimed to bridge theoretical knowledge with real-world problem-solving through an intensive hackathon-style training program. Held in Liège (Belgium) in late 2021, the week-long event challenged students to break from conventional approaches, harness collective intelligence, and develop creative solutions under time constraints. Participants were introduced to the origins and ethos of hackathons, rooted in collaborative problem-solving and competitive innovation, while engaging in workshops and team-based activities designed to foster interdisciplinary thinking. This immersive experience not only emphasized technical and soft skill development but also highlighted the transformative potential of collaborative learning in preparing students for dynamic professional environments.

The project consortium was composed of the University of Warmia and Mazury in Olsztyn (Poland), the University of Vigo (Spain), the Romanian-American University in Bucharest (Romania), the Polytechnic Institute of Porto (Portugal), the Haute École de la Province de Liège (Belgium), and companies Infalia from Greece and Método Consultores from Spain. With financial support from the European Commission, the project culminated in a week-long intensive training program structured as a hackathon, which took place in Liège from September 26 to October 2, 2021.

21.7.1 The UNIFORS Hackathon

To ensure the maximum effectiveness of this intensive learning experience, students were provided with preliminary materials and concepts to prepare them for the upcoming collaborative work. For example, they were introduced to the hackathon concept by explaining to them that it originated in the 1990s in the USA. Its name comes from the abbreviation "hacker" and "marathon." Originally, the "hackathon" concerned the field of computer science, as it was about decoding a computer system as quickly as possible [16].

Participants are also advised that anything is possible, as long as standard approaches are broken, thanks to collective intelligence to come up with innovative ideas and solutions. Several teams would compete in intense sessions to come up with the best solution to a given problem in a limited time, usually 24, 48, or 72 h [17].

Teams would consist of colleagues and include creative personalities from outside the organization. The hackathon always ends with the presentation of the strategies proposed by each team to a jury of experts, the deliberations of the jury, and finally, the announcement of the winning team. One of the main motivations for participants is the prize that the winning team will receive.

In addition to these sessions, the event often hosts several inspiring and fun workshops. In our case, these were activities organized by the students for the eight groups formed at the beginning of the week.

The hackathon demonstrated the value of blending structured preparation with hands-on, experiential learning. The project created a fertile ground for innovation and skill acquisition by equipping students with foundational materials, fostering cross-cultural teamwork, and integrating competitive yet collaborative dynamics. The inclusion of student-led workshops and presentations assessed by experts further enriched the experience, reinforcing the importance of creativity, adaptability, and effective communication in addressing complex challenges. As this initiative demonstrated, such immersive formats not only cultivate technical expertise but also nurture the interpersonal and critical thinking skills essential for modern careers. The lessons learned from this groundbreaking endeavor – spanning pedagogical strategies, participant engagement, and the integration of diverse perspectives – will be explored in detail in the following paragraphs, shedding light on its broader implications for future educational programs and collaborative innovation.

21.7.2 Lessons learned

Students understood that teamwork is important for developing different tasks in an organization. Many students had confidence in the natural or innate skills associated with teamwork – as with many other social skills – and understood that teamwork skills and techniques could be changed. They found that practical exercises, developed in a face-to-face format, were particularly important to integrate different techniques into core knowledge.

Depending on the culture and age of the students, they did not have the same point of view about time management strategies. For example, life–work balance was not considered important for all young people, and most wanted to work hard to achieve a good position in the company as soon as possible.

In the work itself, it is important that different types of time management tools are offered [18] (e.g., scheduler/calendar, planner/organizer, task and checklists, activity logs, macroplanning, anti-agenda), as not everyone feels comfortable with the same tools. Time management principles and methods (e.g., Eisenhower matrix [19]) should only be applied in a practical way, based on one's own experience.

All students claimed that the online course contributed to their theoretical learning, while the face-to-face meetings enabled them to practice and recall key information.

Students mentioned the benefits of the course for better interaction among people, companies, and other stakeholders. They also mentioned that at the beginning, they were not able to see the point of some of the activities, but in the end, everything became clear to them. Some students mentioned that they sometimes felt problems with interpersonal communication, but did not know that it was possible to improve these skills with some exercises. Overall, the course was followed with great interest.

On many occasions, students found it difficult to think outside the box or to provide ideas that do not strictly follow learned concepts. Students need to be confident when using their creativity: they need to feel that failure is allowed and that there are no stupid ideas. Any idea, even the craziest, can be the foundation of an amazing proposal, project, product, or solution. Working in teams, with specific tools to promote creative ideas, they discovered a place where creativity flourishes.

In general, students erroneously associated the characteristics of critical thinking and creative thinking, requiring a clearer separation of these concepts [20]. Students should have been aware that critical thinking is an ongoing process, not an accumulation of facts and knowledge or something that can be learned once and then used in that form forever. If students are aware that critical thinking enables an effective approach to problem-solving, they can easily learn to use certain tools and achieve much better results in their daily practice. After completing the module on critical thinking, students became confident in applying the tools they had learned in practice with excellent results.

Particularly noteworthy was the discussion of methods and techniques of analytical thinking with reference to their usefulness in specific situations [21]. Students gained a broader perspective and the ability to perceive events rationally, as well as the ability to interpret and infer correctly from a large amount of information, facts, and data. Students tended to accept the first solution that they encountered and had difficulty explaining what guided them to choose a particular solution. They need to understand that analytical thinking is nothing without the ability to argue effectively and the ability to present solutions and ideas convincingly. Students should have been aware that they should continuously develop their analytical skills throughout their lives. Analytical thinking builds on experience and best practice to avoid mistakes in the future.

Students discovered that soft skills such as analytical thinking, critical thinking, creativity, or time management play an important role in the decision-making process. Despite cultural differences, students discovered that being aware of types of decisions helps and facilitates the development of skills to make the right choices. Some students realized that complex decision-making can be seen as a standardized process with specific steps. Group decision-making differs from individual choices in many ways. Students identified both the advantages and disadvantages of group decisions. Students appreciated the workshops on decision-making, mainly because of the opportunity to participate in the activities and to discuss different points of view.

The experience also confirmed that working in groups improves problem-solving. However, as participants were of a similar age, even though they were from different countries, the solutions that they found were very similar. The students realized that solving the problem too quickly was not appropriate. It was reasonable to take several steps to solve the problem correctly. The students enjoyed the logic versus creativity activity, as they had to think a lot and approach the problem "outside the box," they created a competition to see who would solve it first. The activity was highly appreciated by the students as they were able to work in groups and share different opinions. Students were able to relate some of the methods presented in this course to others explained in previous courses.

In conclusion, students found the whole activity very interesting, especially the exercises to review and practice the different methods presented in the online course.

21.8 Concluding remarks

The integration of soft skills into university education represents the natural response to the evolving demands of the modern workforce and the shifting learning preferences of students. Traditional pedagogical methods, which prioritize passive knowledge transmission, are increasingly inadequate for cultivating competencies such as collaboration, adaptability, critical thinking, and emotional intelligence. Instead, educators must adopt interactive, experiential approaches that mirror real-world challenges and foster active student engagement. By shifting the instructor's role from a lecturer to a facilitator and mentor, universities can create dynamic learning environments where students practice soft skills through simulations, group projects, and reflective exercises. This transformation enhances skill acquisition and empowers students to take ownership of their learning paths, recognizing that soft skills are not innate traits, but competencies developed through deliberate practice and feedback.

The design of learning content plays a relevant role in this process. Content must be contextually relevant, updated to reflect industry trends, and delivered through diverse formats – from multimedia resources to gamified activities – to cater to different learning styles. Case studies, real-world scenarios, and partnerships with industry professionals bridge the gap between theoretical concepts and practical application, making learning both relatable and impactful. Furthermore, continuous feedback mechanisms ensure that curricula remain responsive to student needs and societal changes. By embedding flexibility and adaptability into course design, educators prepare students to navigate the uncertainties of a rapidly changing global landscape.

According to our experience, diversity and impartiality emerged as foundational principles in soft skills education, as classrooms that embrace diverse perspectives foster empathy, cultural awareness, and inclusive problem-solving, which in turn are essential skills for thriving in multicultural and interdisciplinary environments. Teachers must model impartiality, guiding students to evaluate ideas critically while maintaining respect for differing viewpoints. This balance cultivates environments where students learn to communicate effectively, negotiate conflicts, and collaborate across boundaries, mirroring the complexities of professional and civic life.

The UNIFORS 2020 case study illustrates the effectiveness of immersive, collaborative learning models such as hackathons. Students developed teamwork, creativity, and decision-making skills in a high-stakes context by combining theoretical preparation with hands-on, time-bound challenges. The project highlighted the importance of blending structured frameworks with opportunities for autonomous exploration, enabling students to apply soft skills in authentic, interdisciplinary settings. Lessons from this initiative confirm that soft skills flourish in environments that encourage risk-taking, iterative learning, and peer-to-peer interaction.

In conclusion, the successful teaching of soft skills in higher education demands a holistic, student-centered approach. It requires reimagining classroom dynamics, embracing technological innovation, and fostering inclusive cultures that value both diversity and critical inquiry. As universities strive to equip students for lifelong professional and personal success, the integration of soft skills into curricula must remain a priority. By aligning pedagogical strategies with real-world demands, educators can ensure that graduates possess not only technical expertise but also the interpersonal and adaptive capacities needed to lead, innovate, and thrive in an interconnected world.

References

[1] Pellegrino, J. W., and Hilton, M. L. (Eds.). (2012). *Education for Life and Work: Developing Transferable Knowledge and Skills in the 21st Century*. National Academies Press. DOI:10.17226/13398.

[2] Goleman, D. (1995). Emotional *Intelligence: Why It Can Matter More Than IQ*. Bantam Books.

[3] Brookfield, S. D. (2017). *Becoming a Critically Reflective Teacher* (2nd ed.). Jossey-Bass.

[4] Crawley, E. F., Malmqvist, J., Östlund, S., Brodeur, D. R., and Edström, K. (2014). *Rethinking Engineering Education: The CDIO Approach* (2nd ed.). Springer. DOI:10.1007/978-3-319-05561-9.

[5] Bennett, M. J. (2013). *Basic Concepts of Intercultural Communication: Paradigms, Principles, and Practices* (2nd ed.). Nicholas Brealey Publishing.

[6] Wang, L., and Duan, X. (2025). Generational diversity and team innovation: the roles of conflict and shared leadership. *Frontiers in Psychology*, 15, 1501633. DOI:10.3389/fpsyg.2024.1501633

[7] Liu, J., Gong, X., Xu, S. *et al.* (2024). Understanding the relationship between team diversity and the innovative performance in research teams using decision tree algorithms: evidence from artificial intelligence. *Scientometrics* 129, 7805–7831. DOI:10.1007/s11192-024-05183-0

[8] Kolb, D. A. (2015). *Experiential Learning: Experience as the Source of Learning and Development* (2nd ed.). Pearson Education.

[9] Laurillard, D. (2012). *Teaching as a Design Science: Building Pedagogical Patterns for Learning and Technology*. Routledge. DOI:10.4324/9780203125083.

[10] Karweit, D. (1993). Contextual learning: A review and synthesis. *Educational Reform and Vocational Education* (pp. 53–84). Document number: TD/TNC 63.405, US Department of Education.

[11] Taylor, A. D., and Hung, W. (2022). The effects of microlearning: A scoping review. *Educational Technology Research and Development*, 70(2), 363–395. DOI:10.1007/s11423-022-10084-1.

[12] Clow, D. (2013). An overview of learning analytics. *Teaching in Higher Education*, 18(6), 683–695. DOI:10.1007/s10648-014-9284-0.

[13] Dubiel, A., Kamińska, D., Zwoliński, G., Ramić-Brkić, B., Agostini, D., and Zancanaro, M. (2025). Virtual reality for the training of soft skills for professional education: trends and opportunities. *Interactive Learning Environments*, 1–21. DOI:10.1080/10494820.2025.2450634.

[14] Mittmann G., Zehetner V., Hoehl S., Schrank B., Barnard A., and Woodcock K. (2023). Using augmented reality toward improving social skills: scoping review. *JMIR Serious Games 2023*, 11, e42117. DOI:10.2196/42117.

[15] Fraser, B. J. (2013). Classroom learning environments. In *Handbook of Research on Science Education* (pp. 103–124). Routledge. DOI:10.4324/9780203824696.

[16] Komssi, M., Pichlis, D., Raatikainen, M., Kindström, K., and Järvinen, J. (2015). What are Hackathons for? *IEEE Software*, 32(1), 60–67. DOI:10.1109/MS.2014.78.

[17] Nolte, A., Pe-Than, E. P. P., and Herbsleb, J. D. (2020). How to hackathon: socio-technical tradeoffs in brief, intensive collocation. *Proceedings of the ACM on Human-Computer Interaction*, 4(CSCW), 1–26. DOI:10.1145/2818048.2819946.

[18] Bamford, D. R., and Greatbanks, R. W. (2005). The use of quality management tools and techniques: a study of application in everyday situations. *International Journal of Quality & Reliability Management*, 22(4), 376–392. DOI:10.1108/02656710510591219.

[19] Bratterud, H., Burgess, M., Fasy, B. T., Millman, D. L., Oster, T., and Sung, E. (2020). The sung diagram: revitalizing the Eisenhower matrix. In *Diagrammatic Representation and Inference: 11th International Conference, Diagrams 2020*, Tallinn, Estonia, August 24–28, 2020, Proceedings 11 (pp. 498–502). Springer International Publishing. DOI:10.1007/978-3-030-54249-8_43.

[20] Fatmawati, A., Zubaidah, S., and Mahanal, S. (2019, December). Critical thinking, creative thinking, and learning achievement: How they are related. *Journal of Physics: Conference Series*, 1417(1), 012070. DOI:10.1088/1742-6596/1417/1/012070.

[21] Byrnes, J. P., and Dunbar, K. N. (2014). The nature and development of critical-analytic thinking. *Educational Psychology Review*, 26, 477–493. DOI 10.1007/s10648-014-9284-0.

Chapter 22

Digital thinking as an intensive methodology in the use of soft skills

Rui Raposo[1] and Mário Vairinhos[1]

22.1 Introduction

To stay competitive in today's technology-focused environment, organizations of all kinds need to continually innovate. Digital technology advancements have changed human interaction methods at personal, societal, and product-service levels, which produce both new opportunities and challenges for those working in design, development, and strategic fields. Digital design thinking has become a strong methodology to connect creative processes with empathy and problem-solving to provide meaningful solutions centered around user needs.

The essence of design thinking is more than a linear process or a collection of tools because it represents a mindset focused on human-centeredness and involves collaboration, experimentation, and iteration. Design thinking draws from disciplines including design, neurocognitive science, and organizational behavior to motivate practitioners to examine issues from different viewpoints while questioning underlying assumptions and developing prototypes quickly to evaluate concepts before allocating substantial resources. Though originally linked with physical product design, its use has spread broadly into digital experience development where intangible elements like usability, accessibility, and emotional engagement become key indicators of success.

This chapter provides an overview of design thinking applied to the creation of digital services and products, exploring its principles, processes, and applications while highlighting its significance in fostering innovation in an ever-evolving and shifting digital landscape.

This chapter is organized into five sections, starting with a preliminary reflection on the conceptual evolution from Design Thinking to digital thinking. Next, a selection of the main theoretical and conceptual models with potential application to digital thinking for soft skills is presented, including a critical analysis with the main advantages and disadvantages summarized. Finally, before the

[1]DigiMedia Research Center, Department of Communication and Art, Universidade de Aveiro, Portugal

conclusions, the main challenges that these approaches face in the contemporary world mediated by digital technologies and new emerging paradigms are discussed.

22.2 From design to digital thinking

The roots of what can be seen as design thinking date back to the mid-twentieth century [1], when designers started to move past just aesthetic aspects and toward a broader approach that incorporated problem-solving and user experience. The roots of design thinking can be linked to the contributions of John E. Arnold and Herbert A. Simon, who promoted more organized methods for creative problem-solving. The term ended up gaining greater notoriety in the 1990s through the efforts of David Kelley, founder of the design consultancy IDEO*, and Tim Brown CEO of also IDEO. They proposed a methodology that combined issues linked with empathy, creativity, and rationality to address complex problems and, by doing so, come up with innovative solutions [2]. Work done by Herbert Simon, who introduced the concept of "bounded rationality" [3] and emphasized the importance of decision-making in design tasks, and by Donald Schön, who studied the reflective practices of designers [4] helped to further develop the concept of design thinking. Stanford University's d.school (Hasso Plattner Institute of Design) popularized the concept in the early 2000s by outlining its stages which include the following main actions: empathize, define, ideate, prototype, and test, as depicted in Figure 22.1 – and promoting its use across a wide array of industries and contexts of use. Since then, design thinking has become an essential framework in modern design education and practice. Buchanan's work on "wicked problems" further highlights the complexity placed by design challenges, emphasizing the need for iterative and exploratory approaches [5]. Over time, design thinking has evolved into a structured but flexible framework applicable to a wide range of challenges, from product development to service design.

While design thinking may have brought us up here, its evolution into digital design thinking has yet to define the limits of its evolving horizons and the unpredictable potential of digital technologies. The phrase "We shape our tools and there after our tools shape us" [6] traces a perfect picture of our relationship with many of the digital technologies that have made their way into our daily activities.

Digital design thinking may be broadly defined as the application of design thinking principles to the ideation, creation, improvement, and optimization of digital products, services, and experiences. The theoretical support of digital design

Figure 22.1 Stages of design thinking

*https://www.ideo.com/

thinking merges knowledge from a variety of disciplines, such as cognitive sciences, human–computer interaction, programming, UX design, and organizational behavior. Because of its non-analogue nature, digital design thinking must consider characteristics of digital environments, such as rapid technological evolution, interoperability, decision-making supported by data, and the need and potential for scalability and sustainability. According to Kumar, successful digital design thinking is based on matching solutions with user needs while ensuring they meet business objectives and societal expectations [7]. This match is achieved through the accumulation of digital soft skills, which encompass a blend of technical, strategic, and interpersonal competencies. Dorst emphasizes the importance of problem-framing in design thinking, arguing that how a problem is defined significantly influences the quality of the solution [8]. This perspective is particularly relevant in digital contexts, where problems may often involve more than one stakeholder and a multitude of variables with different types of dependencies.

The already mentioned foundational elements of design thinking, empathy, prototyping, iteration, and co-creation, are also inherently tied to digital design thinking and the development of digital soft skills. Empathy in digital contexts, for instance, involves understanding not only users' explicit needs but also their implicit desires and behaviors, which may emerge through interactions with technology [9]. Similarly, prototyping in digital design often leverages tools like Figma[†] or Adobe XD[‡] to create interactive representations of concepts that can be tested and quickly refined [10]. These activities require professionals to develop proficiency in both technical tools and human-centered approaches, thereby enhancing their digital soft skills.

Digital soft skills may be understood as the non-technical competencies required to navigate, understand, interact, and succeed in digital environments. These skills, which may greatly vary from person to person, include adaptability, communication, collaboration, problem-solving, creativity, and ethical awareness and principles, among others [9]. Digital design thinking provides an ideal framework for developing these skills because it encourages professionals to engage deeply with users, experiment and develop iteratively, and seek interdisciplinarity and ideally transdisciplinarity. For example, co-creation workshops bring together diverse stakeholders, fostering inclusivity, generating diverse ideas, and building trust among participants [11]. Such collaborative settings help professionals cultivate their interpersonal and communication skills that are very useful when working in multidisciplinary teams.

Moreover, the iterative nature of digital design thinking also promotes adaptability and resilience, digital soft skills essential components. Embracing failure as part of the learning process teaches you to remain flexible and persistent even when confronted with uncertainty [12]. This is the sort of mindset that comes in handy in digital contexts, where technologies evolve rapidly, and solutions must continuously adapt to changing user needs and market conditions.

[†]https://www.figma.com/
[‡]https://adobexdplatform.com/

In the context of digital services, products, and experiences, digital design thinking offers a unique framework useful for addressing their characteristics. Unlike the case of tangible goods, digital content, digital artefacts, experiences and the essence of communication flows exist in a virtual space constantly changing, always tending to obsolescence, and dosed by user expectations. Results must guarantee seamless interoperability across platforms, while keeping up with security, privacy, and ethical standards. Interdisciplinary teams made up of designers, developers, data scientists, marketers, and business analysts, despite bringing distinct expertise and perspectives to the table must find common ground on which to communicate and collaborate and build something new. In such environments, digital design thinking serves as both a unifying language and a structured yet flexible approach useful for overcoming ambiguity or strings of thought or action sometimes so specialized no one else in the team understands what they are talking about or looking for. So, how do we go about this challenge of dealing with the many times unpredictable nature of digital technologies, their use and their acceptance?

Digital design thinking may be approached as revolving around five key phases (Figure 22.1), not necessarily sequential or even mutually exclusive.

22.2.1 Empathize

The first phase, **empathize**, involves researching and achieving a deeper understanding of users' needs, motivations, and critical issues and challenges through quantitative and qualitative data collection. This moment, as all the others, is essential because it ensures that solutions are grounded on hard discussable data rather than based on assumptions or preconceived notions. By stepping into users' shoes, designers can possibly uncover latent desires and unmet needs that may not be apparent at first. In a digital context, the user provides more than what is collected through typical data-collection tools such as questionnaires, interviews, focus groups, etc. The user's interaction patterns, attention span, and biometric feedback are also sources of data and information that may be collected and analyzed to understand the users and better design digital solutions for their organizational, group or individual needs.

22.2.2 Define

The second phase, define, focuses on articulating the problem in a clear, practical way using the insights gained during the first phase. Stakeholders work together to create and elaborate statements that capture the essence of the challenge while also allowing for creative inquiry. Effective problem formulation lays the groundwork for later actions, guiding the team to important results without prematurely reducing possibilities.

22.2.3 Ideate

Ideation is the third phase, in which brainstorming sessions unleash communal creativity. Encouraging thought-provoking ideas while deferring judgment creates an environment receptive to breakthrough thinking. Techniques like mind mapping and reverse brainstorming aid in the generation of a diverse range of potential solutions, some of which may appear unrealistic at first look but may inspire feasible alternatives after more consideration.

22.2.4 Prototype

Prototyping is the fourth phase in which ideas take physical form. Digital mockups, clickable wireframes, and functional prototypes are all examples of artifacts that allow teams to iteratively visualize ideas, find problems, and modify functionality. Medium- to high-fidelity prototyping also helps cross-functional teams communicate more effectively, bridging understanding gaps and coordinating efforts towards common goals.

22.2.5 Test

Finally, testing enables the validation of prototypes under close to real-world conditions, providing valuable feedback for improvement. User testing provides insight into the connection between the solutions proposed and the users' actual needs, uncovering areas requiring adjustment or enhancement. Importantly, testing is not a one-time event but an iterative ongoing process open to continuous improvement.

22.2.6 Design thinking in a digital world

Digital design thinking embodies several guiding principles that help to place the user exactly at the center of the whole process. The idea of repositioning the user as the center of technology design is at the heart of contemporary approaches to software development. An obvious example of the influence of design thinking on digital product development methodologies is the approach known as Human-Centered Design (HCD), which is based on the concept of experience, introducing the contribution of psychology to bring technology closer to the user [12]. Closely related is User-Centered Design (UCD), which focuses on designing products that prioritize the needs, goals, and preferences of end users throughout the development process [13]. This approach emphasizes iterative testing and user feedback to ensure that solutions remain aligned with user expectations. At HCD, the process begins with the user and understanding their context, then focuses on their problems and requirements, and finally interacts with design solutions. Also, in other approaches to experience design, such as goal-directed design [14], we can find such similarities and inspiration. This design methodology places at the forefront of the design of digital products and services an understanding of the motivations and attitudes of technology users. To this end, it mobilizes instruments such as personas or the creation of context scenarios, involving digital storytelling techniques, to build an entire framework for promoting empathy with the design team.

Collaboration forms another cornerstone of design thinking, recognizing that innovation rarely occurs in isolation. Cross-pollination of ideas across disciplines establishes links that lead to novel solutions unattainable through soloed efforts.

Furthermore, embracing failure as part of the learning process empowers teams to experiment boldly, knowing that setbacks provide valuable lessons for future success.

In the digital era, design thinking plays a vital role in the development of products and services. While trying to digitize operations, improve customer experience and exploit emerging technologies, including artificial intelligence, block chain and Internet of Things (IoT), organizations are dealing with complexity like never before. In digital ecosystems with such volatility, uncertainty and rapid pace of change, traditional waterfall methodologies find it difficult to survive. Agile practices address some of these issues by promoting flexibility and responsiveness, but, in some cases, they do not prioritize user experience or strategic plans. Design thinking complements agile methodologies by attributing purpose, ensuring that technological advancements serve human needs instead of defining them. This is not always the case as demonstrated in current phenomena like smartphone dependency, Fear of Missing Out connected with social media and other forms of digital dependency.

If a bank wanted to develop a mobile app for its customers, its concerns would probably focus on the following activities. If the development team in charge of the task decided to apply design thinking principles, it would probably begin by engaging with a sample of random customers to collect and understand their banking habits, preferences, expectations and frustrations when dealing with the service. This would enable the collection of costumers' experience insights that might reveal potential concerns about data security, difficulty dealing with or finding account details, or even the desire for personalized advice tailored to their needs. With the data and information collected, the team could go beyond replicating what the costumers already do online or at the bank in person and would be able to redirect their focus on issues like improving trust in their relationship with costumer, simplifying navigation on the app, and providing other personalized services. During the ideation phase, the team could play around with ideas and concepts like gamification or even the inclusion of some artificial intelligence (AI) agents and chatbots that could provide innovative features to the app and differentiate it when compared to apps from other banks or financial institutions. Agile prototyping and iterative testing cycles would help align service goals with user expectations and avoid the need to conduct in-depth redesign and revision further on down the line.

If we decide to do the same exercise in another context, like healthcare and wellbeing, we may consider challenges like improving patient care management using telemedicine digital platforms. Digital design thinking approaches would guide the organization in mapping patient behaviors and interaction patterns with the service provider, identifying critical points, causes and possible solutions, and involve patients, caregivers, and clinicians in activities for co-creating solutions for the problems identified. Such an approach not only addresses operational inefficiencies but also enhances overall patient satisfaction and health outcomes.

Despite its advantages, design thinking is not without limitations when considering its application in the digital world. The exponential growth of the number of users, apps, trends, and legal implications demands substantial time investment upfront for research and exploration, which may conflict with tight deadlines or resource constraints. In a blink of an eye, a new idea may not be that new and be branded as just another app, or something that looks a lot like a service or product already out there. Additionally, its open-ended nature requires strong facilitation skills to keep teams focused, productive and sustainable. Finally, integrating design thinking into established organizational cultures, still grasping what digital transformation means to them, often necessitates cultural shifts that challenge entrenched ways of working.

To overcome some of the abovementioned challenges, teams should combine digital design thinking approaches with other complementary frameworks. For instance, pairing digital design thinking with lean startup methodologies enables possibilities like faster experimentation and market validation, while incorporating design sprints may accelerate progress within compressed timelines. If the organizations add training programs and mentorship initiatives to the equation to build internal capacity, team members will be able to acquire digital skills and develop a mindset adequate for conducting digital design thinking projects in a more effective manner.

A quick look around confirms that we currently live in an era where everyone if hyper-connected, where everything tends to be automated, and where experiences are aimed to be tailored and very personalized. This reality confirms that the relevant role played by digital design thinking approaches in shaping digital services and products will only tend to grow and probably become ubiquitous, as is the technology it deals with. Its emphasis on empathy, collaboration, and experimentation provides organizations and their collaborators with a mindset and a set of skills useful for anticipating trends and delivering solutions that enhance people's lives and experiences in cases where digital media plays an active role. Promoting and nurturing an individual and an organizational culture fueled by curiosity and openness may aid businesses in harnessing the full potential of digital design thinking as a means to drive sustainable innovation and competitive advantage when dealing with products, services, and needs with a digital essence.

22.3 Systematic literature review

Digital design thinking has emerged as a critical approach for addressing the complexities of designing digital products and services in an increasingly interconnected world. The following sections aim to contribute to a better understanding of some of the theoretical foundations, practical applications, and current challenges tackled in the field of digital design thinking.

Table 22.1 summarizes the key findings from the literature reviewed on digital design thinking, explicitly outlining issues such as theoretical foundations, applications, challenges, and competencies.

Table 22.1 Key findings on digital design thinking

Theme	Key finding	References
Theoretical foundations	Design thinking principles (empathy, collaboration, iteration) are foundational for digital innovation	[2,9,15,16]
	Extensions to service design and systems thinking enhance applicability	
	Integration of cognitive processes and problem-solving frameworks supports complex digital challenges	[5,8,17]
Applications	Effective in driving digital transformation and improving customer experiences	[7,18]
	Facilitates innovation through co-creation and user-centered approaches	[10,12]
Challenges	Managing complexity and ambiguity in digital ecosystems remains a significant challenge	[5,8,19]
	Resistance to change, lack of organizational support, and misalignment with business goals hinder adoption	[17,19]
Competencies	Core competencies include empathy, technical literacy, collaboration, adaptability, and ethical awareness	[12–14,20]
	Blending design thinking with other agile methodologies requires specialized skills and tools	[1,10,17]

22.3.1 Methods of design thinking: a comprehensive overview and comparison

Design thinking employs a variety of methods to address complex challenges in the development of digital services and products. From the sample of approaches identified in the literature review conducted, eight methods are further discussed in this chapter because of their effectiveness and widespread use: Empathy Mapping, Journey Mapping, Prototyping, Co-Creation Workshops, Framework for innovation, IDEO's HCD, Stanford's d.school Design Thinking Process, and Systemic Design. Each method encompasses characteristics that make them, according to our opinion, probably more adequate for different stages of the design process and types of challenges. Below, we explore these methods, comparing their applications, advantages, and limitations. The comparison is summarized in a table at the end of this section. This chapter then explores a new method that has burst onto the scene in the last couple of years and is rightfully called.

22.3.1.1 Empathy Mapping

Empathy Mapping is a collective tool employed in the empathize stage of digital design thinking to obtain a more profound understanding of users' emotions, thoughts, and experiences. Designers can gain a deeper insight into users' needs and challenges by visually depicting what they say, think, do, and feel [7]. This approach is especially useful in the initial phases of digital initiatives, where

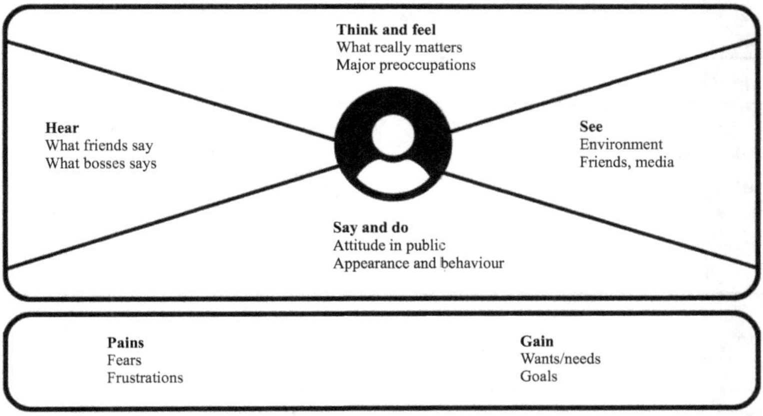

Figure 22.2 Empathy map

Table 22.2 Advantages and limitations of empathy map

Advantages	Limitations
Encourages team alignment around user insights Facilitates empathy-building through visualization Simplifies complex data into digestible formats	Relies heavily on accurate and comprehensive user research May oversimplify nuanced user behaviors if not executed carefully

collecting qualitative information about target users is essential. Its effectiveness relies, however, on thorough and precise research, and it may tend to oversimplify complex behaviors. Figure 22.2 depicts its graphical tool.

Table 22.2 summarizes the main characteristics of the empathy map, organized as its advantages and limitations when applying it.

22.3.1.2 Journey Mapping

Journey Mapping provides a holistic view of the user experience by visualizing the entire customer journey across touchpoints. It helps identify friction points, opportunities for improvement, and areas requiring optimization [7] and is ideal for evaluating existing services or planning new ones. Journey Maps are commonly used in sectors like e-commerce, banking, and transportation to map out deconstruct, understand, and optimize user and experience flows and to possibly enhance satisfaction. Work published by Stickdorn and Schneider [9] has already demonstrated the value of Journey Mapping in uncovering unmet needs and guiding strategic decisions [7]. Its characteristics are summarized in Table 22.3.

Table 22.3 Advantages and limitations of the journey map

Advantages	Limitations
Provides a comprehensive overview of the user experience	Can become overly complex for lengthy or multi-channel journeys
Helps prioritize improvements based on impact and effort	Requires significant upfront research to capture all relevant details
Encourages cross-functional collaboration	

Table 22.4 Advantages and limitations of prototyping

Advantages	Limitations
Reduces risk by identifying issues early in the design process	High-fidelity prototypes can consume considerable time and resources
Enables rapid iteration and adaptation	Over-reliance on prototyping may delay decision-making
Engages stakeholders and users in the design process	

22.3.1.3 Prototyping

Prototyping consists of developing low, medium, or high-fidelity models of products or services to rapidly and iteratively test concepts. Prototypes can vary from basic drawings to interactive digital representations, enabling designers to confirm concepts prior to investing resources in large-scale development. Instruments such as Figma and Adobe XD aid in quick prototyping, allowing designers to confirm their assumptions prior to investing resources in extensive development [1]. While prototyping mitigates risk and attracts stakeholder involvement, high-fidelity prototypes can demand significant resources, and depending too much on them may hinder prompt decision-making. Prototyping plays a crucial role in both the prototype and test stages of digital design thinking and is extensively used in software and hardware design, as well as service innovation, to enhance functionality and collect user input. Principles of interaction design, as outlined by Cooper *et al.* [21], are also essential when creating effective prototypes that align with user expectations and behaviors. Table 22.4 illustrates the key consequences of prototyping.

22.3.1.4 Co-Creation Workshops

Co-Creation Workshops bring together different stakeholders (i.e., users, designers, developers, and business leaders) with the purpose of collaboratively generating ideas and co-developing solutions. These workshops promote inclusivity and end up generating diverse ideas while helping the participants build trust among the participants [10]. They are effective for ideation and concept development activities, especially in contexts that need input from multiple perspectives. They are often employed in community-driven participative projects, corporate innovation

Table 22.5 Advantages and limitations of Co-Creation Workshops

Advantages	Limitations
Fosters inclusivity and shared ownership of solutions	Success depends on skilled facilitation and participant engagement
Generates diverse ideas through cross-disciplinary collaboration	Time-consuming and resource-intensive compared to other methods
Builds trust and rapport among participants	

initiatives, and public sector reforms. However, their success depends on skilled facilitation and participant engagement, making them time-consuming and resource-intensive compared to other methods. A paper by Sanders and Stappers [12] does, however, outline the importance of co-creation in empowering users and driving meaningful innovation [10]. Table 22.5 highlights the advantages and limitations of this technique.

22.3.1.5 Framework for Innovation

The Framework for Innovation, an upgrade of the Double Diamond Framework, developed by the UK Design Council, visualizes the design process as two overlapping diamonds representing divergent and convergent thinking phases. The first diamond focuses on exploring problems (divergence) and defining them (convergence), while the second addresses generating solutions (divergence) and implementing them (convergence). This method, as did the Double Diamond framework, encourages deep exploration before settling on solutions, providing a clear structure for dealing with uncertainty [22]. Its predecessor, the Double Diamond, deemed by some as rigid compared to more flexible methodologies and requires strong facilitation to manage transitions between phases. Its revision has included iteration instances within and among the two structuring diamonds and is kicked off with a challenge and ends in an outcome (Figure 22.3). It also applies

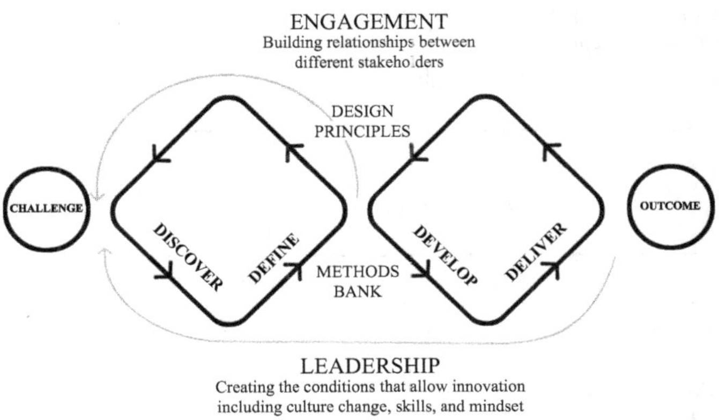

Figure 22.3 Framework for Innovation

design principles such as being people-centered, communicating visually and inclusively, collaborating and co-creating and yet again and as a sort of mantra, iterate, iterate, iterate. This method is particularly effective for tackling ambiguous or "wicked" problems where the challenge itself is sometimes unclear. It is widely used in product development, service design, and organizational transformation projects.

Table 22.6 summarizes the consequences of applying the framework for innovation.

22.3.1.6 IDEO's HCD

IDEO's HCD methodology places people at the core of the design process and emphasizes empathy, iteration, and agile prototyping. The HCD model was further elaborated by Brown and Katz [23], who emphasized the need to balance desirability, feasibility, and viability when designing human-centered innovations. It includes three main phases: Inspiration, Ideation, and Implementation [11]. Something Brown and Katz called the three spaces of innovation HCD is well-suited for projects requiring deep user engagement, such as healthcare services and social impact initiatives, but can be resource-intensive due to extensive user research and testing. Other characteristics are provided in Table 22.7.

22.3.1.7 Stanford's d.school design thinking process

Stanford's d.school design thinking process includes five stages: Empathize, Define, Ideate, Prototype, and Test, being the classical definition of Design

Table 22.6 Advantages and limitations of Framework for Innovation

Advantages	Limitations
Provides a clear structure for dealing with uncertainty	May feel rigid compared to more flexible methodologies
Encourages deep exploration before settling on solutions	Requires strong facilitation to manage transitions between phases
Balances creativity with practicality through its dual-convergence approach	
Enables and promotes iterations throughout the process	

Table 22.7 Advantages and limitations of HCD

Advantages	Limitations
Strong focus on empathy and user needs.	May feel rigid compared to more flexible methodologies
Proven track record in delivering impactful solutions	Requires strong facilitation to manage transitions between phases
Encourages collaboration across disciplines	

Table 22.8 Advantages and limitations of d.school process

Advantages	Limitations
Clear and accessible framework for beginners	Can be overly simplistic for complex, multi-layered problems
Emphasizes iterative cycles of learning and refinement	Requires discipline to avoid skipping critical steps
Fosters cross-functional collaboration	

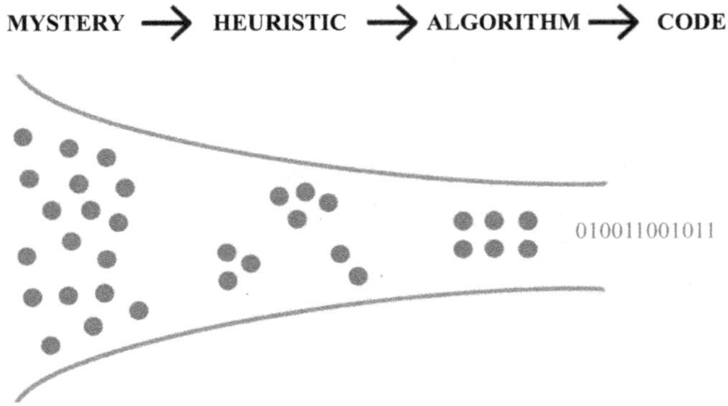

Figure 22.4 Martin's funnel of knowledge

Thinking. Its non-linear nature accommodates the unpredictable nature of digital development, encouraging flexibility and collaboration [16,24]. While accessible and adaptable, the d.school process can oversimplify complex problems if steps are skipped or rushed (Table 22.8).

Its versatility makes it suitable for a wide range of projects, from digital product design to organizational strategy development (i.e., the funnel of knowledge, Figure 22.4).

22.3.1.8 Systemic Design

Systemic Design builds upon conventional design thinking to tackle complex, interconnected issues by incorporating principles of systems thinking. It is suited for tackling systemic challenges such as climate change, urban development, and public health digital transformation [18] because it considers interconnections and long-lasting effects. It combines systems thinking concepts with design methods to develop comprehensive solutions that account for interconnections and lasting effects. Systemic Design is especially important in areas that need systemic actions. Nonetheless, its intricacy and resource demands might hinder adoption in the absence of expertise.

Table 22.9 Advantages and limitations of systemic design

Advantages	Limitations
Addresses root causes rather than symptoms of problems	Complexity can make it difficult to implement without expertise
Promotes sustainability and resilience in solutions	Long timelines and high resource requirements may deter adoption
Encourages collaboration among diverse stakeholders	

Research by Buchanan lays the theoretical foundation for Systemic Design, highlighting its potential to tackle wicked problems [4], which limitations and advantages are listed in Table 22.9.

22.3.2 Summary of digital design thinking methods

Based on what has already been presented, it is easy to see that different approaches are better at different stages of the digital design thinking process or according to the challenges or contexts of use presented. Empathy Mapping is great for understanding your target users, whereas Journey Mapping is suitable for understanding the complete journey of the users. Prototyping allows us to experiment quickly, and Co-Creation Workshops unleash collective creativity and trust. At the same time, frameworks such as IDEO's HCD, Stanford's d.school process and Systemic Design provide high-level guidance for moving through the entire digital design

Table 22.10 Summary of methods

Method	Stage of Digital Design Thinking	Key strengths	Limitations	Ideal use case
Empathy Mapping	Empathize	Encourages empathy; simplifies complex data; aligns teams	Dependent on quality of research; potential for over-simplification	Early-stage user research and persona creation
Journey Mapping	Define	Holistic view of user experience; identifies friction points	Complexity increases with multi-channel journeys; requires thorough research	Evaluating or planning user-centric services
Prototyping	Prototype and test	Reduces risk; enables rapid iteration; engages stakeholders	High-fidelity prototypes can be resource-intensive; delays final decisions	Validating concepts and gathering feedback
Co-Creation Workshops	Ideate and define	Fosters inclusivity; generates diverse ideas; builds trust	Dependent on facilitation skills; time-consuming and costly	Collaborative problem-solving and innovation

(Continues)

Table 22.10 (Continued)

Method	Stage of Digital Design Thinking	Key strengths	Limitations	Ideal use case
Framework for Innovation	Entire process	Provides clear structure; encourages exploration and convergence	May feel rigid; requires strong facilitation	Complex, ambiguous problems
IDEO's HCD	Entire process	Focuses on empathy and user needs; proven track record	Resource-intensive; less effective for technical problems	Human-focused challenges
Stanford's d.school DT	Entire process	Flexible and accessible; fosters iteration and collaboration	Can oversimplify complex problems; skipping steps under-mines effectiveness	Versatile use cases
Systemic design	Entire process	Addresses root causes; promotes sustaina-bility; considers interdependencies	Complex and resource-intensive; requires expertise	Large-scale, interconnected challenges

thinking experience. Table 22.10 summarizes much of what was discussed regarding each method, the stage of the digital design thinking process where each makes sense to be adopted, their key strengths, limitations and ideal use case.

22.4 Bridging design thinking methods with digital design challenges and professional competencies

In the field of digital design innovation, the aforementioned approaches (Empathy Mapping, Journey Mapping, Prototyping, co-creation workshops, framework for innovation, IDEO's HCD, Stanford's d.school Design Thinking Process, Systemic Design) are not simply tools for consideration, but methodologies that are born out of consideration of how to confront the challenges presented by building digital products, services and experiences. The digital landscape included volatile variables such as fast-changing technologies, changing user expectations, cross-platform compatibility, and the necessity of seamless and efficient integration of data-driven insights. To effectively apply these methods in digital contexts, professionals must possess a blend of technical, creative, and interpersonal competencies that enable them to manage ambiguity, nurture collaboration, and deliver impactful results that centered on the users and their relationship with digital technologies.

22.4.1 Challenges in digital design thinking

A quick overview of the literature available reveals that digital design thinking is a dynamic field with robust theoretical underpinnings and practical applications in all and any field that encompasses the use of digital media, services and products as one of its defining variables. Recent advancements emphasize its role in digital transformation, customer experience enhancement, and systemic problem-solving. A successful digital transformation requires aligning innovation with organizational strategy and fostering a people-centric approach, as emphasized by Robu and Lazar [25]. However, challenges such as complexity, organizational resistance, and skill gaps persist and have room for multilevel and multi-goal improvements. By cultivating the necessary competencies and leveraging interdisciplinary approaches, practitioners can harness the full potential of digital design thinking to drive meaningful innovation in an ever-evolving technological landscape. Focusing on scalable implementations, long-term impact assessments, iterative approaches to the development and deployment of solutions, and the integration of emerging technologies like AI and IoT into design thinking frameworks are but a few concerns that are not only trending, but also fundamental to ensure mindset and attitude as dynamic, flexible and organic as the digital landscape.

The progress of digital technology demands iterative design processes capable of accommodating constant change. Digital products, services and experiences cannot afford to freeze up and must evolve in tandem with emerging technologies. Furthermore, the complexity of the user experience has increased exponentially, and users are used to interacting seamlessly across a multitude of devices and platforms (i.e., mobile applications, websites, social media, wearables, voice assistants, smartphones). Users expect consistent, intuitive, and, crucially, accessible and inclusive experiences across all of them, so it is only normal that designers should therefore be aware of and understand what these expectations demand.

While data-driven insights are invaluable in informing design decisions, a critical balance must be struck. The allure of quantitative analytics must not overshadow the importance of qualitative user research. A reliance solely on metrics risks creating solutions that optimize for measurable outcomes at the expense of genuine human needs. Moreover, digital solutions are frequently required to scale rapidly, often globally. This scalability must be coupled with robust performance and security considerations. Equally important is the principle of sustainability. Digital design must embrace long-term thinking, minimizing environmental impact and promoting ethical practices. Finally, successful digital projects invariably involve multidisciplinary teams. Everyone, from designers and developers to data analysts, marketers, and business analysts should find ways to collaborate and communicate within these heterogeneous teams. The diversity of viewpoints, therefore, must be bridged through effective communication and a common understanding of project goals to channel these perspectives towards meaningful outcomes.

We may ask how each of the digital design thinking methods addresses these digital challenges. Empathy mapping, for example, proves valuable in understanding how users perceive and interact with technology. By uncovering pain points related to usability, accessibility, and emotional engagement, it allows designers to create truly user-centered digital experiences. This requires competencies in conducting user interviews, applying surveys, and contextual inquiries, coupled with the ability to interpret qualitative data and translate it into objective design requirements. Journey mapping then aids the visualization of multi-channel user interactions, revealing gaps in the interaction or digital flow and contributing to the improvement of identified workflows. Practitioners must possess strong analytical skills to correlate and synthesize data from diverse sources, integrating quantitative metrics with qualitative insights, and map, visualize and explain these complex journeys effectively.

Before starting large-scale development, prototyping, a crucial component of design thinking, enables rapid concept testing, assumption confirmation, and functionality improvement. It is essential to have practical experience with front-end technologies together with a solid understanding of prototype tools and interface design principles. Co-creation workshops bring together stakeholders from many industries to collaboratively develop innovative digital solutions. This is particularly relevant when developing platforms that require input from end users as well as technical experts. In this situation, having strong facilitation skills and familiarity with agile methodologies are competencies worth having.

In the case of methods like the Framework for Innovation professionals must be flexible in applying this framework to varying project scopes and timelines, demonstrating strong problem-framing abilities. Even in the face of fast technological development, IDEO's HCD methodology makes sure that digital products are firmly grounded in human needs. Its focus on storytelling works especially well for explaining complicated digital ideas to stakeholders who are not technical, therefore it seems sense that empathy-building strategies and storytelling abilities are crucial for its adoption and application.

With an emphasis on collaboration and iteration, Stanford's d.school Design Thinking Process, provides a flexible road map for addressing digital challenges. Because digital progress is unpredictable, its non-linear nature fits in nicely. Strong cross-functional collaboration abilities and an experimental and resilient mindset are crucial. Finally, Systemic Design tackles the interdependence of digital ecosystems, empowering businesses to create sustainable and scalable solutions. This is especially important for large-scale projects. Understanding policy frameworks and being able to think systemically are essential for seeing interdependencies, foreseeing possible repercussions, and trying to guarantee medium to long-term viability.

As seen, a certain set of fundamental skills is required for the effective use of design thinking approaches in the context of use that involve digital solutions of some nature. The most important of them are clearly empathy and a user-centered approach and focus. Effective digital design thinking is based on a clear grasp of user demands, expectations, behaviors, and emotions. To overcome their own innate prejudices and presumptions, professionals in the field must deliberately develop the capacity to seek out different viewpoints from team members, users and the specialized literature. Technical literacy is equally important, as well as hands-on practical know-how in the use of digital tools, platforms, and technology, which will guarantee that designs are practical and useful in addition to being aesthetically appealing and engaging. This includes knowing coding languages, being skilled at software prototyping, and being able to use data analytic tools efficiently.

Collaboration and communication skills are crucial since all teams articulate ideas, listen, and give or receive feedback when required. That is why everyone has gotten through, and all team members contribute to the task at hand regardless of their area of specialization. Digital design thinkers equally need to cross disciplines and interpret complex technical language to make it acceptable to all the stakeholders as well as the team members. In the case of digital innovation, disruption is constant and so is the need to be agile; therefore, it is important to be comfortable with challenges and failures and understand that these are opportunities for growth. Lastly, problem framing and critical thinking is fundamental in determining what the right problem to solve is because context is far more important than solutions and for designers, differentiating between the symptoms and the real problem is critical to ensure that the most effective solutions are offered.

Finally, ethical awareness cannot be compromised, for designers must be responsible for considering the ethical consequences of their work. Such responsibility involves ensuring privacy protection, fostering inclusivity, and actively attempting to do the least amount of harm possible. These skills, if cultivated and practiced thoughtfully, give digital design thinkers the ability to balance the users' relation with the digital world and create meaningful and responsible products, services and experiences.

Table 22.11 Summary of methods and their challenges

Method	Challenge addressed	Key competencies required
Empathy Mapping	– Understanding user perceptions and interactions with technology – Identifying usability and accessibility pain points	– Conducting user interviews and surveys – Interpreting qualitative data – Building empathy with users
Journey Mapping	– Visualizing multi-channel user interactions – Identifying gaps in the digital ecosystem	– Analyzing large datasets from various sources – Synthesizing quantitative and qualitative data – Visualization skills
Prototyping	– Testing ideas quickly and refining functionality before full-scale development	– Proficiency in prototyping software (e.g., Figma, Adobe XD) – Familiarity with interaction design principles
Co-Creation Workshops	– Facilitating collaboration between technical experts and end-users – Generating innovative solutions	– Strong facilitation skills - Agile methodology knowledge – Cross-functional collaboration skills
Framework for Innovation	– Dealing with uncertainty in digital innovation – Encouraging thorough exploration of problems	– Flexibility in applying the framework - Strong problem-framing abilities

(Continues)

Table 22.11 (*Continued*)

Method	Challenge addressed	Key competencies required
IDEO's HCD	– Ensuring digital products remain rooted in user needs – Communicating complex concepts to stakeholders	– Storytelling skills – Empathy-building techniques – Participatory design expertise
Stanford's d.school DT	– Providing a flexible roadmap for unpredictable digital development – Emphasizing iteration and collaboration	– Experimentation mindset - Resilience to failure – Cross-functional collaboration skills
Systemic Design	– Addressing interconnectedness in digital ecosystems – Designing scalable and sustainable solutions	– Systems thinking capabilities - Knowledge of policy frameworks and regulatory environments – Ethical awareness

As shown in Table 22.11, each design thinking method restated has in fact dealt with challenges that needed to be solved and what key competencies were required to tackle the challenges addressed.

22.5 Conclusions

We are living in a time when, once again in our recent history, profound changes are approaching in the way we work, communicate and engage with the digital world. New technological paradigms are beginning to emerge with a transformative potential for society that we cannot ignore. Technologies that we are aware of, but that now seem to be emerging in a more permanent and mature way, with Virtual and Augmented Reality, AI, have gradually entered our lives, transforming the way we socialize, communicate and produce wealth.

Digital Design Thinking approaches, which have proven their quality in the recent past, are now viewed by various areas of knowledge and economic activities as instruments that are not only desirable but also necessary for all young students and those entering the job market, allowing them to develop skills and tools for problem-solving, in the search for creative and innovative solutions.

References

[1] Kimbell, L. (2011). Rethinking design thinking: Part I. *Design and Culture*, 3 (3), 285–306.

[2] Brown, T. (2009). *Change by Design: How Design Thinking Transforms Organizations and Inspires Innovation*. New York: Harper Collins.

[3] Simon, H. A. (1972). Theories of bounded rationality. *Decision and Organization*, 1(1), 161–176.

[4] Visser, W. (2010). Schön: Design as a reflective practice. *Collection*, (2), 21–25.
[5] Buchanan, R. (1992). Wicked problems in design thinking. *Design Issues*, 8 (2), 5–21.
[6] Culkin, J. M. (1967). A schoolman's guide to Marshall McLuhan. *The Saturday Review*, 51–53, 70–72.
[7] Kumar, V. (2012). *101 Design Methods: A Structured Approach for Driving Innovation in Your Organization*. New York: Wiley.
[8] Dorst, K. (2011). The core of 'design thinking' and its application. *Design Studies*, 32(6), 521–532.
[9] Stickdorn, M., and Schneider, J. (2012). *This Is Service Design Thinking: Basics, Tools, Cases*. New York: Wiley.
[10] Liedtka, J. (2018). Why design thinking works. *Harvard Business Review*, 96 (5), 72–79.
[11] Preece, J., and Rogers, Y. (2015). *Interaction Design: Beyond Human–Computer Interaction*. Chichester: Wiley.
[12] Sanders, E. B.-N., and Stappers, P. J. (2012). *Convivial Toolbox: Generative Research for the Front End of Design*. Amsterdam: BIS Publishers.
[13] Interaction Design Foundation – IxDF. What is User Centered Design (UCD)? Interaction Design Foundation – IxDF. [Online]. Available: https://www.interaction-design.org/literature/topics/user-centered-design.
[14] Kelley, D., and Kelley, T. (2013). Creative confidence: Unleashing the creative potential within us all. *Crown Business.*
[15] Cross, N. (2023). *Design Thinking: Understanding How Designers Think and Work.* London: Bloomsbury Publishing.
[16] Gonen, E. (2019) "Tim Brown, Change by Design: How Design Thinking Transforms Organizations and Inspires Innovation (2009)," *Markets, Globalization & Development Review*, 4(2), 8, https://doi.org/10.23860/MGDR-2019-04-02-08 Available at: https://digitalcommons.uri.edu/mgdr/vol4/iss2/8.
[17] Lawson, Bryan. (2006). *How Designers Think – The Design Process Demystified*. Cambridge: Cambridge University Press.
[18] Sangiorgi, D., Lima, F., Patrício, L., Joly, M.P., and Favini, C. (2019). A Human-Centred, Multidisciplinary, and Transformative Approach to Service Science: A Service Design Perspective. In: Maglio, P.P., Kieliszewski, C.A., Spohrer, J.C., Lyons, K., Patrício, L., and Sawatani, Y. (eds) *Handbook of Service Science, Volume II*. Service Science: Research and Innovations in the Service Economy. Cham: Springer, https://doi.org/10.1007/978-3-319-98512-1_7.
[19] Tschimmel, K. (2012). Design Thinking as an effective Toolkit for Innovation. In *ISPIM Conference Proceedings* (p. 1). Bergen: The International Society for Professional Innovation Management (ISPIM).
[20] Verganti, R. (2009). *Design Driven Innovation: Changing the Rules of Competition by Radically Innovating What Things Mean*. Brighton, MA: Harvard Business Press.

[21] Cooper, A., Reimann, R., Cronin, D., and Noessel, C. (2014). *About Face: The Essentials of Interaction Design* (4th ed.). New York: Wiley.

[22] Verganti, R., Dell'Era, C., and Swan, K. S. (2021). Design thinking: Critical analysis and future evolution. *Journal of Product Innovation Management*, 38(6), 603–622.

[23] Brown, T., and Katz, B. (2011). Change by design. *Journal of Product Innovation Management*, 28(3), 381–383.

[24] "Design Thinking Bootleg", Stanford d.school., [online] Available: https://dschool.stanford.edu/resources/design-thinkingbootleg.

[25] Robu, D., and Lazar, J. B. (2021). Digital transformation designed to succeed: Fit the change into the business strategy and people. *Electronic Journal of Knowledge Management*, 19(2), 133–149.

Chapter 23
Summary

Itziar Goicoechea[1]

Throughout this book, we have explored the importance of soft skills in both the workplace and personal life, and how their development is crucial for long-term success. In an increasingly interconnected and collaborative world, technical skills, though essential, are no longer enough to thrive. The ability to communicate effectively, collaborate in teams, resolve conflicts, adapt to change, and manage time efficiently, among others, has become a necessity for every professional.

As a conclusion, a summary of each chapter will be provided to emphasize the most relevant aspects of soft skills.

23.1 Introduction

This first chapter introduces the growing importance of soft skills in higher education, especially in engineering. While students must master the technical skills specific to their field, soft skills are crucial for securing employment and advancing professionally.

The text presents the main competency frameworks established by the European Commission: DigComp (digital competencies), GreenComp (sustainability), LifeComp (personal and social competencies), and EntreComp (entrepreneurship), as well as the classification provided by The EU Skills Panorama and The Occupational Information Network (ONET)*. In this classification, soft skills are divided into three main categories:

1. Methodological skills including time management, problem-solving, decision-making, learning strategies, planning, and digital skills.
2. Social skills including interpersonal communication, teamwork, conflict management, and intercultural understanding.
3. Cognitive skills including analytical, critical, reflective, and creative thinking.

With all this information, a table has been created to organize the relationships among various European Union competence frameworks and the soft skills discussed in the book.

[1]Departamento de Deseño na Enxeñaría, Universidade de Vigo, Spain

This chapter also emphasizes the importance of integrating these skills into engineering programs, as they complement traditional technical training and are increasingly in demand by employers. The book is presented as a tool to help both professors and students develop these essential competencies.

23.2 Why soft skills?

In today's world, characterized by technological development and Industry 4.0, soft skills have become of fundamental importance alongside traditional technical competencies. This chapter explores their relevance from multiple perspectives:

1. **Context and Need.** The Fourth Industrial Revolution has transformed the work environment, where human capital is the element that connects technology with reality. Although technical competencies remain essential, soft skills such as communication, teamwork, and adaptability have become equally crucial for professional success.
2. **Role of Universities.** Universities and training centers must recognize this need and adapt their educational programs to include the development of soft skills. This involves not only teaching technical content but also implementing methodologies that promote interaction, critical thinking, and empathy among students. Continuous education and personalized learning are key to preparing future professionals for an ever-changing work environment.

   Educational institutions are adapting to this new reality by
   - Implementing more flexible and practical programs,
   - Developing collaborations with companies and organizations,
   - Incorporating digital technologies into teaching,
   - Emphasizing continuous and personalized learning.
3. **Personal Benefits.** Soft skills benefit individuals in three main areas:
   - Professional development improves employability and job performance,
   - Interpersonal relationships facilitate communication and teamwork,
   - Self-actualization increases self-awareness and the ability to manage personal development.
4. **Importance for Industry 4.0.** Companies seek professionals who not only master tools and technologies but also possess interpersonal competencies such as effective communication, teamwork, and conflict resolution. These skills are essential for fostering a collaborative and adaptive work environment where innovation and creativity can thrive.

In the context of automation and digitalization, soft skills become more relevant because

- Routine jobs are being automated,
- Greater adaptability to change is required,
- Collaboration in multicultural and virtual teams is increasingly common, and

- Effective communication and leadership are fundamental for managing complex projects.

In conclusion, the combination of technical and soft skills is essential for success in today's job market. Fostering these competencies not only enhances individuals' employability but also contributes to the growth and sustainability of organizations in an increasingly competitive market.

23.3 What do companies demand? What do universities offer?

In 2018, the UNIFORS 2020 project, funded by Erasmus+ of the European Union, conducted surveys in six European countries to identify the most valued soft skills in the labor market. In 2024, these surveys were repeated to assess the evolution of skill demands in the current work environment, after the COVID-19 pandemics. The surveys targeting companies focused on the competencies they consider essential for their employees' professional success, revealing a high demand for skills such as teamwork, critical thinking, and digital literacy.

At the same time, surveys were conducted with students to understand their perceptions of the crucial skills for their professional development and the challenges they face in acquiring them. While companies emphasized the importance of social and digital skills, students valued problem-solving and creativity, expressing frustration over the lack of training opportunities in these areas.

More recently, in 2025, an analysis of online job advertisements (OJAs) showed that 51.4% of postings required soft skills, highlighting the importance of flexibility and adaptability. The OJAs also indicated that cognitive skills, such as critical thinking and problem-solving, are increasingly in demand, underscoring the need for educational institutions to align their programs with labor market expectations.

These findings emphasize the growing need for educational institutions to adapt their training programs to effectively prepare students for an ever-evolving work environment.

23.4 Time management

Time management is a vital skill that significantly influences both personal and professional aspects of life. It enables individuals to prioritize tasks effectively, minimize procrastination, and allocate their time wisely, leading to enhanced productivity and reduced stress levels. By mastering time management, one can achieve a more balanced lifestyle, ensuring that work commitments do not overshadow personal well-being.

Implementing effective techniques is key to successful time management. Methods such as the Pomodoro Technique, which encourages focused work sessions followed by short breaks, the Eisenhower Matrix that helps prioritize tasks based on urgency and importance, and SMART goals that ensure objectives are Specific, Measurable, Achievable, Relevant, and Time-bound, are all valuable strategies. These techniques not only streamline task execution but also foster a sense of accomplishment and motivation.

In addition, developing essential skills such as planning, organization, time estimation, and self-confidence is crucial for maximizing productivity. Utilizing both digital tools, like task management apps and calendars, and traditional methods, such as to-do lists and planners, can significantly enhance task organization and deadline adherence. Ultimately, effective time management leads to improved efficiency, better decision-making, and a higher quality of life, allowing individuals to thrive in both their personal and professional endeavors.

23.5 Decision-making

Decision-making is a critical process that influences various aspects of life, from everyday choices to strategic decisions in professional environments. It begins with the clear identification of the problem that needs to be addressed, which sets the direction for finding effective solutions. Once the problem is defined, gathering relevant information becomes essential to understand the circumstances and explore possible alternatives.

Decisions can be categorized into various types, including strategic, tactical, operational, rational, and intuitive. Each type has its own context and methodology, indicating that there is no universal approach. For instance, strategic decisions often require in-depth analysis and a long-term vision, while operational decisions may be more immediate and based on practical experience.

The decision-making process involves evaluating the identified options by considering factors such as feasibility, cost, time, and potential consequences. This critical analysis is vital for selecting the best alternative. Additionally, decisions can be made individually or in groups. Group decision-making can enhance the process by incorporating diverse perspectives and expertise, leading to more innovative solutions. However, it can also introduce challenges, such as slower decision-making and the risk of conformity, which may stifle creativity.

Dawson identifies five situations where group decision-making is particularly beneficial: when the team can develop more solutions, when external viewpoints or expertise are needed, when ethical standards need to be elevated, when group support is necessary for the expected decision, and when there is uncertainty about the acceptance of one's own ideas.

Ultimately, once a decision is made, implementing the corresponding action and evaluating the results is crucial. The ability to learn from past decisions, both successful and unsuccessful, is fundamental for continuous improvement. In summary, decision-making is an art that combines analysis, intuition, and collaboration, and mastering it can lead to greater success and satisfaction in both personal and professional realms.

23.6 Problem-solving

Problem-solving is a fundamental skill classified by the European Union within the competencies of communication, collaboration, and creativity. It is defined as the process of developing and implementing solutions to practical, operational, or conceptual problems in various work contexts. This chapter highlights its importance for both students and educators, emphasizing its impact on professional development by helping to identify goals, assess progress, and verify outcomes. The ability to solve problems is essential not only in the academic realm but also in the workplace, where challenges are constant and require effective solutions.

The problem-solving process is divided into several key stages. First, it is crucial to identify and define the problem clearly and precisely, as a good definition is the first step toward an effective solution. Next, alternative solutions are generated, fostering creativity and critical thinking. Then, the best options are evaluated and selected, considering factors such as feasibility, cost, and potential impact of each solution. To facilitate this process, tools such as pros and cons lists, situational analysis, and techniques like the "5 Whys" and Pareto analysis are suggested, which help break down complex problems into more manageable parts.

Teamwork should also be encouraged, as collaboration can enrich the problem-solving process by bringing in diverse perspectives and experiences. Keeping the stages of idea generation and evaluation separate allows for a more effective and creative approach to finding viable solutions, preventing premature judgment from limiting creativity.

Finally, the chapter underscores the importance of continuous practice in problem-solving, suggesting that accumulated experience enhances the ability to face future challenges. Adaptability and a willingness to learn from mistakes are essential for personal and professional growth, making problem-solving a key competency in today's world. In a constantly changing work environment, the ability to solve problems effectively not only improves productivity but also fosters a more innovative and collaborative workplace.

23.7 Learning strategies

The development of soft skills is essential in today's professional landscape, as these interpersonal and communication competencies are increasingly valued by employers. To cultivate these skills, as well as technical ones, a continuous learning approach is necessary, allowing individuals to adapt to a constantly changing work environment. In this context, learning strategies become key tools for optimizing the educational process and fostering student autonomy.

Metacognition, which refers to the ability to reflect on one's own learning process, is fundamental for students to evaluate their knowledge and progress. By developing this skill, learners can identify their strengths and weaknesses, enabling them to adjust their study methods effectively. However, it is important to recognize that common myths about learning can hinder the effectiveness of these strategies, such as the belief that intelligence is fixed or that learning must be a linear process.

Another key factor is the relationship between motivation and success; as students achieve their goals, their motivation tends to increase, creating a positive cycle that fosters greater engagement with learning. To maximize this effect, it is recommended to establish a learning plan that includes distributed and varied practices, which not only improves information retention but also enriches content understanding. By integrating these strategies, students can become more effective and autonomous learners, prepared to face future challenges.

23.8 Planning

Planning constitutes a fundamental methodological competence recognized in the European Union's Skills Panorama, being especially crucial for team performance and project management in engineering.

In the context of team effectiveness, planning emerges as a critical process that develops in the early stages of team development. This process involves two main elements: gathering information from members and stakeholders and evaluating this information to formulate effective strategies. It acts as an essential transition mechanism that facilitates the identification of tasks, goals, and resources, becoming a recurring process that allows teams to adapt to new circumstances and challenges.

From the project management perspective, planning defines the project's direction by identifying fundamental requirements and constraints. Its main results materialize in the management plan and project documents, characterized by being an iterative process that requires continuous adjustments according to new information and changing environmental conditions.

Within the European Union's competency framework, planning is integrated into three fundamental frameworks: it is relevant for lifelong learning (LifeComp), crucial for digitalization (DigComp), and fundamental for sustainability (GreenComp). To develop these planning skills, various teaching methods are employed, including design thinking teamwork, game-based learning, team-based learning, problem-based learning, and project-based learning, with the latter being the most effective, specifically in the field of engineering.

The successful implementation of these methods requires key elements such as a proactive instructor role, the use of support tools and artifacts, a clear structure and guidance during the process, and continuous evaluation of skills development. However, this process faces significant challenges, such as the need for proactive guidance from instructors, the considerable time and effort required to guide teamwork, and the lack of standardized criteria for evaluating teamwork skills. Nevertheless, these challenges come with important opportunities, such as the implementation of evidence-based approaches, the development of standardized frameworks for assessment, and the integration of technological collaboration tools.

In conclusion, planning reveals itself as an essential competence in today's world, characterized by its digitalized nature and focus on sustainability.

Engineering education provides an ideal environment to cultivate these skills through active and evidence-based learning methods, thus preparing future engineers to address the complex global challenges they will face in their professional practice.

23.9 Digital skills

"Digital competences" have gained significant relevance in the current context, as they are fundamental for driving transformation in various areas of society and the professional realm. These competences encompass a set of skills, knowledge, and attitudes that enable individuals to interact effectively and responsibly with digital technologies. The European Union has recognized this importance and has developed frameworks such as DigComp 2.2, which provides clear guidelines for assessing and progressively developing these competences in different contexts.

The DigComp 2.2 framework focuses on several key areas, including information and data literacy, communication and collaboration, digital content creation, online safety, and problem-solving. These areas are essential for citizens not only to become consumers of information but also to be active creators and critics in the digital environment. Likewise, a specific framework for the educational sector, known as DigComEdu, has been created, targeting students, teachers, and educational organizations, emphasizing the need to integrate these competences into teaching and learning.

A crucial aspect of digital competences is the ability to access and critically evaluate information. In a world where misinformation and fake news are prevalent, it is vital for individuals to develop skills to discern the veracity of the information they consume and share. This involves not only seeking reliable sources but also understanding the algorithms that influence search results and how these can bias the information presented.

Protecting devices and personal data is another essential component of digital competence. With the rise of cyber threats, it is fundamental for users to implement security measures, such as using strong passwords, two-factor authentication, and protection software. Equally, managing online privacy has become critical, as users must be aware of the amount of personal information they share and how it can be used.

In the end, health and well-being in digital environments are aspects that should not be overlooked. Excessive use of devices and constant exposure to social media can have negative effects on mental and physical health. Therefore, it is important to set limits on screen time and promote healthy habits that encourage a balance between digital life and real life.

Clearly, digital competences are essential today, not only for professional development but also for active and responsible participation in society. Acquiring these competences enables individuals to navigate effectively in an increasingly digitized world, contributing to their personal well-being and that of the community.

23.10 Team working

Teamwork is an essential skill in both professional and social environments, where individuals collaborate interdependently to achieve a common goal. An effective team is characterized by clear and open communication, a shared focus on objectives, and a supportive atmosphere that fosters diverse perspectives. As a team forms, it goes through several phases: from initial formation, where relationships are established, followed by a second phase of conflict (storming), where disagreements and differences arise. Next is the norming phase, where clear norms and roles are defined. The fourth phase is called performing, and this is where the team truly excels, working cohesively and productively. As a final point, we have the adjourning phase, which occurs when the team disbands after achieving its goals.

However, it is crucial to recognize that emotions play a fundamental role in team dynamics. Empathy, assertiveness, and good self-esteem are vital for maintaining a positive and collaborative environment. Similarly, each team member may take on different roles, as described in the Belbin model, which helps maximize the group's performance and effectiveness.

Nevertheless, teamwork also faces challenges. It is important to identify and address "hitchhikers" (members who do not contribute) and "couch potatoes" (passive members) who can hinder the team's progress. In summary, teamwork is not just about achieving goals, but also about building strong relationships and creating an environment where each member feels valued and motivated to contribute to collective success.

23.11 Interpersonal communication

Chapter 11 focuses on the importance of interpersonal communication for engineers and professionals. This skill is not only essential for collaboration and project management but also influences career advancement. It emphasizes that effective communication goes beyond the mere transmission of information; it involves a deep understanding of the motivations and emotions underlying the exchange. Therefore, skills such as empathy, active listening, and the ability to tailor messages are fundamental, especially in interdisciplinary teams working on projects with significant social impacts.

This chapter also mentions that the success of a project often depends on how engineers interact with clients, colleagues, and stakeholders. The ability to explain technical concepts in an accessible manner and build trusting relationships is crucial. Various communication contexts within the engineering field are discussed, such as teamwork, conflict resolution, and professional networking.

However, there are challenges that can obstruct effective communication, such as the use of technical jargon, the complexity of information, cultural differences, and time constraints. To overcome these barriers, several strategies are proposed, including active listening, clear expression of ideas, harmonious conflict resolution, and cultural sensitivity.

The chapter concludes by highlighting that the development of communication skills is an ongoing process that requires reflection and practice. Engineering students are encouraged to engage in extracurricular activities, internships, and courses that reinforce these competencies, which will not only enhance their professional performance but also contribute to their personal growth.

23.12 Conflict management and negotiation

This chapter addresses two fundamental topics in project management: conflict management and negotiation, both considered critical professional skills for success in technical and engineering environments.

Regarding conflict, this text defines it as a natural phenomenon in organizational social interactions that occurs at multiple levels. Three main types of conflict within teams are identified: relationship conflict (arising from interpersonal differences), task conflict (disagreements over work-related ideas and opinions), and process conflict (disagreements over logistics and delegation).

An important concept introduced is that of "conflict profiles," which represent how these different types of conflict manifest in combination within teams. Research has identified three main profiles: task conflict-dominant, minor/moderate general conflict, and dysfunctional conflict. The most effective teams tend to exhibit the task conflict-dominant profile, with minimal levels of interpersonal and process conflict.

For effective conflict management, the text emphasizes prevention over resolution. It recommends focusing on the team formation stage, establishing goal alignment, collaborative planning, and the use of team charters. Common barriers to conflict resolution are also identified, including cognitive biases, power dynamics, and emotional barriers.

Regarding negotiation, it is presented as a joint decision-making process between two or more parties to reach mutually acceptable agreements. Key concepts introduced include Zone of Possible Agreement and Best Alternative to a Negotiated Agreement. The text highlights five conflict management styles based on concern for oneself and concern for others: integrating, accommodating, dominating, avoiding, and compromising.

Finally, the chapter warns about the "groupthink" phenomenon, where the desire for consensus can override the critical evaluation of alternatives, leading to poor decision-making. The importance of developing negotiation skills through practical methods such as role-playing and authentic team projects in academic settings is emphasized.

This knowledge is essential for preparing current and future engineers to effectively handle conflicts and negotiations in their professional environments.

23.13 Intercultural understanding

This chapter examines the significance of intercultural communication in today's globalized world. A key concept discussed is Hall's Iceberg Model, which illustrates that only about 20% of a culture is visible, while the remaining 80% lies beneath the surface, encompassing deeper values and beliefs.

The growing importance of intercultural communication is attributed to factors such as the Internet, increased global travel, and workforce mobility. While language differences present an obvious challenge, deeper cultural barriers – such as variations in communication styles and underlying values – pose greater complexity.

Erin Meyer's Culture Map model is introduced, outlining eight key dimensions of cultural differences: Communicating (low vs. high context), Evaluating (direct vs. indirect feedback), Persuading (principles vs. applications first), Leading (egalitarian vs. hierarchical), Deciding (consensual vs. top-down), Trusting (task- vs. relationship-based), Disagreeing (confrontational vs. avoiding), and Scheduling (linear vs. flexible time).

This chapter also explores culture shock, the experience of uncertainty and confusion when adapting to a new cultural environment. This process is described through four stages: honeymoon, crisis, recovery, and adjustment. In the same vein, reverse culture shock is addressed, highlighting the challenges of readjusting to one's home culture after an extended period abroad.

Common barriers to intercultural communication include stereotypes, discrimination, and xenophobia. A reference is made to Unia, a Belgian institution that identifies 19 discrimination criteria, ranging from racial and linguistic factors to social origin.

Practical strategies for effective international communication are emphasized, including maintaining an open mindset, recognizing personal cultural biases, ensuring clear communication, and respecting local customs. A cultural checklist for international presentations is also provided, covering aspects such as language formality, content structure, and expected behaviors.

Ultimately, the chapter underscores that successful intercultural communication extends beyond language proficiency; it requires cultural awareness, adaptability, and respect for diverse perspectives. This idea is encapsulated in Henry Miller's quote: "One's destination is never a place, but a new way of seeing things."

23.14 Analytical thinking

Analytical thinking is an essential skill that allows individuals to break down complex problems into more manageable parts, facilitating understanding and resolution. This systematic approach is based on logic and data analysis, making it a crucial tool in both educational and professional settings. Through a structured process, analytical thinking guides individuals in identifying problems, formulating hypotheses, collecting relevant data, and analyzing the information obtained.

One of the most important characteristics of analytical thinking is its ability to avoid cognitive biases that can distort the perception of reality. Common errors, such as premature closure, can lead to incorrect conclusions if the problem is not approached critically and openly. Therefore, it is crucial to adopt an approach that fosters curiosity and exploration, allowing for multiple perspectives to be considered before reaching a conclusion. This process not only improves the quality of decisions but also promotes an environment of continuous learning.

Additionally, the use of formal methods, such as ontological theory and contextual logic, can significantly enhance knowledge organization and decision-making. These tools provide a framework that helps structure thought and clarify the relationships between different elements of a problem. Ontological theory, for example, allows analysts to better understand the categories and relationships that exist within a data set, while contextual logic aids in interpreting information based on the context in which it is presented.

Analytical thinking also encourages collaboration and teamwork, as it enables individuals to share their perspectives and approaches to problem-solving. By integrating different viewpoints, more creative and effective solutions can be generated. In summary, analytical thinking is not only vital for problem-solving but also a skill that can be developed and refined over time, contributing to personal and professional growth. Its constant practice can lead to greater effectiveness in decision-making and improved performance in various areas of life.

23.15 Creative thinking

Chapter 15 delves into the concept of creative thinking, emphasizing its relevance in various aspects of life, both professional and personal. Often, creativity is associated with innate talent, but the text challenges this notion, arguing that creativity is, in fact, a skill that can be cultivated and refined over time.

Two fundamental types of thinking are presented in the creative process: convergent thinking, which focuses on finding the most effective solution to a specific problem, and divergent thinking, which seeks to generate a wide range of ideas and possibilities. This distinction is crucial for understanding how to approach challenges more effectively.

The chapter also details the four stages of the creative process:

1. Preparation: where information is gathered, and the problem is defined.
2. Incubation: a period of reflection where ideas can mature subconsciously.
3. Illumination: the "eureka" moment when a new idea or solution emerges.
4. Verification: the final stage where the idea is evaluated and implemented.

To facilitate the development of creative thinking, several techniques are introduced, such as SCAMPER, which encourages modifying existing ideas, and the Six Thinking Hats, which promote different perspectives in decision-making.

Ultimately, a comparison is drawn between creative thinking and Design Thinking, emphasizing that the latter not only seeks innovative solutions but also prioritizes empathy and collaboration in the problem-solving process. This holistic approach highlights the importance of understanding the needs of others to generate ideas that truly resonate and are effective.

23.16 Critical thinking

Critical thinking is an essential mental process that enables individuals to analyze and evaluate information, as well as question its validity and relevance. This approach is crucial in various contexts, from education to professional environments and everyday life, as it helps individuals make informed and well-founded decisions.

By developing critical thinking skills, people can enhance their ability to solve problems, identify biases and fallacious arguments, and formulate more accurate judgments.

To cultivate critical thinking, it is fundamental to practice consistently and employ learning methods that encourage reflection and deep analysis. This can include debates, case studies, and the use of open-ended questions that challenge assumptions. Likewise, educational technologies, such as online platforms, simulations, and interactive resources, have become valuable tools for facilitating the development of these skills, allowing students and professionals to explore different perspectives and scenarios.

In an increasingly complex world filled with information, critical thinking becomes even more relevant. It not only helps individuals navigate the abundance of available data but also empowers them to contribute to a more informed and engaged society. By fostering critical thinking, individuals are equipped to face contemporary challenges with confidence and clarity, promoting constructive dialogue and more effective decision-making.

23.17 Reflective thinking

Reflective thinking has emerged as a fundamental tool in the learning process, especially in engineering education. This approach allows students to deepen their acquired knowledge, relate previous concepts to new information, and identify areas for improvement in their skills. Research conducted at Kaunas University of Technology examined the implementation of reflective tools in 27 study modules that use the design thinking method.

The design thinking process consists of six phases: understanding, observation, synthesis, ideation, prototyping, and testing. Different reflective tools were implemented in each phase. For example, in the understanding phase, mental and conceptual maps were used; in the observation phase, empathy maps and user journey mapping were employed; while in later phases, SWOT analysis, SCAMPER techniques, and prototype feedback grids were utilized.

The study results revealed that students positively valued these reflective tools, highlighting their usefulness in connecting theory with practice and developing professional skills. However, they also pointed out the need for clearer instructions and concrete examples for implementation. The research concluded that structured reflection, when properly aligned with learning objectives and implemented with clarity, significantly improves the understanding of theoretical concepts and fosters the development of critical skills necessary in professional engineering practice. Furthermore, students emphasized that combining different reflection tools throughout the design thinking process enabled them to develop deeper insights and more innovative solutions to engineering challenges.

23.18 A course sheet for training soft skills

Chapter 18 presents a course proposal focused on the development of soft skills, designed for university students across various disciplines. Recognizing that while these students may possess basic technical knowledge in their respective fields, they often lack essential interpersonal and communication competencies. This course aims to address this gap. The proposal suggests that the course can be implemented as an independent program or integrated into a broader curriculum that includes topics such as project management, ethics, and leadership.

The course is structured to offer a total of 5 European Credit Transfer System (ECTS), distributed across 20 h of direct instruction, 36 h of e-learning, and 69 h of autonomous work, mainly by teams, allowing for flexibility in credit allocation and time commitment. This structure is designed to accommodate the diverse academic requirements of students, providing a comprehensive and customizable learning experience. Course modules cover organizational skills, creativity, problem analysis, and decision-making, each with a focus on practical application and teamwork.

Additionally, the importance of collaboration and interaction between students and instructors is emphasized. Online forums are planned for ongoing discussions, as well as the possibility of face-to-face meetings and consultations via video calls to address specific questions. Course evaluation will be conducted through rubrics that assess both content and collaboration, along with satisfaction surveys to continuously improve the learning experience. In summary, this course not only seeks to equip students with soft skills but also fosters a collaborative and adaptive learning environment that responds to the needs of a diverse student population.

23.19 A collection of cases of use for understanding soft skills

In the rapidly evolving field of engineering, the integration of technical and soft skills has emerged as a critical factor for achieving professional success. While technical skills, such as proficiency in mathematics, programming, and engineering principles, form the foundation of an engineer's expertise, they are no longer sufficient in isolation. Engineers must also cultivate a range of soft skills, including effective communication, leadership, adaptability, and teamwork, to navigate the complexities of modern work environments.

The Future of Jobs Report 2023 highlights the increasing importance of soft skills in the job market, identifying key competencies such as analytical thinking, creativity, and resilience as essential for future success. These skills enable engineers to approach problems from multiple perspectives, innovate solutions, and adapt to changing circumstances. As projects become more interdisciplinary and collaborative, the ability to communicate ideas clearly and work effectively with diverse teams is paramount.

To support the development of these vital skills, various tools and strategies are available for both students and professionals. For example, the Eisenhower Matrix serves as a practical tool for enhancing time management by helping individuals prioritize tasks based on urgency and importance. This method not only improves productivity but also reduces stress by providing a clear framework for decision-making. In addition, engaging in problem-solving dynamics, such as group brainstorming sessions or role-playing exercises, fosters collaboration and enhances communication among team members, which is crucial for tackling complex engineering challenges.

What is more, the ability to work effectively in teams and resolve conflicts constructively has become an indispensable requirement in today's work environment. Engineers often find themselves collaborating with professionals from various disciplines, necessitating strong interpersonal skills to ensure smooth cooperation. As technology continues to advance, and the nature of work becomes increasingly interconnected, soft skills are positioned as a key differentiator in an engineer's career trajectory.

In conclusion, investing in the development of soft skills is not merely an enhancement of one's professional toolkit; it is a strategic imperative that significantly boosts employability and career advancement. By prioritizing the cultivation of these skills, engineers can contribute to a more productive, innovative, and harmonious work environment, ultimately leading to greater success in their careers and the projects they undertake.

23.20 A strategy for assessment of soft skills

The evaluation of soft skills has emerged as a crucial aspect of higher education, particularly in fields such as engineering, where technical expertise alone is no longer sufficient for success. Despite their importance, the abstract and subjective nature of soft skills presents significant challenges in both their definition and assessment, making it difficult for educators to measure student progress effectively.

To tackle these challenges, various evaluation strategies and tools have been developed to provide a more structured approach to assessing soft skills. Among these tools are questionnaires, interviews, and rubrics, which allow educators to measure skill development in a more objective manner. These assessment methods not only help educators gauge student progress but also provide valuable feedback to students regarding their performance, strengths, and areas for improvement. This feedback is crucial for fostering a growth mindset and encouraging students to actively engage in their personal and professional development.

A notable case study, conducted in Galicia (Spain), illustrates the successful implementation of a soft skills program aimed at entrepreneurs in the food sector. This program was designed to address the specific needs of participants by focusing not only on teaching technical skills but also on integrating practical activities that fostered the development of interpersonal skills. By engaging in real-world scenarios and collaborative projects, participants were able to enhance their communication and teamwork abilities. The results of the program showed significant improvements in participants' perceptions of their soft skills, highlighting the effectiveness of a comprehensive approach to education that balances both technical and soft skill development.

Overall, the evaluation of soft skills is fundamental to preparing students for the workforce, as employers increasingly seek candidates who possess a blend of technical knowledge and interpersonal abilities. As educational institutions continue to adapt to the evolving demands of the job market, it is essential to implement assessment methods that recognize and promote the development of these vital skills. By doing so, educators can ensure that graduates are not only equipped with the necessary technical expertise but also possess the soft skills required to navigate the complexities of modern professional environments and face future challenges with confidence.

23.21 Soft skills in university classroom: best practices and lessons learned

In the current context of higher education, teaching soft skills has become a crucial aspect of preparing students for the workforce. However, this process faces several challenges, especially in a digital environment where interactions can be limited. To address these difficulties, it is essential to adopt a more interactive and experiential approach that moves away from traditional lectures and fosters active student participation.

One of the best practices in teaching soft skills is the clear specification of editing requirements and presentation formats. This not only helps students develop professional communication competencies but also establishes clear expectations that are essential for their success. Additionally, implementing clear deadlines and consequences for assigned tasks cultivates time management skills and a sense of responsibility, reflecting real-world project environments where adherence to deadlines is critical.

Educators should act as facilitators, guiding students in their learning process and promoting self-reflection. This involves creating an environment where collaborative learning is the norm, allowing students to work together on projects and activities that develop their interpersonal and communication skills. For instance, using round tables instead of rows encourages natural group discussions and fosters interaction. Moreover, the design of the learning environment can positively influence the educational experience, facilitating collaboration and creative thinking.

A data-driven learning approach is also fundamental. R&D professionals must learn to utilize new technologies and data analysis platforms to predict training demand, like how advertising agencies use data to direct their campaigns. This enables educators to tailor their teaching methods to the changing needs of students.

A notable example of the effectiveness of these methods is the UNIFORS 2020 project, which implemented hackathons as an innovative teaching method. These events not only encouraged collaboration among students but also allowed them to apply their knowledge in practical situations, thereby developing soft skills such as teamwork, problem-solving, and creativity.

In the end, integrating relevant and up-to-date content, along with using diverse formats such as multimedia resources and gamified activities, makes learning more engaging and accessible. Continuous feedback and flexibility in curriculum design ensure that educational programs remain aligned with student needs and industry trends. In summary, teaching soft skills in higher education is essential for preparing students for future challenges and requires a dynamic and collaborative approach that values diversity, inclusion, and innovation.

23.22 Digital thinking as an intensive methodology in the use of soft skill

In today's digital age, design thinking has evolved into what is known as digital thinking, an approach that prioritizes empathy and collaboration in problem-solving. This shift responds to the need to adapt to a constantly changing technological environment, where user expectations are increasingly high, and solutions must be innovative and effective. Digital thinking is based on a structured process consisting of five key phases: Empathize, Define, Ideate, Prototype, and Test, each of which plays a crucial role in creating user-centered solutions.

The first phase, Empathize, involves deeply understanding the needs and desires of users. This is achieved through observation, interviews, and immersion in the user's context, allowing teams to capture not only what users say but also what they feel and experience. In the Define phase, the findings from the empathy stage are synthesized to clearly articulate the problem to be solved, establishing a clear focus for solution development.

The Ideate phase is where creativity comes into play. Here, teams generate a wide range of ideas and potential solutions, fostering an environment where innovation is valued and thinking outside the box is encouraged. Subsequently, in the Prototype phase, tangible representations of the selected ideas are created, allowing teams to experiment and explore different approaches quickly and economically. The last phase, the Test involves evaluating the prototypes with real users, providing valuable feedback to refine and improve solutions before final implementation.

Throughout this process, soft skills are fundamental. Adaptability is crucial, as teams must be willing to adjust their approaches based on feedback and changes in context. Effective communication is equally important, ensuring that all team members and stakeholders are aligned and committed to the common goal. Additionally, critical thinking allows teams to evaluate ideas and solutions objectively, ensuring that informed, data-driven decisions are made.

However, the implementation of digital thinking, which is based on the principles of design thinking, faces several challenges. The complexity of contemporary problems can generate organizational resistance, especially in environments where hierarchical structures are rigid, and innovation is not part of the culture. Furthermore, the need for interdisciplinary skills can hinder collaboration across different areas, potentially limiting the effectiveness of the process. Despite these obstacles, digital thinking emerges as an essential tool for fostering innovation and developing user-centered solutions.

In conclusion, digital thinking is not only a methodology for problem-solving but also represents a paradigm shift in how organizations approach innovation. By integrating soft skills and a user-centered approach inspired by design thinking, companies can successfully navigate a constantly evolving digital world, creating solutions that not only meet current needs but also anticipate future ones. This approach not only enhances the user experience but also positions organizations as leaders in a competitive and rapidly transforming market.

23.23 Conclusions

In this final chapter, after providing a brief summary and conclusions from each of the chapters that make up this book on soft skills, we want to highlight a key message.

Throughout this journey, we have explored a diverse range of essential competencies – communication, leadership, adaptability, teamwork, problem-solving, and more – that go beyond technical knowledge and play a fundamental role in engineering practice. Each chapter has provided valuable insights, practical tools, and real-world applications that demonstrate how these skills are not just complementary but essential for success in both academic and professional settings.

As we conclude, it is important to reflect on the broader impact of soft skills in engineering. The ability to work collaboratively, navigate uncertainty, and approach problems with a human-centered perspective is what truly distinguishes exceptional engineers.

In a world where technology is advancing at an unprecedented pace, the true differentiator of engineers is not just their technical expertise but their ability to communicate, lead, adapt, and collaborate in increasingly complex environments. Engineering, by its very nature, seeks to solve real-world problems, and those solutions depend not only on mathematical or scientific knowledge but also on the ability to understand human needs, make decisions under uncertainty, and work effectively with diverse stakeholders.

To sum up, the development of soft skills is not an optional addition but an essential requirement from the very beginning of academic training. It is not enough to know how to calculate, design, or program; engineers must also learn to negotiate, resolve conflicts, manage time, think critically, and innovate with an ethical and sustainable vision. Preparing well-rounded engineers for the challenges of the future requires a balance between technical rigor and personal and professional growth.

This book is the result of the dedication and expertise of a group of professionals who have generously shared their knowledge and insights to strengthen these essential skills in engineering. To each of them, we extend our deepest gratitude. Their contributions will help shape future generations of engineers who are more resilient, creative, and aware of the impact of their work.

May this book serve as a starting point for fostering a more comprehensive approach to engineering education – one where technical proficiency and human-centered skills go hand in hand. The engineers of the future will not just solve problems; they will change lives. And to achieve this, we need professionals with analytical minds and human hearts.

Index

www.ingramcontent.com/pod-product-compliance
Lightning Source LLC
LaVergne TN
LVHW011828200926
848484LV00002B/339

* 9 7 8 1 8 3 7 2 4 0 1 7 3 *